Distillation Symposium 2014

Topical Conference at the 2014 AIChE Spring Meeting and 10th Global Congress on Process Safety

New Orleans, Louisiana, USA
30 March – 3 April 2014

ISBN: 978-1-63439-065-1

Printed from e-media with permission by:

Curran Associates, Inc.
57 Morehouse Lane
Red Hook, NY 12571

Some format issues inherent in the e-media version may also appear in this print version.

Printed by Curran Associates, Inc. (2014)

For permission requests, please contact AIChE
at the address below.

AIChE
3 Park Avenue
New York, NY 10016-5991

Phone: (203) 702-7660
Fax: (203) 775-5177

www.aiche.org

Additional copies of this publication are available from:

Curran Associates, Inc.
57 Morehouse Lane
Red Hook, NY 12571 USA
Phone: 845-758-0400
Fax: 845-758-2634
Email: curran@proceedings.com
Web: www.proceedings.com

TABLE OF CONTENTS

(1a) Possible vs. Practical: Engineers Must Lead the Development of Practical Technologies 1
William Banholzer
(28a) Can Well-Configured Hot Vapor Bypass Condenser Controls Become Unstable? 2
Henry Z. Kister and Daryl Hanson
(28b) Moving Gamma Scanning Forward – Quantitative Analysis of Tray Capacity 29
Lowell Pless
(28c) How Sensitive is Your Amine Plant to Operating Conditions? 31
Ralph Weiland, Jesse Santos, Attilio J. Praderio, Nilia Maharaj and Michael Schultes
(28d) "MEGA Tower" Design Considerations 41
Daniel R. Summers, Mark W. Pilling and Daryl Wiesman
(28e) Collaborations Between FRI and OSU Chemical Engineering 42
James R. Whiteley, T.J. Cai and Clint P. Aichele
(28f) "Ecmd™ Trays: A Quarter Century of Excellence" 48
Bob Miller, Kevin J. Richardson and Michael Marsh
(28g) Bibliography of Mr. Mike Resetarits 49
Tony Cai and Simon Chambers
(48a) Change Your Board Operator to a Process Manager with State-Based Control 54
Tom Nolan and Dustin Beebe
(48aa) Huaa – When Learning Is Not Enough 69
Mike Bearrow
(48ac) Autocad - Smart Grid – Agriculture - Architecture and the United States Government 73
Ethenia Scott
(48ad) Production, Characterization and Catalytic Studies of Biobased Carbon Materials 74
Qiangu Yan, Jilei Zhang and Zhiyong Cai
(48ag) Uncertainty in Sour Gas Viscosities Estimation, What Is the Effect on Your Reservoir Inflow and Tubing Performance 75
Adel Elsharkawy
(48ah) Chromonic Nanocarriers for Chemotherapeutics: Size Distribution and Control Release Studies 76
Rahul Misra and Sanat Mohanty
(48ai) Double Containment Piping Solutions for Safety and Environmental Concerns 77
Patrick Fedor and Darin Johnson
(48b) DTP Process: On-purpose Propylene Production Technology 84
Kazunori Honda, Atsushi Okita, Jumpei Takahashi, Koji Oyama, Nobuyasu Chikamatsu, Mitsuo Morita and Shuji Obayashi
(48d) Self-Cleaning "Bernoulli" Type Filters Used in Onshore and Offshore Applications 89
Artur W. Krueger
(48f) Effects of Support on Sulfur Tolerance and Regeneration of Pt Catalysts Measured By Ethylene Hydrogenation and EXAFS 90
Jorge Pazmino, Chuansheng Bai, Jeffrey T. Miller, Fabio H. Ribeiro and W.N. Delgass
(48h) Improved Operational Efficiency and Reliability through Insulation Materials Selection 91
Steven Coppella
(48ij) Compliance with EPA Boiler MACT Standards: Mercury-In-Fuel Gas 92
Patrick Laine
(48k) Young Professional Simulation Tutorial 93
Naomi Hua and Mike Donahue
(48m) Development Of Polymeric Sulfonic Acid Composite Membranes For Fuel Cell Applications 97
Jimoh Adewole, Abdullah S. Sultan, Amir Al-Ahmed and S. M. Javid Zaidi
(48o) Facility Siting for Major Projects – Implementation of Consequence Analysis/Quantitative Risk Analysis, a Project Development Lifecycle Framework 105
Mohammad Faruq Haider
(48p) Global Energy and Transportation and Mobil Oil 118
Ethenia Scott
(48r) Characterization of Iron Phthalocyanine As the Active Material for Lithium Batteries 119
Sarwan S. Sandhu, Joseph P. Fellner and David Anneken
(48s) Design of a Free-Fall Reactor for Fast Pyrolysis of Waste Plastics 120
Pravin Kannan and Ahmed AlShoaibi

(48t) Biosorptive Dehydration of Ethanol/Water Azeotropes Using Compound Starch-Based Adsorbent 121
Wenping Wang, Jinsheng Sun, Xijia Cao and Guangxin Liu
(48u) Purification of 2-Amino-1-Phenylethanol Enantiomers By a New Technique Combining Distillation and Crystallization 123
Lie-Ding Shiau and Hou-Guo Teng
(48w) Fabrication of Low Cost Insulating Material from Kaolin Clay for Construction Purposes 124
Naim Faqir, MA Al-Harthi, Hamad AbdulWahhab, Mazen Alshaaer and Reyad Shawabkeh
(48x) Thermodynamic Study of Binary PAH (Anthracene + Phenanthrene) Solid Mixtures 125
James W. Rice, Jinxia Fu, Emma Sandström and Eric M. Suuberg
(48y) How Confident Are You That Your Major Accident Risks Are Under Control? 126
Ellis Graeme and Robert Smith
(48z) Sil Determination: Shortcomings with the Use of LOPA 127
Alan King
(61a) TRU-SCAN and TRU-CAT Scanning of a Structured Packing Operating at DEEP Vacuum 129
Simon Chambers, Lowell Pless and Ron Carlson
(61b) Performance Characteristics of an Intermediate Area High Performance Structured Packing 137
Zarko Olujic, Thomas Rietfort, Helmut Jansen and Egon Zich
(61c) Absorption Hydraulic and Mass-Transfer Characteristics of Raschig Super-Pak 250 138
Lukas Valenz, Frantisek J. Rejl, Jan Haidl and Vaclav Linek
(61d) Gas-Phase Mass-Transfer Characteristics of Mellapak 250Y Structured Packing Under Distillation Conditions 143
Frantisek J. Rejl
(61e) Nagaoka Packing "Super H Pack" 150
Tadayoshi Nagaoka
(61f) Enhancing Multi-Product Fractionator Efficiency Using Properly Designed Vapor and Liquid Distribution Devices 154
SounHo Lee
(61g) A Submerged Particle Model for Liquid Holdup in Packed Beds 165
Subbarao Duvvuri, Nurhidayah Mohammad, Mohd.Azmi Bustam and Mohamed I. A. Mutalib
(82a) A New Screening Tool for Designing Energy Efficient Distillation Systems 166
Joshua Huff, Mohit Tawarmalani and Rakesh Agrawal
(82b) More Operable Dividing Wall Columns 168
Gautham Madenoor Ramapriya and Rakesh Agrawal
(82c) Dividing Wall Column as Energy Saving Retrofit Technology 169
Igor Dejanovic, Helmut Jansen and Zarko Olujic
(82d) Process Intensification through Distillation Synthesis, Dividing Wall Columns, and Novel Recycle Methods 170
John Pendergast, Daniel Trauth, Mahesh Sawant and Ravi Dixit
(82e) Parallelization Opportunities in Distillation Calculations 176
Ronald W Bondy, David Bluck and Srividya Gummadi
(82f) Batch Distillation Strategy for Higher Purity Products 191
Zbigniew T. Fidkowski
(119a) Troubleshooting and Rectification of a Giant C3 Splitter Tower Problem Part 1: Troubleshooting 201
Henry Z. Kister, Randy Miller, Brian Clancy-Jundt and Daniel R. Summers
(119b) Troubleshooting and Rectification of a Giant C3 Splitter Tower Part 2: Rectification 203
Daniel R. Summers, Henry Z. Kister, Brian Clancy-Jundt and Randy Miller
(119c) A High Capacity Re-Tray That Wasn't - a Practical Lesson in Counter-Intuitive Thinking 205
Sean Hennigan
(119d) Evaluation and Testing of a Two Phase Feed Distributor 206
Veronica R. Saunders, Eric Clavelle and Andre Bernard
(119e) Steam Stripping of Aniline Process Wastewater -- a Tutorial, Review, and Critique 223
Robert G. Kunz
(119f) Kettle Reboiler Hydraulics 251
Mike Resetarits, Tony Cai and Lowell Pless
(146a) Distillation Fundamentals 253
M.R. Resetarits, A.N. Vennavelli and A. Y. Ogundeji
(146b) Trays for Distillation and Absorption 267
Tony Cai, Mike Resetarits and A. Y. Ogundeji
(146c) Structured Packings for Distillation and Absorption 274
Simon Chambers and A.N. Vennavelli

(146d) Liquid Extraction Processes & Equipment 281
Mike Resetarits, A.Y. Ogundeji and T.J. Cai
(146e) Acid Gas Absorption Fundamentals 288
Clint P. Aichele, James R. Whiteley, Yash Tamhankar and Mike Resetarits
(146f) Radioisotope Applications for Obtaining Detailed Process 298
Andy Burleigh and Lowell Pless
Author Index

354012 Possible vs. Practical: Engineers Must Lead the Development of Practical Technologies

Abstract:

Chemical Engineering is need today more than ever. There is a strong societal desire for sources of energy and feedstocks that can sustainably support a high standard of living for all. Thermodynamic, material and economic challenges stand in the way. Most of the general public does not have the basic understanding of the energy flows, chemical transformations, and the scale of the systems that support our current way of life. These concepts are fundamental to chemical engineering. Chemical engineers should be on the vanguard of separating hype from truly practical options yet occationally we let society down. This talk will present several examples where the fundamentals of chemical engineering can be used to access the practicality of proposed solutions to energy and water issues, as well as explain why the venture capital model struggles in the materials market.

Can Well-Configured Hot Vapor Bypass Condenser Controls Become Unstable?

Henry Z. Kister, Fluor, Aliso Viejo, CA

and

Daryl Hanson, Valero, San Antonio, TX

Presented at the Distillation Topical Conference, AIChE Spring Meeting, New Orleans, Louisiana, March 30 - April 3, 2014

Can Well-Configured Hot Vapor Bypass Condenser Controls Become Unstable?

Henry Z. Kister, Fluor Corp., Aliso Viejo, CA and
Daryl W. Hanson, Valero Energy, San Antonio, TX

Our survey of tower failures (1, 2) identified the hot vapor bypass as the most troublesome pressure and condenser control method in distillation. About one third of the pressure and condenser control malfunction case histories reported in the literature were problems with hot vapor bypasses, most in refineries. Most problems are due to poor configuration of hot vapor bypass piping, which evolves from poor understanding of its principles. These principles have been in the literature for decades, described by excellent papers by Whistler (3), Hollander (4), and later by the classic article by Chin (5) and in some texts (6, 7). When configured correctly, the experience is that hot vapor bypasses are usually non-troublesome (3-7).

Recently, we encountered some situations that even though the hot vapor bypass schemes used correct configurations, problems were experienced. To gain better understanding we applied field measurements and testing as well as hydraulic analysis which led to successful solutions. This paper describes our experiences, findings, and the lessons learnt for troubleshooting and design of hot vapor bypass schemes.

HOW HOT VAPOR BYPASS CONTROLS WORK

A hot vapor bypass is one of the flooded condenser control schemes used for total condensers (no vapor product). In these methods, the condenser area is partially flooded by condensate. The flooded tubes do not contact the vapor and perform little condensation. Column pressure is controlled by manipulation of the flooded area. Raising condenser liquid level floods additional tubes, which reduces the condensation area, raising tower pressure. Conversely, lowering the liquid level in the condenser exposes more tubes, which increases condensation area, and subsequently lowers tower pressure. Although the flooded area performs little condensation, it serves the vital purpose of subcooling the condensate before it leaves the exchanger.

Fig.1 shows a correctly configured pressure control by hot vapor bypass. Unlike other flooded condenser control schemes, in which the condenser is elevated above the reflux drum, in the hot vapor bypass scheme the condenser is at ground level, while the drum is elevated, often mounted on the lowest platform. So the liquid level in the condenser is 10 to 20 ft below that in the drum (which may be horizontal or vertical). The condensate is lifted by the vapor pressure difference between the condenser liquid surface, which is at the bubble point temperature, and the drum liquid surface, which is colder due to the condensate subcooling.

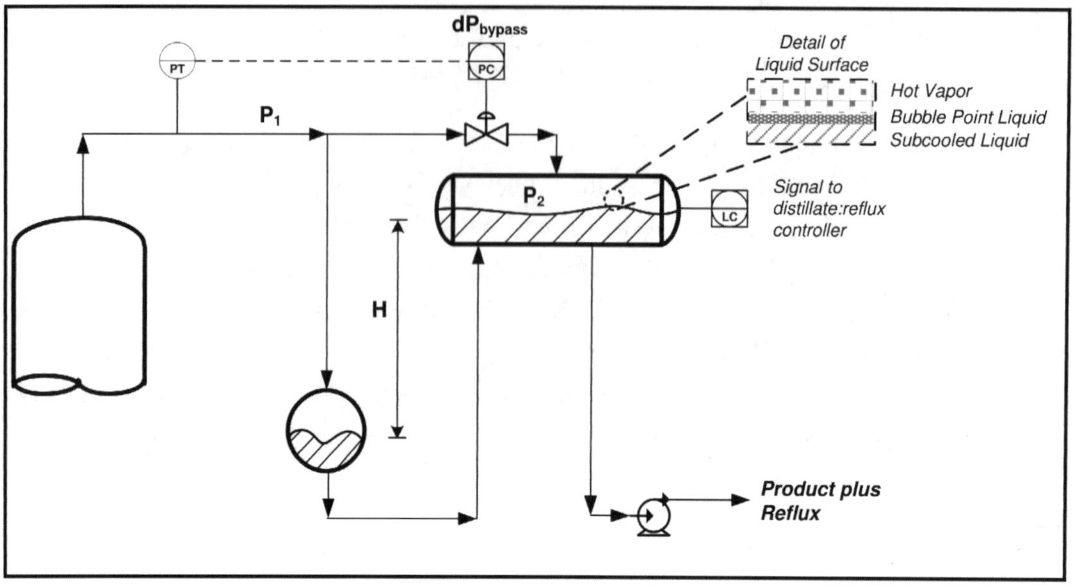

Figure 1: Typical Hot Vapor Bypass Controls

Even a few degrees of subcooling can often raise the condensate 50 to 100 ft, even more. To limit the lift to the desired height (i.e., to the reflux drum), hot vapor from the tower overhead is used to heat the liquid surface in the drum. The vapor pressure in the drum is set by the liquid surface temperature, not the subcooled temperature. Liquid is a good thermal insulator. The surface temperature, and therefore the vapor pressure in the drum, is determined by the heat balance between the hot vapor inflow and the heat outflow from the surface to the bulk liquid and from the vapor space to the atmosphere. Opening the hot vapor bypass valve heats the drum liquid surface, thereby reducing the lift, while closing the valve allows the liquid surface to cool, thereby raising the lift.

The lift is manipulated to control the tower pressure. Opening the bypass valve heats the drum liquid surface, raising its vapor pressure, and therefore the drum pressure. This pushes liquid from the drum into the condenser, flooding more tubes, reducing condensation, and therefore raising the tower pressure. Conversely, closing the valve cools the drum liquid surface, reducing its vapor pressure and therefore the drum pressure, and sucking liquid from the condenser to the drum. This exposes additional condensing area and therefore lowers tower pressure.

The popularity of the hot vapor bypass scheme in large total condensers stems from its major capital savings. Locating large condensers at ground level eliminates the need for massive condenser support structures, piping cooling water to elevations, and provides easy access for maintenance. The piping is simple, the control valve is small, and the response is fast (1, 7). These advantages can translate into handsome savings in steelwork, platforms, and maintenance. These savings can be major in large installations, especially where a battery of condensers rather than a single exchanger is used. Like other flooded condenser schemes, this arrangement delivers subcooled liquid for the reflux and product pumps, maximizing their available NPSH and lowering platform height requirements.

CORRECT AND INCORRECT HOT VAPOR BYPASS CONFIGURATIONS

The hot vapor bypass arrangement, however, has several disadvantages, and as stated, its operation can be troublesome (1, 2, 5 - 12), mostly due to poor configuration of hot vapor bypass piping. The main lessons learned by the industry and the good practices to overcome the traps, as reported in the literature, are:

- Correct piping is mandatory. Bypass vapor must enter the vapor space of the reflux drum (Figure 1). The bypass should be free of pockets where liquid can accumulate; any horizontal runs should drain into the reflux drum. Liquid from the condenser must enter the reflux drum well below the liquid surface. The bottom of the drum is the most suitable location. If the entry nozzle is at the top of the drum, the liquid line needs extending to discharge near the bottom of the drum. Any other liquid streams entering the drum, like the reflux pump minimum flow recycle, must also enter at the drum bottom. These recommendations, first published in the literature sixty fifty years ago (3, 4), have been strongly endorsed by recent sources (2, 5, 6, 25). Figure 2a shows one case (8) in which violation of this practice led to severe pressure fluctuations, inability to control tower pressure, and a capacity bottleneck. Figure 2b shows the fix that eliminated the problem. References (2, 4, 6, 13, 14, 25) report other experiences in which incorrect piping led to instability, poor control, and hammering.

- Operation may be troublesome if the drum liquid surface is agitated; a case of instability due to inadvertent agitation has been reported (4, 5). Agitation may be due to high-velocity impingement of the hot vapor jet on the liquid surface, due to currents introduced by improper liquid inlet design, or due to atmospheric cooling condensing vapor at the ceiling of the drum, with the condensate dripping down.

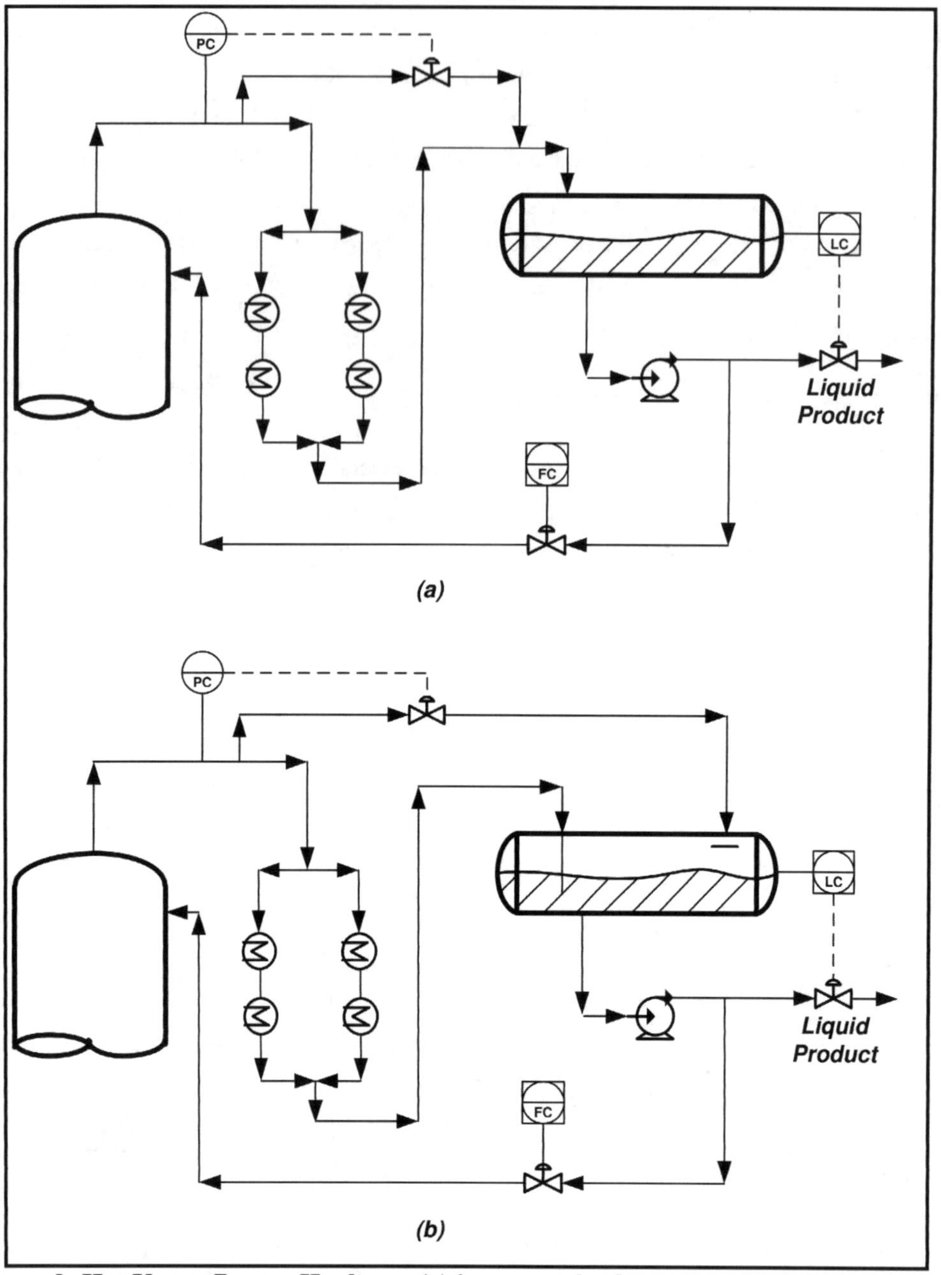

Figure 2: Hot Vapor Bypass Hookups: (a) incorrect, leads to pressure fluctuations; (b) modified, good pressure control. (From ref. 8. Reprinted courtesy of *Chemical Engineering*.

- Like other flooded condenser control methods, hot vapor bypass controls are suitable only for total condensers. The liquid leg between the condenser and drum prohibits venting of any non-condensables. To handle small amounts non-condensables, such as those trapped in the condenser during startups or upstream upsets, a vent (not shown on Figure 1) is required from their accumulation point(s). The condenser vent can be directed to the vapor space of the drum or to elsewhere. If a vent line is absent, instability and capacity bottlenecks may result, as described in one published case history (15).

- Since the success of the hot vapor bypass technique is contingent on maintaining a small vapor pressure difference, the reflux drum vapor space may need to be insulated to minimize interference from rainstorms (4-6). This issue was reported to be more pronounced with narrow boiling-range mixtures (4, 5), and at high pressure (4), where small temperature changes have a large effect on the split of overhead flow between the condenser the bypass. At the other extreme, the incidence of Rayleigh fractionation with wide-boiling mixtures (heavy components condensed out without mixing with the remaining mixture), can also interfere with this control system (3).

- Because of the liquid leg between the accumulator and the drum, this technique can suffer from interaction between the drum and the condenser liquid level (6, 10-12) and from U-tube oscillations (6, 11, 12). To minimize the interaction, the pressure controller should be tuned much tighter than the drum level controller (10, 12). This can be an issue if the reflux drum is small, and the level controller needs to be tuned fast to avoid overflow or loss of level. This is uncommon; we have encountered this situation only once, but another case was reported in the literature (16).

- Leakage of vapor through the bypass valve at the closed position can substantially reduce condenser capacity as demonstrated in one reported case history (17).

- Undersizing of the bypass control valve may lead to inability to keep the tower pressure up in cold winter days. Occasionally this is also experienced at low rate operation due to highly subcooled condensate. In some cases, there may be a tendency to suck liquid from the condenser to the drum or destabilize tower pressure. The effects of undersizing the valve are most severe when the drum is not insulated. In one case (24), poor pressure control due to undersizing was improved by installing a throttling valve in the liquid line from the condenser to the drum. In another case (26), pressure fluctuations and instability due to excessive subcooling at low rates, plant startups, and cold nights was countered by controlling the cooling water flow rates to prevent excessive subcooling.

- The amount of subcooling and vapor bypass rates can only be determined empirically, and design is difficult. Simplified sizing procedures are available (4, 19), but these are based on heating all the subcooled liquid to its bubble point, and therefore grossly conservative. A recent interesting method (26) models the drum surface as a heat exchanger, but this method is based on variation of surface temperature along the drum length, which has not been observed in well-functioning drums.

- There is an inverse response problem (8); when pressure rises, it closes the bypass valve, which initially increases pressure until the condenser level begins to change.
- In some cases, the hot vapor bypass control valve is manipulated by the drum pressure instead of the tower pressure, as in a case described in Ref. 24. Dynamically, this control is inferior because the vapor volume in the drum is much smaller than in the tower and more variable in response to ambient changes.

In summary, this method has some major advantages, but its operation can be troublesome.

Our discussion below starts with assuming a well-configured, well-designed hot vapor bypass scheme that does not violate the above guidelines. Nonetheless, some of the above guidelines will be re-visited in our discussion below.

HYDRAULICS OF HOT VAPOR BYPASS SCHEMES

For the correctly-configured hot vapor bypass arrangement in Fig.1, assuming negligible line pressure losses, a pressure balance gives

$$P_1 - P_2 = H + \Delta P_{cond} \quad (1)$$

Where

P_1 is the pressure at the junction between the line to the condenser and the condenser bypass, psia:
P_2 is the pressure inside the reflux drum vapor space, psia;
H is the head difference between the reflux drum liquid level and the condenser liquid level, psi;
ΔP_{cond} is the condenser pressure drop, psi;

The density used to calculate the head differential H is the difference between the liquid and vapor densities, due to the presence of a static leg of vapor in the condenser inlet line. The vapor density is based on P_1 and the condenser inlet temperature. The liquid density is best approximated as the density of the subcooled liquid leaving the condenser.

Eq. 1 states that the driving head required to pump the condensate into the drum is supplied by the pressure difference between the condenser liquid surface and the drum liquid surface. P_1 is the vapor pressure of the liquid surface in the condenser, and is the bubble point pressure at the condensing temperature. P_2 is the vapor pressure at the reflux drum surface. For P_1 to be higher than P_2 , the temperature at the drum surface needs to be lower than that at the condenser surface. Due to the subcooling of the drum liquid, P_2 is lower than P_1 .

It does not take much subcooling to largely lower the vapor pressure in the drum. Table 1, derived from physical property data for hydrocarbons (20), lists vapor pressure and liquid head variations with temperature for three common applications for hot vapor bypasses:

Table 1.

Effect of Temperature on Vapor Pressure for Typical Hot Vapor Bypass Applications (Basis: API Data Book, Ref. 20)

Application	Pressure, psia	Temp, °F	Vapor Pressure change rate, psi/°F	Liquid Density, lb/ft^3	Liquid Head change rate, ft/°F
C_3- C_4 Splitter	305 psia	126°F	3.2 psi/ °F	28.3	15 ft/°F
(72%/28% LV Propylene/Propane)	218 psia	99°F		30.2	
Debutanizer	150 psia	175°F	1.5 psi/ °F	31.2	7 ft/°F
(Butane)	76 psia	125°F		33.7	
Dehexanizer	40 psia	220°F	0.4 psi/ °F	36.0	1.6 ft/°F
(Hexane)	15 psia	158°F		38.3	

Table 1 shows that for a C_3 - C_4 Splitter, a 1°F subcool can lift the liquid up 15 feet. With a debutanizer, it takes just over 2°F subcool to lift liquid to the same height. With the dehexanizer it takes a bit more (10°F subcool), but even then, it is well-within the subcooling capability of the condenser. Typical subcools range from 5°F to 50°F.

In typical hot vapor bypass schemes, the liquid level in the reflux drum is about 10 to 20 ft above the liquid level in the condenser. So subcooling of less than 10°F is sufficient in most practical applications. With the actual subcooling being much higher than 1°F to 10°F, the vapor pressure difference will tend to pump the condensate to a much higher elevation than needed.

The hot vapor provided by the bypass in Fig. 1 regulates the vapor pressure in the drum. It keeps the liquid surface temperature at a value that will give the desired liquid head H in Eq. 1. The hot vapor condenses onto the drum liquid surface and heats it up. At steady state, the valve opening induces sufficient vapor to maintain the drum liquid surface temperature at the value corresponding to the desired vapor pressure P_2 That satisfies Eq. 1. There is a net heat flow from the hot liquid surface to the subcooled liquid underneath. This heat flow, as well as the atmospheric heat losses, need to be matched by condensing the hot vapor.

The pressure difference in the bypass can be written in terms of the bypass control valve pressure difference, ΔP_{bypass} , giving

$$P_1 - P_2 = \Delta P_{bypass} \quad (2)$$

Combining Eq. 1 and Eq. 2 gives

$$\Delta P_{bypass} = \Delta P_{cond} + H \quad (3)$$

When the line friction losses are significant, Eq. 3 needs to include them, giving

$$\Delta P_{bypass} + \Delta P_{bypass\ line} = \Delta P_{cond} + \Delta P_{liq\ line} + \Delta P_{vap\ line} + H \quad (3A)$$

Where

$\Delta P_{liq\ line}$ is the pressure drop in the liquid line from the condenser to the reflux drum, psi

$\Delta P_{vap\ line}$ is the pressure drop in the tower overhead vapor line to the condenser downstream of the point where the hot vapor bypass splits off, psi.

$\Delta P_{bypass\ line}$ is the pressure drop in the hot vapor bypass vapor line (other than the control valve), psi.

For clarity, the following discussion will be based on Eq.3. However, all the following equations can be expanded to include the line pressure drop terms as shown in Eq. 3A.

In hot vapor bypass schemes, the vapor pressure differences directly manipulate the flooded height in the condenser. Eq. 1 states that the vapor pressure difference is balanced by the liquid head lift plus the condenser pressure drop. To raise pressure in the tower, additional tubes in the condenser need to be flooded, so the liquid head lift H needs to be lowered. To achieve this, the vapor pressure differential $P_1 - P_2$ needs to be reduced (Eq. 1). With P_1 constant, this is achieved by opening the hot vapor bypass to heat the liquid surface in the drum, which raises the drum vapor pressure P_2. Conversely, to reduce tower pressure, the liquid level in the condenser is lowered, so the liquid head lift H is raised. This is achieved by closing the bypass valve. The reduction in hot vapor flow allows the liquid surface in the drum to cool, lowers P_2 , and the larger $P_1 - P_2$ difference sucks liquid from the condenser into the drum, thus exposing more condenser area for condensation.

HYDRAULIC IMBALANCES

Eq. 3 states that when the liquid head and the condenser pressure drop are small, the bypass pressure drop needs to be small. Under these conditions the hot vapor bypass control system becomes vulnerable to hydraulic imbalances. To the best of our knowledge, this vulnerability has not been previously reported.

At low heads and small pressure drops,the valve tends to open. If the valve pressure drop exceeds the right hand side of Eq. 3 even at 80 – 90% percent opening, the valve will tend to widely open and go off-control. To match the large pressure drop on the left hand side of Eq. 3, the liquid head H will tend to rise. In some situations it will keep rising until it sucks the liquid leg right out of the condenser.

Imbalance can also develop because different factors govern the valve pressure drop and the valve opening. The valve pressure drop is largely governed by the drum surface temperature. The valve opening is governed by the tower pressure controller. The two may vary independently and interact. Such interaction can be initiated by a disturbance that causes sudden cooling to the liquid surface in the drum. The drum pressure quickly drops, while the tower pressure, and therefore, the valve opening, remain temporarily constant. The bypass pressure drop rise renders the left hand side of Eq. 3 higher than the right hand side.

The higher pressure differential increases the vapor flow through the bypass. Also, per Eq. 3, H rises, exposing condenser tube area and causing the tower pressure to fall. The tower pressure fall opens the valve, further increasing the hot vapor flow. If the combined increase in hot vapor flow is sufficiently large to re-heat the liquid surface and quickly reinstate the drum pressure, the drum will return to steady state following the bump. In contrast, if the increase in flow does not balance the fall in drum pressure fast enough, or if it interacts in a manner that aggravates the initial surface disturbance, H in Eq. 3 will keep rising, in some situations until it sucks the liquid leg right out of the condenser. This is illustrated by Cases 1 and 2 in this paper.

A good solution for both imbalance issues is to increase the pressure drop at the outlet of the condenser. Fig. 3 shows the addition of such a throttling valve as described in Cases 1 and 2.

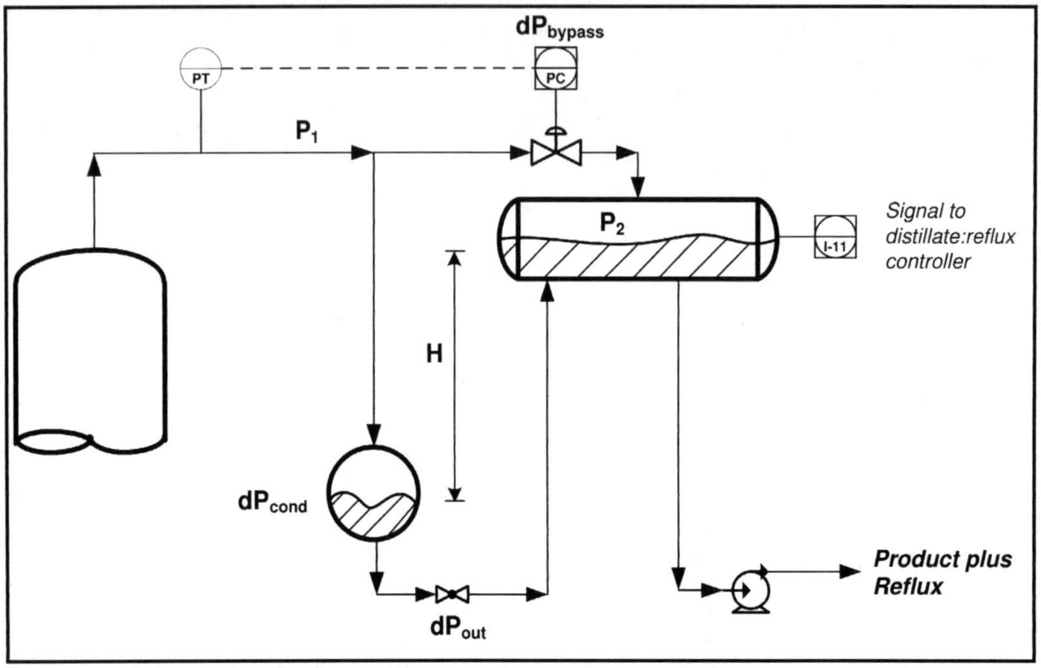

Figure 3: Addition of a Throttling Valve at the Condenser Outlet Reduces the Bypass Valve Vulnerability to Hydraulic Imbalances

The addition of this valve changes adds a term ΔP_{out} to Eq.1, converting it to Eq. 4

$$P_1 - P_2 = H + \Delta P_{cond} + \Delta P_{out} \tag{4}$$

Where ΔP_{out}, is the throttling valve pressure drop, psi

Combining Eq. 4 with Eq. 2, which remains unchanged, gives

$$\Delta P_{bypass} = \Delta P_{cond} + \Delta P_{out} + H \tag{5}$$

Eq. 5 can be expanded to include the line friction terms in Eq. 3A, giving

$$\Delta P_{bypass} + \Delta P_{bypass\ line} = \Delta P_{cond} + \Delta P_{liq\ line} + \Delta P_{vap\ line} + \Delta P_{out} + H \qquad (5A)$$

The additional term, ΔP_{out}, permits larger pressure drop through the valve (ΔP_{bypass}). The additional pressure drop retards the sudden mass movement of liquid from the condenser to the drum upon a strong pull from a disturbance at the drum liquid surface. In case of the control valve being close to full opening, the additional pressure drop helps to return the valve to its normal operating range. Finally, the additional pressure drop counters the tendency for U-tube oscillations mentioned earlier.

Another hydraulic imbalance can develop when tower overhead is condensed by an elevated air condenser followed by a ground-level cooling water condenser (Fig. 4). This system generally works well when the liquid level is in the cooling water condenser. However, during cold or wet ambient conditions, at low rates,, or when the condensers are clean, the air condenser may easily provide all the needed condensation area. The pressure controller will open the bypass, pushing liquid up into the air condenser. With the liquid level in the air condenser, Eq. 1 changes to

$$P_1 - P_2 = \Delta P_{cond,\ air} + \Delta P_{cond,\ water} - h \qquad (6)$$

and Eq. 3 changes to

$$\Delta P_{bypass} = \Delta P_{cond,\ air} + \Delta P_{cond,\ water} - h \qquad (7)$$

For this scenario to work, the sum of the condensers pressure drops need to be well above the liquid head h. This is unlikely because the feed to the water condenser is all liquid, which gives a low condenser pressure drop. Most likely, the air condenser pressure drop will not be high enough to keep ΔP_{bypass} in Eq. 7 within the valve control range, and the bypass valve will open fully. Even full opening of the valve will not provide enough heat to keep the liquid level down in the drum. The drum will fill up and control will be lost.

The simplest solution is to make sure that the liquid level is always in the water consenser by reducing the air condenser duty during winter. This is achieved by turning off fans, closing louvers, and controlling the air condenser outlet temperature by manipulating the fan motor speed (best) or the pitch of the blades. Alternatively, but not commonly practiced in this scenario, additional pressure drop (e.g., a throttling valve) can be added in the condensate line from the cooling water condenser outlet or in the condenser inlet line.

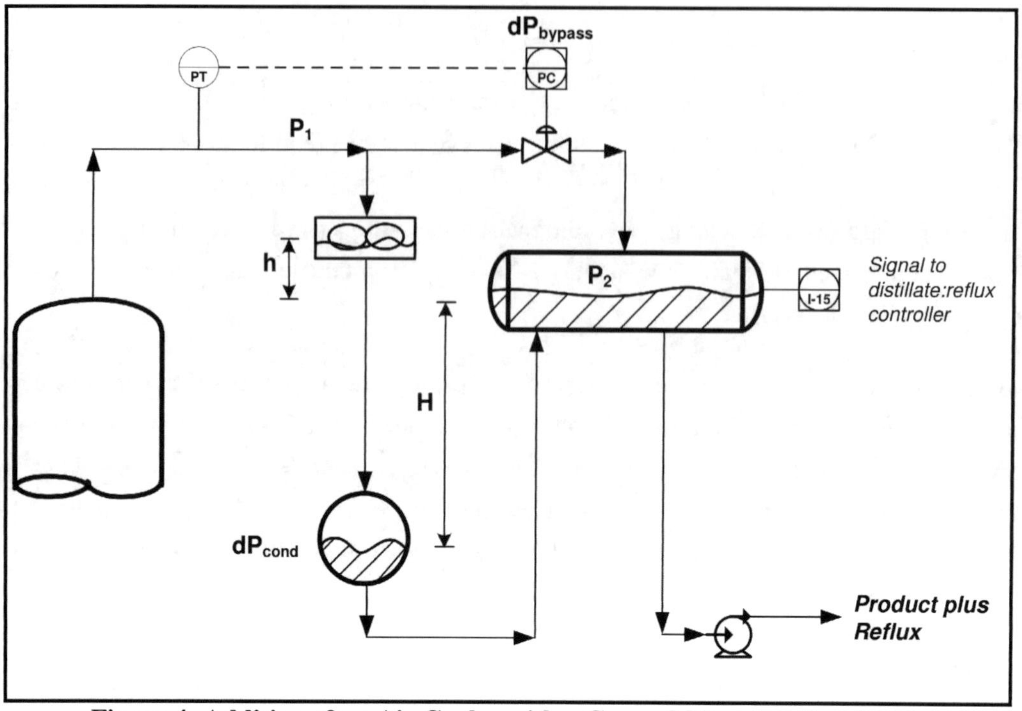

Figure 4: Addition of an Air Cooler with a Ground Mounted Condenser

STEADY STATE HEAT TRANSFER

At steady state, the heat supplied by the hot vapor bypass balances the heat flow from the drum liquid surface to the drum bulk liquid plus any heat losses from the vapor space of the drum, i.e.

$$Q_{HVB} = Q_{cond} + Q_{loss} \tag{8}$$

Where

Q_{HVB} is the total heat supplied by condensation of vapor supplied by the hot vapor bypass, Btu/h

Q_{cond} is the heat transferred from the drum liquid surface to the drum bulk (subcooled) liquid, Btu/h

Q_{loss} is the heat loss from the exposed drum wall area above the liquid surface to the atmosphere, Btu/h

As long as the liquid surface in the reflux drum is steady and unagitated, most of the heat flow from the surface into the subcooled liquid, Q_{cond}, is by conduction. In reality, convection and bulk movement raise the heat transfer from the surface to the subcooled liquid. All the heating, cooling, and vapor pressure adjustment processes take place at the liquid surface. The vapor space and the liquid surface are hot, at the condensing temperature corresponding to P_2 . This hot temperature does not travel deep below the liquid surface (see illustration in Fig.1). The vapor film heat transfer coefficient for condensation at the liquid surface is high (26). Since most process liquids are good thermal insulators, most of the temperature gradients occur within a thin hot liquid layer. According to Whistler (3), the hot liquid layer at the surface can be as thin as 1 inch. Experience with thermal scans confirms that the hot layer is only a few inches thick with a well-designed system. Below this hot surface crust the liquid remains at its subcooled temperature.

The heat transferred from the surface to the bulk liquid by conduction and convection through quiescent liquid, $Q_{cond,}$ is given by:

$$Q_{cond} = h_{QHL}\, A_S\, (T_{surface} - T_{subcool}) \tag{9}$$

Where

h_{QHL} is the heat transfer coefficient for conduction and convection through the hot liquid layer, Btu/hr per ^oF per ft^2 of drum surface area

A_S is the area of the liquid surface in the reflux drum, ft^2

T is the temperature, ^oF, with the subscripts surface denoting at the drum liquid surface and subcool denoting the subcooled bulk liquid.

The heat supplied by the hot vapor bypass also needs to provide the heat lost from the wall of the drum above the liquid surface to atmosphere, Q_{loss}. Assuming the drum liquid surface is at about the same temperature as the drum vapor space (usually a good assumption), the loss of heat from the wall of the drum is given by

$$Q_{loss} = h_{atm}\, A_{DW}\, (T_{surface} - T_{ambient}) \tag{10}$$

Where

h_{atm} is the heat transfer coefficient for convection and radiation from the exposed drum wall area above the liquid to the atmosphere, Btu/hr per ^oF per ft^2 of drum surface area

A_{DW} is the area of the drum wall above liquid surface, ft^2

T is the temperature, ^oF, with the subscripts surface denoting at the drum liquid surface and ambient denoting the ambient temperature outside the drum.

Combining Eq. 9 with Eq. 10 permits comparison between the two mechanisms

$$Q_{cond} / Q_{loss} = [h_{QHL} / h_{atm}] [A_S / A_{DW}] (T_{surface} - T_{subcool}) / (T_{surface} - T_{ambient}) \quad (11)$$

Due to the drum curvature, the wall area above the liquid surface A_{DW} in horizontal drums exceeds the liquid surface area A_S. In vertical drums, the length usually is about 3 times the diameter, so unless the liquid level is near the top, the wall area will again exceed the liquid surface area. In either case, the area ratio term A_S / A_{DW} in Eq. 11 is less than 1. In most condensers, the ambient temperatures, especially during a cold winter night, can be much lower than the subcooled temperature, so the temperature difference ratio term in Eq. 11 is less than 1, often by a large factor. At the same heat transfer coefficient, the atmospheric heat losses usually exceed the heat flow from the liquid surface to the drum liquid, often quite significantly.

For uninsulated vessels, the ambient losses heat transfer coefficient h_{atm} varies widely with the ambient conditions. For the temperature range in Table 1, typical heat transfer coefficients for ambient heat loss range from (18, 21) about 2-3 Btu/hr per ^oF per ft^2 in still air, to 5-7 Btu/hr per ^oF per ft^2 on a windy day. On a rainy day, the coefficient can rise as high as 15 -20 Btu/hr per ^oF per ft^2. With still liquid at the drum surface, the heat transfer coefficient of conduction plus convection at the hot liquid layer in the drum is of the order of 2-10 Btu/hr per ^oF per ft^2. It follows that for uninsulated drums, the ambient heat losses exceed the heat flow from the surface to the bulk liquid even under favorable ambient conditions. On a stormy day, the ambient heat losses can be more than an order of magnitude higher than the heat flow from the surface to the bulk liquid.

Insulating the drum can reduce the ambient heat losses by an order of magnitude compared to even still air. Lieberman (18) found that for well-insulated process vessels (about 3 – 4 inches of fiberglass or mineral wool insulation with no large exposed metal surfaces) the ambient losses heat transfer coefficient is 0.25 to 0.3 Btu/hr per ^oF per ft^2 .

There are three implications of this comparison:

- For uninsulated vessels, ambient changes, especially sudden rain or thunderstorms, can generate major instability in hot vapor bypasses. The problem is most severe in winter and in cold climates. This echoes previous recommendations (4-6) to insulate the reflux drums when using hot vapor bypass controls.

- Sizing the bypass should be based on the total heat losses $Q_{HVB} = (Q_{cond} + Q_{loss})$ on a cold and rainy winter night. Equations 9 and 10, with the appropriate heat transfer coefficients, and based on the coldest conceivable ambient and subcooled temperatures, can be used to avoid gross oversizing of hot vapor bypass valves. For conservative design, the value calculated can be multiplied by a safety factor of 2 or even 3.
- For insulated vessels, a much smaller bypass is required.

LOSS OF CONDENSER HEAT TRANSFER CAPACITY

Lieberman (17, 22) reports cases in which leakage through a hot vapor bypass, even when fully shut, largely lowered condenser heat transfer capacity. In one case (17), blocking-in the bypass raised condenser capacity by 50%, circumventing the need to use a larger condenser that was already purchased.

The above discussion can provide valuable insight to the cause of the problem reported by Lieberman. As stated earlier, the heat supplied to an uninsulated drum, and therefore the size of the control valve, is primarily dictated by the ambient heat losses during periods of cold, rainy (or snowy) weather. Case 2 below describes a case where the bypass valve was not large enough to provide enough vapor when cooling water was cold.

During periods of calm and warm weather, the heat losses may be an order of magnitude less than during cold, stormy periods. During these warm and calm periods, the control valve will tend to be oversized, operating almost shut. A nearly-shut control valve promotes leakage and unstable control. Per the above discussion, the key to prevent control valve oversizing is adequately insulating the drum.

The leaking valve that limited condenser capacity in Lieberman's case was a butterfly valve (17). This type of valve is prone to leak, especially when oversized. Lieberman highlights the importance of designing the bypass to prevent excessive leakage. The valves used for hot vapor bypasses should be tight shut-off. As an additional measure, a small manual bypass with a throttling valve, as well as isolation block valves in the hot vapor bypass, will permit servicing the control valve should it leak. A check for leak can be performed by opening the manual bypass, shutting the control valve, and using a surface pyrometer to check whether the hot vapor bypass remains hot.

SOME OPERATING EXPERIENCES

CASE 1. Condenser outlet throttling mitigates instability

Installation In this column, the hot vapor bypass was correctly configured (as in Fig.1). The condenser liquid level was 10-12 ft below the liquid level in the drum. The drum was not insulated. The control usually worked well when the tower overhead flow rate (reflux plus product) was less than 25 KBPD. Beyond 25 KBPD, the drum would often suddenly fill up. The drum level controller would increase reflux rate. The drum level would then dive, the reflux rate plummet, and the reflux pump cavitate. The only way to restore stability was to cut back reflux. The problem was more severe in summer. This restricted the column capacity.

A similar instability at times occurred at lower rates. Figure 5 shows the operating charts for such an event. During this event the reflux plus product rates were at 17 KBPD. At about 10:40 PM the hot vapor bypass differential pressure jumped from the normal 4.5 psi to 7 psi in about 1 minute, and to 12 psi within 5 minutes. Over the same 5 minutes the tower pressure declined only slightly from 289.5 psig to 288.5 psig, raising the valve opening from 50% to 80%, while the drum liquid level jumped from 40% to 80%. Over the next 15 minutes, the pressure drop, valve opening and drum liquid level returned to their pre-event values, while the column pressure slightly rose to 290.5 psig. Throughout the event there was little change to the boilup, reflux and product flow rates and to the condenser outlet temperature.

Testing Pressure gauges were installed and measured dP across the condenser. Field measurements taken during an instability at high reflux and product rates showed that the condenser pressure drop rapidly fell (due to reduced head) and condensate temperature quickly rose as the liquid slugged from the condenser to the drum, suggesting that the condensate seal in the condenser could have been broken.

Hydraulic analysis We elected to perform our hydraulic analysis for the event in Figure 5, in which the reflux and product flow rates were low and the condensate seal was not broken, as evidenced from the condenser outlet temperature remaining unchanged. For the pressure balance, we applied Eq. 3A which included the line losses, which in this case were significant. Calculated pressure drops are in Table 2. Table 2 shows a 1.1 psi discrepancy, probably due to inaccuracy in the measurement of the pressure drop across the control valve or in the calculations.

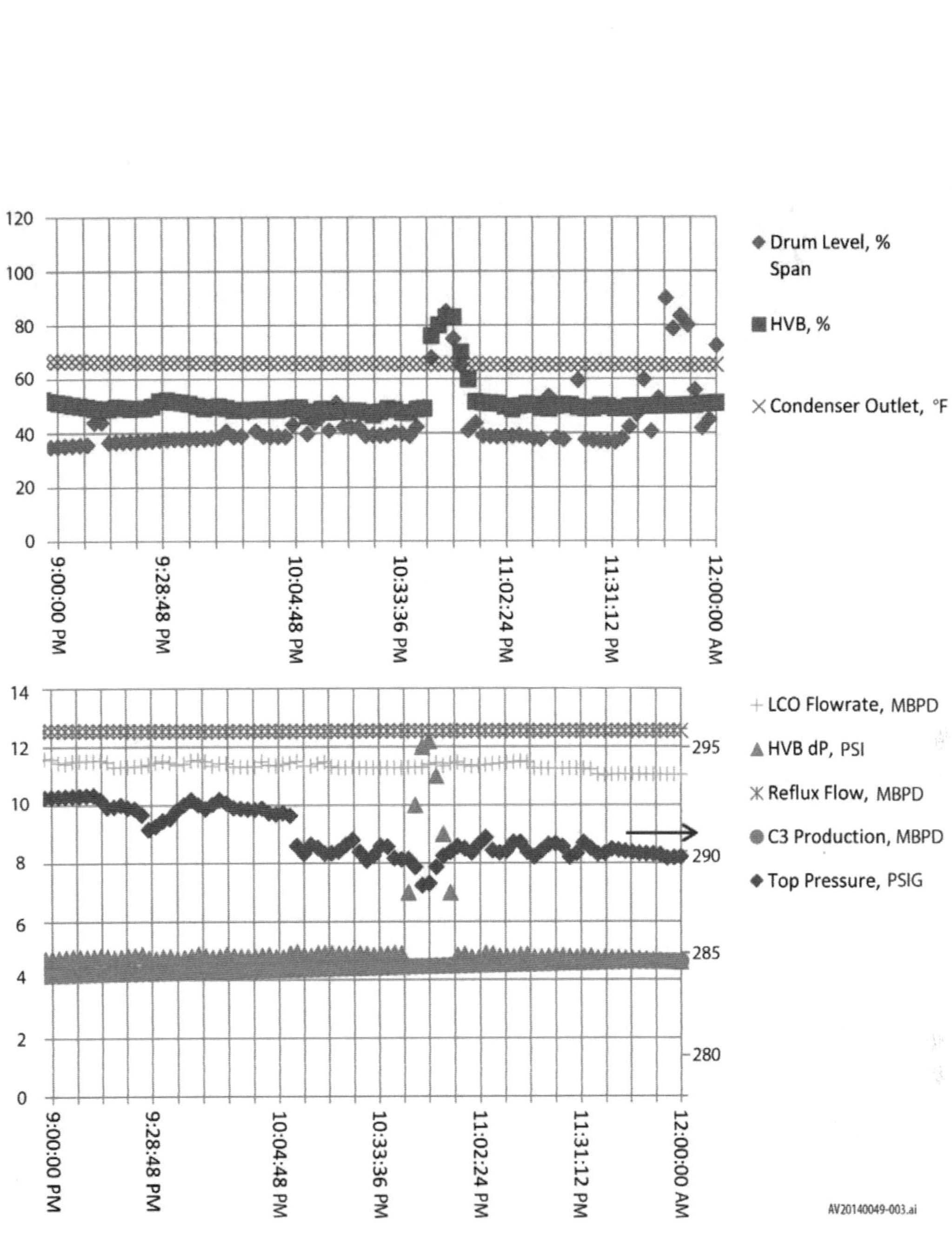

Figure 5. Tower Instability Event Operating Chart

TABLE 2. PRESSURE BALANCE FOR CASE 1

Term	Left Hand Side of Eq. 3a	Right Hand Side of Eq 3A
ΔP_{bypass} psi	4.5	
$\Delta P_{bypass\ line}$ psi	0.1	
ΔP_{cond} psi		0.5
$\Delta P_{liq\ line}$ psi		0.4
$\Delta P_{vap\ line}$ psi		0.5
H, Liquid head, psi		2.1
Total, psi	4.6	3.5

Figure 6 is a plot of flow versus valve opening at various valve pressure drops based on the control valve characteristics and the physical properties for the vapor. Figure 6 shows that opening the valve from 50% at 4.5 psi pressure drop to 80% at 12 psi pressure drop increased the vapor flow rate through the valve from 5,500 lb/h to 31,000 lb/h. Over the 5 minutes event, with approximately a linear increase in valve opening and pressure drop per Figure 6, the additional vapor that entered the drum (above the normal rate) was about 1000 lb.

The drum is 9 feet ID and just over 20 feet long (tangent to tangent). To raise the drum liquid level from 40% to 80% on the level transmitter takes 21,000 lb. Adding this quantity of liquid to the drum over 5 minutes is a tremendous movement of liquid, increasing the quantity of liquid entering the drum by a factor of 3, from 17 KBPD to about 50 KBPD. The key to understanding the mechanism is to determine where this huge amount of liquid can come from.

Mechanism From the control valve calculation above, the additional amount of vapor entering the drum would account for about 1000 lb. This may be augmented by condensation of the vapor present in the drum vapor space, which is replaced by the liquid during the event. Condensation of the drum vapor will account for another 2000 lb. This leaves 18,000 lb to be explained.

Another explanation may be the onset of heavy rain. We do not have concise records of the ambient conditions at the refinery at the time of the event, but records from the nearby New Orleans airport show light rain throughout the evening, with periods of easing off and others of intensifying. Ambient temperature was about 63 °F and wind was about 4 miles per hour occasionally gusting to 20 miles per hour.

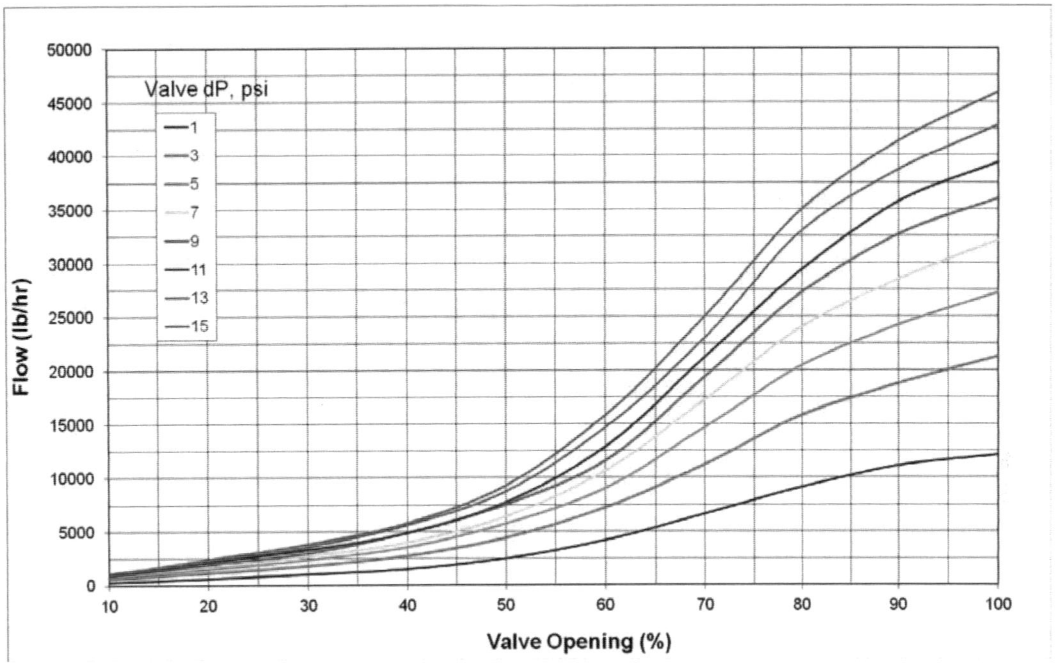

Figure 6. Valve Vapor Flow vs. Control Valve Opening

The incremental amount of vapor condensation in the drum upon heavy rain can be calculated from Eq. 10. As stated earlier, ambient heat transfer coefficients for uninsulated drums vary from about 2-3 Btu/hr per °F per ft^2 at dry, windless conditions to about 15-20 Btu/hr per °F per ft^2 during rainy and windy conditions. Based on a change in heat transfer coefficient of 15 Btu/hr per °F per ft^2 upon the onset of a sudden heavy rain we calculate an increase in condensation rate of about 3000 lb/h. This would diminish to about 2000 lb/h when the drum liquid level rose to 80% as the drum shell area exposed for condensation, A_{DW} in Eq. 10, is diminished. Overall, a heavy rain accounts for additional liquid generation of 200 lb, which does little to explain the 18,000 lb liquid movement.

One conceivable source remains: liquid from the condenser sucked into the drum. There are two condenser shells in parallel, each 46 inches in diameter and 20 ft long. Allowing for the tube volume, the condensers hold approximately 8,000 lb of liquid when full. We estimate that during the low rate operation of Fig. 5 (17 KBPD reflux plus product), and at the relatively cold ambient temperatures, only about 25% of the condenser heat transfer area performed condensation. So the condensers contained only about 6,000 lb of liquid. By itself, this falls short of explaining the large liquid movement of 18,000 lb.

However, here there is a large augmenting factor here. As liquid is sucked out of the condensers, additional tube area is exposed, increasing condensation. Emptying all the liquid from the condensers would quadruple the condensation area. The liquid generated by raising the condensation area from the 25% at the beginning of the event to 100% five minutes later is calculated at about 16,000 lb. This liquid together with the 6,000 lb drained from the condenser exceed the 18,000 lb increase in drum liquid. This means that some liquid remained in the condensers, retaining their liquid seal and the subcooled outlet temperature, as has been observed in Fig.5.

The intensified condensation rate dropped the tower pressure. To re-instate the tower pressure the hot vapor bypass valve opened rapidly as the pressure fell. Eventually the hot vapor caught up with the disturbance. The drum liquid surface heated up, the drum vapor pressure rose, the differential pressure across the valve declined, the liquid returned to the condenser, column pressure went back up, and normal operation resumed.

During the event in Fig.5, the low rate operation and the low cooling water temperature (as can be inferred from the low condenser outlet temperature) led to a relatively large liquid inventory in the condensers and sufficient condensation capacity to cushion against totally emptying the condenser. Had the rates been higher, or the ambient temperature warmer, the liquid inventory in the condensers would have been smaller, and it would have been much easier to lose the condenser liquid seal. This was the experience described earlier at high reflux plus product rates and especially during summer. These were also the operating conditions during our tests. Losing heat transfer upon loss of condensate seal is a common problem in condensing sides of reboilers (2, 6, 23).

Initiation Up to this point, we followed that movement of liquid produced by a strong suction. However, we did not address the cause of this suction. To move 18,000 lb of liquid in 5 minutes takes a very strong suction force that needs to be identified.

Figure 5 shows a fast rise in control valve differential pressure, i.e., of the difference between the tower and the drum pressures. Since the tower pressure changed only slightly (Fig. 5), the big change was in the drum pressure, 8 psi over five minutes. The drum pressure is the vapor pressure of the liquid surface in the drum. An 8 psi vapor pressure reduction corresponds to cooling of the liquid surface of the drum by only 2.5°F (Table 1). The problem now is to identify the source of a 2.5°F cooling of the drum surface temperature.

One possibility is heavy rain. However, as was stated earlier, the heavy rain would have increased the condensation rate only by 2000 - 3000 lb/h, easily offset by the additional 25,000 lb/h flow rise through the hot vapor bypass during the same period. Further, there were many other events (including the tests) that took place under dry conditions.

Another possibility is hot vapor impingement ruffling the liquid surface. However, at the beginning of the event the vapor velocity was about 6 ft/s, giving a $\rho_V V^2$ of 100 lb/ft s^2, too low to ruffle the liquid surface 5 feet below (where ρ_V. is vapor density, lb/ft^3, and V is vapor velocity at the drum inlet nozzle, ft/s). Also, this vapor flow rate was commonly used during operation without any problem.

During the first minute of the event, the valve pressure drop rose from 4.5 psi to 7 psi, corresponding to a drop in drum surface temperature of 1 °F. At the drum pressure of 285 psig the liquid surface was at 125 °F, while the bulk subcooled liquid in the drum was at about 66 °F. As little as 1.5 percent of the subcooled liquid reaching the drum liquid surface will explain a temperature fall of 1 °F and a consequent pressure fall of 2.5 psi.

This brings us to the liquid entry. The liquid enters via a 6" nozzle at the bottom of the drum, discharging upwards at 5.7 ft/s at the beginning of the event, a velocity exceeding the good design practice of 4 ft/s max. Normally, this would not have been an issue, but here, with the liquid level was only 3.7 ft above, the initial momentum may have carried some of the subcooled liquid jet it to the surface, possibly piercing the hot liquid surface. Some of this may even occur during normal operation, with some surface cooling by the rising jet balanced by some of the hot vapor.

Instability may initiate upon step-up of the subcooled liquid reaching the surface. Fig. 5 shows that just prior to the pressure differential rise, the tower pressure fell by about 0.5 psi and the drum level rose by about 2-3%. Both indicate a step-up in condensation rate, possibly initiating by heavy rain hitting the drum vapor space. The drum pressure fell, sucking liquid from the condenser. A level rise of 2% is equivalent to about 1,000 lb. Sucking 1,000 lb from the condensers would raise the exposed condensation area by around 50%, quickly reducing tower pressure. The bypass pressure drop remained constant at that time, meaning that both the drum and tower pressures fell by the same 0.5 psi. The additional liquid flow, about 1,000 lb in two minutes, would increase the drum liquid inlet velocity from 5.7 ft/s to 7 ft/s. The intensified jet raised more subcooled liquid onto the liquid surface, generating a "fountain effect" that cooled the surface. The drum pressure fell, sucking more liquid from the condenser, the jet and the fountain intensified, the surface further cooled, more liquid was sucked from the condenser, and so on. This process is self accelerating. At the pressure differential peak, the liquid rate tripled, the liquid jet velocity exceeded 15 ft/s, high enough for the jet to break right through the liquid surface, even at the higher level of 80%. .

At the pressure differential peak, the hot vapor bypass flow reached 31,000 lb/h, the vapor inlet velocity 34 ft/s, and the $\rho_v V^2$ 3,300 lb/ft s^2, while the level rose (to 80%) to 2.1 ft below the vapor inlet. Somewhere along the event, vapor impingement on the liquid surface would have begun raising more subcooled liquid to the surface, augmenting the instability.

Solution The low friction pressure drop in the condenser and its inlet vapor and outlet liquid lines, a total of 1.4 psi (Table 2), offers little hydraulic resistance to counter the self-accelerating fountain effect. Increasing the pressure drop in the lines leaving the condenser can mitigate the self-accelerating process. As the liquid flow from the condenser is sucked into the drum, the additional friction pressure drop rapidly increases. The line pressure drop increases with the square of the flow, and severely limits that increase in liquid flow to the drum, giving the hot vapor a chance to catch up. The throttling valve should be in the liquid line and more than 10 line diameters away from the drum liquid inlet to avoid increasing the liquid inlet velocity and the fountain effect.

To overcome, a manual throttling valve was installed in the line from the condenser to the drum, as in Figure 3. The dP across the valve is measured locally, and maintained at 2 psi. The operators check the differential pressure across the valve and adjust it as needed. This completely eliminated the problem. The modified system can now run the reflux plus product up to even 40 KBPD during both summer and winter with no instability.

Epilogue Adding the throttling valve raised the hot vapor bypass pressure drop from about 1.4 psi to 3.4 psi. The throttling valve slows the rate of drainage from the condenser, making it more difficult to suck the condenser liquid into the drum.

This case brings to light an instability not previously reported in hot vapor bypass systems. The instability develops because different factors govern the valve pressure drop and the valve opening. The valve pressure drop is largely governed by the drum surface temperature. The valve opening is governed by the tower pressure controller. The two may vary independently and interact. Such interaction can be initiated by sudden cooling to the liquid surface in the drum. The vapor pressure at the drum liquid surface quickly drops, while the tower pressure, and therefore, the valve opening, remain temporarily constant. A good solution is to increase the pressure drop in the outlet of the condenser (Fig. 3). A good practice is to have a total friction pressure drop (not including the liquid head) of not less than about 4 to 5 psi.

Prevention can also be possibly achieved by keeping the liquid inlet velocity low, and/or installing a horizontal baffle, or better, a stilling chamber, above the liquid inlet, or extending the liquid inlet into the drum and directing it horizontally. This, however, is solely based on the hydraulic analysis above and has not been incorporated in this case. Another good practice would be also to install a horizontal baffle in front of the vapor nozzle, to disperse the vapor flow and not allow it to impinge on the liquid surface upon intensification.

CASE 2. Tempered water and condenser outlet throttling mitigate instability

This was a sat gas deethanizer, operated at 400 psig at the top. The pressure control in this tower was again by a well-configured hot vapor bypass system, similar to Fig.1. Here too the liquid drum level was a relatively short distance, 6 ft, above the liquid level in the condenser. In this column, the pressure fluctuated. The hot vapor bypass would go wide open when the cooling water was cold. The drum was not insulated. A pressure survey showed that the pressure differential across the hot vapor bypass was only about 1.5 psi. The condenser pressure drop was about 0.75 psi, and so was the liquid head.

Calculations for this case show that the liquid head of about 6 ft is equivalent to about 0.75 psi, and the condenser pressure drop another 0.75 psi. In this system, the rate of liquid head change with liquid surface temperature would be even larger than for the C_3- C_4 Splitter in Case 1, meaning that small temperature changes of the drum liquid surface lead to large changes of the condenser liquid level. There was little hydraulic resistance in the condensate line, so the liquid level in the condenser would easily fluctuate with even slight overshooting of the drum liquid surface temperature. This would cause heat transfer to fluctuate and result in the tower pressure swings.

Another issue in this case is that the control valve was not large enough to supply the heat required when the cooling water was cold and therefore the drum liquid was highly subcooled.

To address the problem, a booster pump was added to recycle some of the cooling water from the condenser outlet back to the inlet similar to the system depicted in Case 28.8, p.387 of "Distillation Troubleshooting" (2). This helped, similar to the improvement reported by Souza (26). An even larger improvement was achieved by installing a throttling valve at the condenser process outlet as in Fig. 3 and Case 1, and controlling a small differential pressure (about 1.5 psi) across it. No more pressure fluctuations after this.

CASE 3. Winter instability with air and cooling water condensers in series

This naphtha splitter tower (Fig. 4) had an air condenser in series with a cooling water condenser. The reflux drum was insulated. The pressure control worked well most of the year, but became jerky during winter, especially on colder days, making it difficult to keep a stable pressure in the tower.

The principles involved were addressed in the section on hydraulic imbalances. Most of the time, the liquid level was in the cooling water condenser, so the controls worked like they should. During colder days in winter, the air condenser was capable of supplying more than the total condensation duty, so the liquid level was in the air condenser. When this happened, the control became unstable as described earlier.

The problem was solved by turning off fans and shutting louvers on the air condenser, and setting up the air condenser outlet temperature control (that manipulated the speed of the fan). This lowered the duty in the air condenser and made sure that the liquid level always remained in the water condenser.

ACKNOWLEDGEMENT

The authors wish to thank Chad Jones (presently at Motiva) for his assistance in gathering the field data for this paper. The authors also thank Mark Murphy from Fluor for his control valve characteristics analysis and Dr. Walter Stupin from Fluor for his invaluable input to the hydraulic analysis.

REFERENCES

1. Kister, H. Z., "What Caused Tower Malfunctions in the Last 50 Years?", Trans. IChemE, Vol 81, Part A, p. 5, January 2003.
2. Kister, H. Z. "Distillation Troubleshooting", Wiley – Interscience, NJ, 2006.
3. Whistler, A. M., "Locate Condensers at Ground Level," Pet. Ref. 33(3), 1954, p. 173.
4. Hollander, L. "Pressure Control of Light-Ends Fractionators," ISA J. 4(5), 1957, p. 185.
5. Chin, T. G., "Guide to Distillation Pressure Control Methods," Hydrocarbon Proc. 58(10), 1979, p. 145.
6. Kister, H. Z., "Distillation Operation", McGraw-Hill, NY, 1990.
7. Smith, C. L., "Distillation Control – An Engineering Perspective", Wiley, NJ, 2012.
8. Kister, H. Z., and J. F. Litchfield, "Distillation: Diagnosing Instabilities in the Column Overhead", Chem. Eng., p.56, September 2004.
9. Buckley, P. S., W. L. Luyben, and J. P. Shunta, "Design of Distillation Column Control Systems", Instrument Society of America, Research Triangle Park, NC, 1985.
10. Shinskey, F. G., "Distillation Control for Productivity and Energy Conservation", 2d ed., McGraw-Hill, New York, 1984.
11. Lupfer, D. E., "Distillation Control for Utility Economy", Proceeding of the 53rd Annual Convention, Gas Processors Association, March 25-27, Denver, Colorado, p.159, 1974.
12. Nisenfeld, A. E., and R. C. Seemann, "Distillation Columns", Instrument Society of America, Research Triangle Park, North Carolina, 1981.
13. Schneider, D. F., and M. C. Hoover, "Practical Process Hydraulic Considerations", Hydroc. Proc., p. 47, August 1999.
14. Duguid, I., "Take this Safety Database to Heart", Case MS16, Chem. Eng., p. 80, July 2001 (includes database at www.che.com/CEEXTRA).
15. Sloley, A. W., "Simple Methods Solve Exchanger Problems", Oil & Gas Journal, p. 85, April 20, 1998. Durand, A. A., "Sizing hot Vapor Bypass Valve", Chem. Eng., August 25, p.111, 1980.
16. Laird, D., and J. Cornelisen, "Control-System Improvements Expand Refinery Processes", Oil & Gas J., p.71, September 25, 2000.
17. Lieberman, N. P., "Troubleshooting Process Operations", 4th Ed., PennWell Books, Tulsa, OK, 2009.
18. Lieberman, N. P. "Process Equipment Malfunctions", McGraw-Hill, NY, 2011.
19. Durand, A. A., "Sizing hot Vapor Bypass Valve", Chem. Eng., August 25, p.111, 1980.

20. The American Petroleum Institute, "API Technical Data Book", 7th. Ed., The American Petroleum Institute and EPCON International Publishing Services, Washington, DC, 2005.

21. Kern, D. O., "Process Heat Transfer", McGraw-Hill, NY, 1950.

22. Lieberman, N. P. "Troubleshooting Process Plant Control" Wiley, New Jersey, 2009.

23. Kister, H. Z., and T. C. Hower, Jr., "Unusual Case Histories of Gas Processing and Olefins Plant Columns," Plant/Operations Prog. 6(3), 1987, p. 151.

24. Hartman, E. L., and T. Barletta, "Reboiler and Condenser Operating Problems", PTQ, p. 47, Summer 2003.

25. Sloley, A. W., "Effectively Control Column Pressure", Chem. Eng. Progr., p. 38, January 2001.

26. Souza, L. L. G., "Model Devised for Plant Hot-Gas Bypass Systems", Oil & Gas J., 1.118, Sept 6, 2010.

349989 Moving Gamma Scanning Forward – Quantitative Analysis of Tray Capacity

Abstract:

Improving process operating margins can include maximizing production of higher value products and/or minimizing production costs. Process plant operations depend heavily on fractionation towers to produce products meeting customer specifications. Tower gamma scanning is well established in the process industries as a qualitative tool to help troubleshoot fractionation towers. Using new measurement detector technology and innovative analysis of gamma scanning data quantitative information is now available about the useful capacity of trayed fractionation towers.

For trayed towers a new methodology, FrothView™, has been developed and tested whereby the total froth height on a tray is measured from gamma scan data using new detector technology and software interpretation. The "inspiration" for this development was taken from a presentation given by Mr. Mike Resetarits, Technical Director of Fractionation Research Inc. (FRI), at the 2011 A.I.Ch.E. Spring Meeting titled "On Distillation Tray Entrainment". The author remembered from the presentation that the degree of entrainment is very difficult to quantify. If the degree of entrainment cannot be reliably quantified then it is very difficult to use entrainment in tray correlations particularly to measure tray capacity.

Most, if not all, tray capacity correlations use a calculated tray froth height or level. Up to now there are few practiced methods for measuring a liquid or froth level on a tray from gamma scan data. However; no matter the practitioner the conventional methods allocate a tray's aerated liquid height somewhere in the middle of the layer. This method is adequate for diagnosing tray damage – whether a tray is capable of holding liquid or not. But beyond this determination, conventional tray liquid level measurement has no correlation to any other operating variable.

A couple of years ago Tracerco had the opportunity to scan a trayed pilot plant tower under varying operating conditions, from weeping to flooding. One objective of this study was to find a method to quantify the scan results to provide a more objective diagnosis of the tray's hydraulic condition. The experimental observations, tray model predictions, and total froth height measurements from gamma scan data using a new detector technology had excellent agreement.

Case studies will demonstrate the beneficial difference of measuring total tray froth heights from gamma-scan data versus the conventional method of measuring the mean tray liquid height.

How Sensitive Is Your Treating Plant to Operating Conditions?*

Ralph Weiland[1], Jesse Santos[2], Attilio Praderio[3], Nilia Maharaj[4], and Michael Schultes[5]

[1]Optimized Gas Treating, Inc., 12337 Jones Rd., Houston, TX 77070
[2]INEOS GAS/SPEC Technology Group, INEOS Americas LLC, 10777 Westheimer, Suite 210, Houston, TX 77042
[3]LNG Technology Development, Conoco Phillips Company, 3000 Post Oak Blvd, Houston, TX 77056
[4]Atlantic LNG Company of Trinidad and Tobago, Port of Spain, Trinidad
[5]Raschig GmbH, Mundenheimer Straße 100, D-67061 Ludwigshafen am Rhein, Germany

Abstract

Normal expectations in operating amine treating plants for acid gas removal is that small changes in operating conditions will result in correspondingly small responses in plant performance. However, such expectations are not always well founded. To establish credibility for the process simulator used in the design of a new LNG plant (ProTreat®), we begin by comparing its mass transfer rate-based simulation results with performance data from an LNG plant operated by Atlantic LNG Company of Trinidad and Tobago (Atlantic LNG).

Attention is then turned to another LNG project currently under study. For this system under consideration, our focus is on the CO_2 removal system which consists of a single, large absorber serviced by two regenerators in parallel. Initially the plant would process about 1,400 MMscfd of gas containing about 16 mol% CO_2. A sensitivity analysis leads to some interesting observations regarding potential susceptibility of treating performance to departures of certain operating conditions from design values, and provides reasons for this sensitivity.

Background

There are a number of well-accepted limits on values of plant operating parameters. Examples are corrosion considerations for carbon steel which usually limit rich amine CO_2 loadings below about 0.4 – 0.45 moles of CO_2 per mole of amine, and upper limits on maximum line velocities to prevent the direct scouring of surfaces and the removal of protective films, particularly sulfide layers. Tower internals have natural limitations on gas and liquid rates beyond which jet flooding or downcomer backup and choke flooding of trays, or packed column flooding, occur. These are natural hydraulic capacity limits. Finally, solvent capacity is itself limited by temperature and acid gas partial pressures.

Corrosion, temperature, and acid gas pressures can all limit rich solvent loadings, i.e., solvent capacity, and limitations on tower hydraulic capacity restrict throughput. These limitations restrict performance but without causing over-sensitivity of performance to small changes in the values of operating parameters. However, the normal small-response-to-a-small-stimulus paradigm changed with the advent of very fast-reacting solvents such as piperazine for CO_2 removal.

The development and availability of highly precise simulation tools, especially tools such as the ProTreat® simulator which is based on the direct calculation of *mass transfer rates*, has encouraged the design and construction of new plants with less and less design margin. It has also allowed engineers to

* Paper presented at AIChE Spring meeting, New Orleans, LA, 31 March 2014.

assess precisely the effect of operating parameters on performance, and it has revealed the existence of operating cliffs or points of instability on the performance map[1]. Although regions of increased sensitivity have been predicted even for moderately fast-reacting CO_2-MEA absorbers, when piperazine is used to promote MDEA, reactions rates become very large indeed, and unexpectedly high sensitivity of performance has been observed. None of this has been predictable using the more traditional ideal-stage simulation tools even when they have been modified with efficiencies and even with attempts to force reaction kinetic rates into what is fundamentally an equilibrium model. What enables the mass transfer rate model to reveal what has for so long been hidden behind the façade of the ideal stage?

A real absorber contains a certain number of actual trays or a depth of a specific packing intended to promoted contact across the interface between the counter-currently flowing vapor and liquid phases. The flows are always to some extent turbulent, and the turbulence level depends on the tray or packing design, and the fluid properties and flows. Turbulence affects the absorption rate because it affects the mass transfer coefficients that prevail within the phases. In parallel with heat transfer coefficients in various types and designs of heat exchanger, mass transfer coefficients for a wide range of tray and packing types have been well correlated with design details, flow parameters, and properties. In other words, the mass transfer characteristics of tower internals are well understood and well established—there is no guesswork. Of course, absorption rates also depend on concentration *differences* between the phases. The separation actually achieved by a real tray or a certain depth of real packing depends directly on the absorption rate of the component as dictated by the mass transfer characteristics of the internals. The model is completely integrated with the real-world equipment and how it is operated. The ideal stage concept, on the other hand, replaces every important detail with the single, simplifying assumption of equilibrium between the phases. Indeed there are no composition differences at all between phases, and therefore there is really no reason for a separation! Regardless of applying efficiencies or finessing the equilibrium assumption in any other way, the ideal stage assumption eliminates reality from the calculations, leaving a model that is not only unable to perceive critically important process detail, but which is also vulnerable to gross error.

Case Studies

This article consists of two case studies. In the first study, the question, "How closely can one predict the performance of an LNG plant's CO_2 removal section?" is answered by comparing performance predictions with operating data from one of the LNG trains at Atlantic LNG in Port Fortin, Trinidad, the sixth largest LNG facility in the world in annual LNG production. The second case study is an analysis of certain aspects of the CO_2 removal section of a project currently under consideration. The case study examines the sensitivity of the design to operating and design parameters.

Atlantic LNG

Diglycolamine Agent® (DGA®) is used at 41 wt % to treat inlet gas with the composition shown in Table 1. The process flow diagram (Figure 1) is fairly conventional, although mention should be made of the fact that a substantial portion of the reflux water from the stripper overhead condenser is mixed with process makeup water and returned to the top of the absorber rather than to the regenerator. At the top of the absorber, four wash trays recover DGA Agent from the treated gas. There are 20 contacting trays in the absorption section.

The rich amine is flashed to remove hydrocarbons, cross-exchanged with hot lean amine, and sent to the regenerator. The regenerator contains three wash (reflux) trays and 17 stripping trays. The reboiler is energized using hot oil.

Table 1 Composition of Gas to CO_2 Absorber

CO_2 (mole %)	0.34
Methane (mole %)	95.5
C2+ (mole %)	4.1
Nitrogen (mole %)	< 0.1

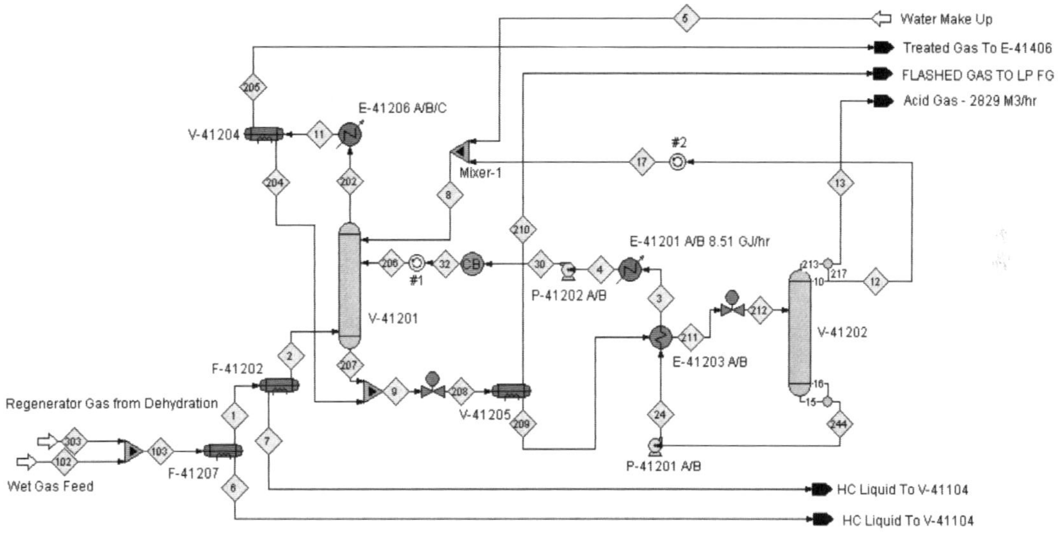

Figure 1 Simplified PFD for the Atlantic LNG Train-4 CO_2 Absorption System

Absorber temperature and composition profiles indicate that the tower is mass transfer rate controlled. Simulated temperatures of various streams throughout the plant are compared with measured data in Table 2. Stream numbers refer to those shown in Figure 1.

Table 2 Actual vs. Simulated Stream Temperatures

Stream	Actual Temp (°C)	ProTreat (°C)	Difference (°C)
103	24.8	24.9	-0.1
202	32.5	33.9	-1.4
206	43.3	44.1	-0.8
207	24.2	23.5	0.7
211	81.8	81.8	0.0
213	111.9	111.8	0.2
13	44.5	44.5	0.0
24	126.7	126.0	0.7

Except for the treated gas temperature, the simulated temperatures of all other streams match plant measurements to better than 1°C; the treated gas differs by only 1.4°C. The measured lean amine loading was 0.04 moles CO_2/mole DGA while the simulated value was 0.036 moles CO_2/mole DGA. These values are almost identical, validating the accuracy of the regenerator simulation. The treated gas CO_2 level was measured at 25 ppmv, whereas, the simulated value was in the range 50–60 ppmv. This is certainly more discrepancy than we are used to seeing. However, the feed gas is known to contain heavy ends, and plant personnel have identified the presence of foaming in the column. Even a small amount of foaming in an otherwise perfectly operating system can increase the vapor-liquid area for mass transfer by 10 or 20%, and this is sufficient to give a predicted value of 25 to 30 ppmv of carbon dioxide.

LNG Project Under Consideration

This plant is in the initial study phases. As such, the name of the project and location will not be disclosed. However, the study phase is the time to perform sensitivity studies and determine (1) that it will be operable over the entire range of expected conditions and (2) whether there are regions in which the operation of the plant might be overly sensitive to one design parameter or another and, if there are, that it operates well away from such regions (or at least has contingencies in place to ensure stable operation can be maintained). Regarding carbon dioxide, the raw gas to the plant under consideration is at the opposite end of the spectrum from the Atlantic LNG facility. The carbon dioxide concentration in the raw gas is about 45 times higher, nominally 16 mole %. Figure 2 shows a simplified PFD. A simplified gas analysis with relevant conditions is provided in Table 2. The absorber has been designed with 3 one-pass valve trays to act as wash trays and assist in the recovery of any vaporized INEOS® GAS/SPEC® CS-1160 solvent from the treated gas. This short wash section swages into the absorption section. The main part of the absorber was designed with two identical 4-metre deep beds of No. 1.5 Raschig Super-Rings®, selected for their excellent hydraulic capacity and very high specific surface area[2,3]. In tests conducted by Fractionation Research, Inc. (FRI), this packing combines high throughput with excellent mass transfer performance. At 60% of flood, the absorber diameter is just over 25 feet, which at 48 barg is a substantial tower shell by any measure.

Table 2 Study Case LNG Feed Gas to the CO_2 Absorber

Temperature (°C)	**46**
Pressure (barg)	54
Volume Flow (kNm3/h)	1 400
Composition (Dry Basis)	
CO_2 (mole %)	16
Methane (mole %)	79
Ethane (mole %)	3
Propane (mole %)	1
Butanes (mole %)	0.5
Other (mole %)	0.5

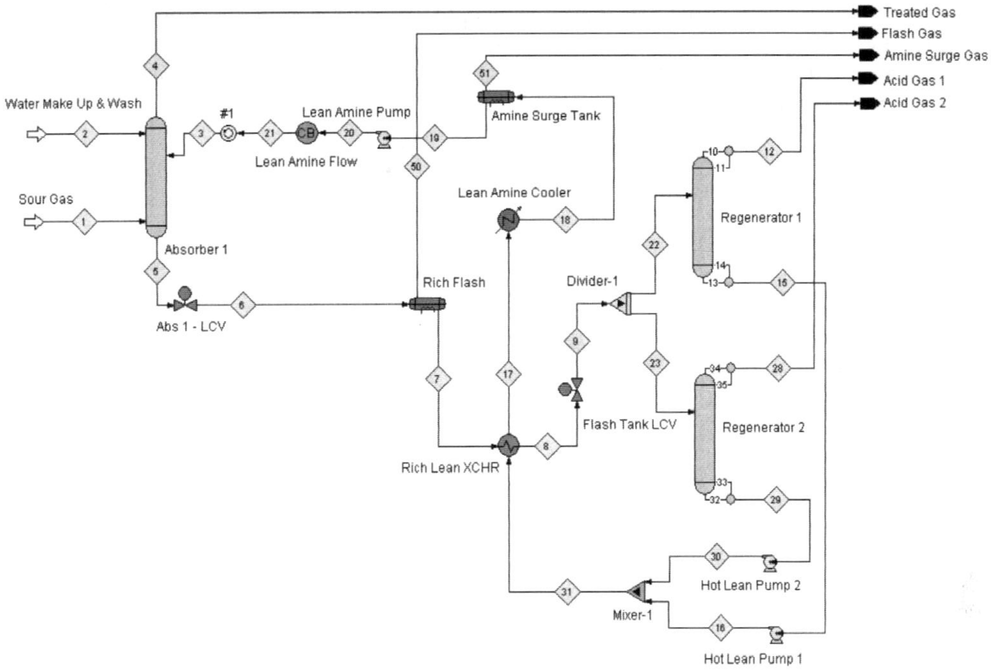

Figure 2 Simplified PFD for the Study Case LNG CO_2 Absorption System

The regenerator's design stripping ratio is 0.8 (stripping ratio is the ratio of water vapor to acid gas in the overhead vapor line going to the condenser). The solvent circulation rate is set to achieve a rich amine loading of 0.5 moles of CO_2 per mole of amine in the solvent. At the design lean amine temperature of 115.8°F into the absorber, the treated gas was simulated to contain 21.4 ppmv of CO_2.

A study of how overall performance might respond to variations in design and operating parameters revealed that, for the most part, the carbon dioxide content of the treated gas was remarkably insensitive, so the design appeared quite solid from an operational standpoint. However, performance was found to be sensitive to (a) the depth of packing in the absorber, and (b) the temperature of the lean amine.

Sensitivity to Absorber Packing Depth

Figure 3 shows the simulated effect of packed bed depth on treating performance as measured by the carbon dioxide level in the treated gas. Note the logarithmic scale on the concentration axis. The design point of an 8-m deep packed bed appears to give a comfortable margin away from the 50 ppmv specified limit for CO_2. However, reducing the bed depth to 7 m, a difference of only a single metre, would result in the treated gas exceeding this specification by at least 20 ppmv versus meeting it with a comfortable 30 ppmv margin. Obviously, for a tight design getting the packing depth right is critical. Indeed, for a truly safe design one must have confidence in the reliability of the simulator. If a simulator that uses ideal stages in any form is used, it must rely on the engineer having high confidence in the accuracy of the HETP value he uses to translate into reality. Unfortunately, even when it exists at all, information on HETP values in amine systems is usually very unreliable. Because the ProTreat® simulator has a real mass and heat transfer rate-basis which allows it to avoid the efficiency and HETP questions altogether, and because it has been finely tuned to a large amount of commercial plant performance data, it is one of the only commercially available simulators that can reliably get it right. The mass transfer rate model deals directly with the real internals in the tower, not with an idealization needing the band aid of an HETP.

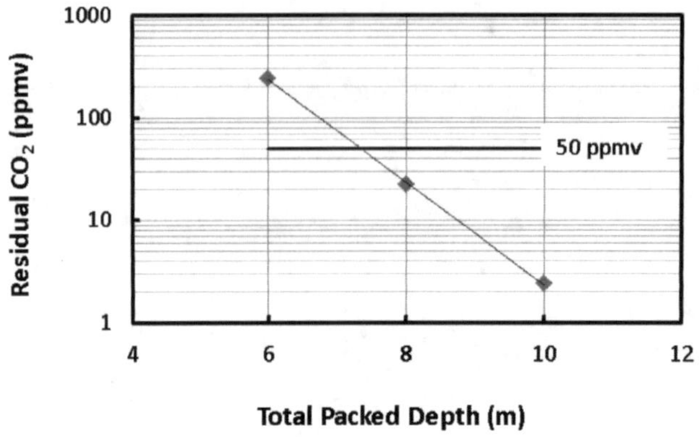

Figure 3 Sensitivity of CO_2 in Treated Gas to Packing Depth in the Absorber

Absorber performance is exponentially sensitive to packed depth, mainly because the absorber's performance in this case is controlled, not by the solvent lean loading, but by the mass transfer itself. The easiest way to understand this is by looking at the CO_2 concentration profile across the absorber, as shown in Figure 4. Detail is made visible by using the logarithm of CO_2 concentration and the temperature profile is shown for reference. The size of the temperature bulge is substantial in the bulge region of the absorber. The carbon dioxide concentration is only slowly changing there because of the high CO_2 backpressure at such temperatures. In other words, the solvent is nearing saturation towards the bottom of the absorber. However, throughout the rest of the absorber the carbon dioxide concentration continues to fall exponentially. Indeed, the CO_2 concentration in equilibrium with the lean amine is less than 1 ppmv so the absorber truly is well away from being lean-end pinched.

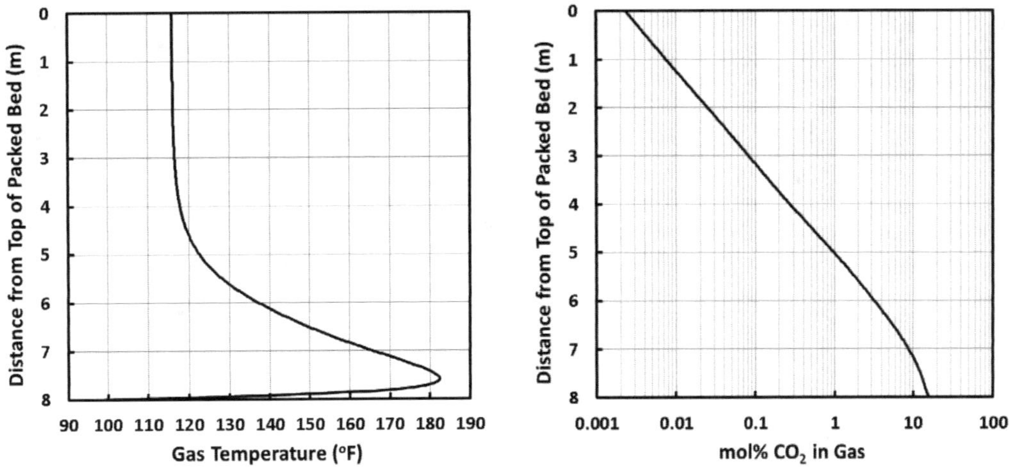

Figure 4 Temperature and CO_2 Concentration Profiles in Gas across the Absorber

Sensitivity to Operating Parameters

If the absorber goes off specification, remedial action must be taken. There are several candidates for control variables, the most obvious being: lean solvent flow rate, lean solvent temperature, regenerator reboiler duty. This absorber is sized for only 60% of flood. Therefore, the solvent rate might be an excellent control variable, at least in terms of tower hydraulic capacity (provided of course that pumps have been adequately sized). The Regenerator reboiler duty, on the other hand, is a poor control variable simply because treating is almost independent of lean loading, (provided only that the loading is low enough). Figure 5 shows the lack of sensitivity of treating to lean loading. As shown in the figure, a 15% increase in reboiler duty results in a lean loading reduction of only 0.0036 loading units and this decreases carbon dioxide in the treated gas by only 2 ppmv.

Lean solvent temperature was the final control variable considered. As the lean amine temperature is increased, one naturally expects to find temperature increases throughout the column in response. A higher temperature solvent is unable to hold as much acid gas as a low temperature one. In other words, it's net loading decreases. This is shown in Figure 6 where the CO_2 level in the treated gas and the net solvent loading are shown side by side. Between 100 and 126°F the rich loading remains constant regardless of the lean amine temperature because the solvent under those conditions is able to absorb virtually all the CO_2 (at least to the 20 ppmv or so level). However, at or above 126°F the rich loading starts to drop with increasing temperature. The ability of the solvent to "load up" with CO_2 is being compromised by it becoming too hot, and what cannot be absorbed must leave in the exiting gas. Thus the treated gas quality starts to suffer very severely. The effect is very noticeable when one is dealing with ppm specifications on the treated gas.

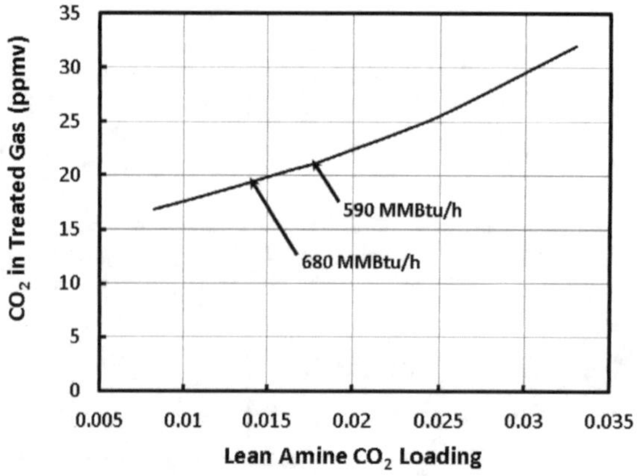

Figure 5 **Insensitivity of CO_2 in Treated Gas to Lean Amine Loading (and Reboiler Heat Duty)**

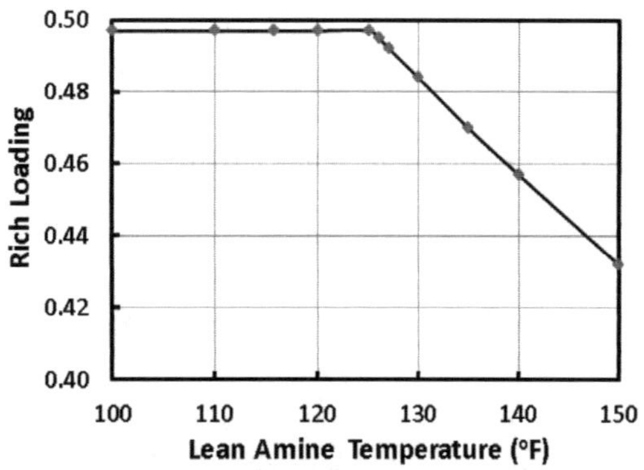

(a) How Rich Loading Varies with Lean Amine Temperature

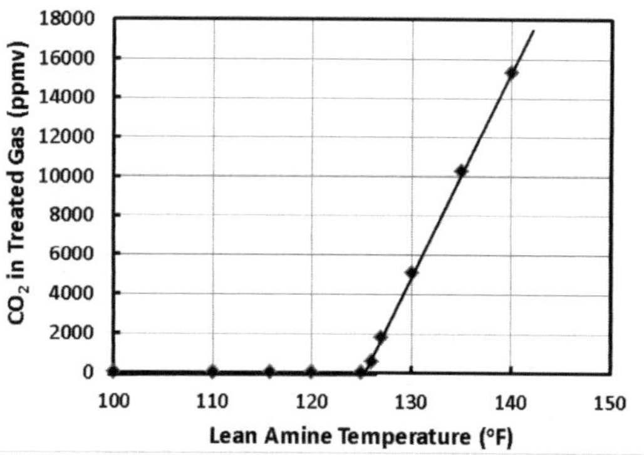

(b) Failure to Make Specification Gas when the Lean Amine is Too Hot

Figure 6 **Net Solvent Loading and Response of CO_2 in Treated Gas to Lean Amine Temperature**

It is instructive to examine how treated gas composition changes with lean amine temperature in more detail. Figure 7 is a replot of Figure 6(b) on a logarithmic basis to magnify the region where treating fails. The design temperature is 115°F (46°C), and as long as the temperature is kept within 10°F (5°C) of the design point, the plant appears to be very stably operable. However, as Figure 7 shows, once the lean amine temperature approaches 125°F (52°C), the absorber will become extremely unstable; indeed, it will become inoperable. As already discussed, the reason for this apparently remarkable sensitivity to an operating condition has to do with the capacity of the solvent—that part of the CO_2 in the raw gas cannot be absorbed because of a solvent capacity limit passes directly into the treated gas. Continuing to decrease the solvent temperature, however, does not result in continued improvement to treated gas quality. Eventually it starts to have a deleterious effect because the solvent viscosity is increasing with lowering temperature and mass transfer resistance to CO_2 absorption starts to increase.

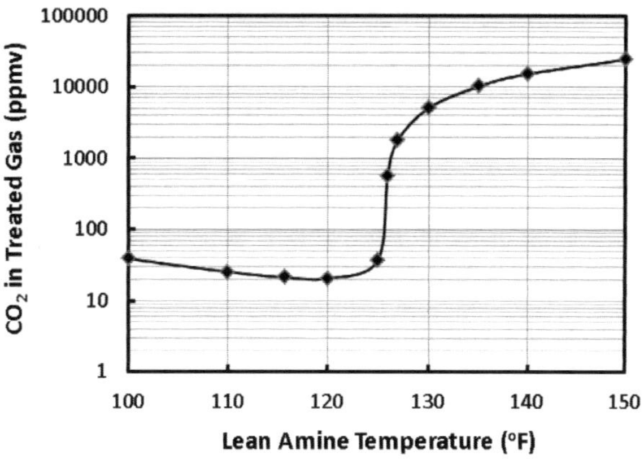

Figure 7 Extreme Sensitivity of CO_2 in Treated Gas to Lean Amine Temperature

The remaining question is: what is the best way to control the absorber if one has to operate with a lean amine that is too warm. The answer is fairly simple: manipulate the variable that will increase the solvent capacity, i.e., its net loading capacity. The choices are limited. Increasing solvent strength is always an option but it is not a control strategy. Producing a leaner solvent would be completely ineffective because lean loading has only a tiny effect on net loading capacity; in no sense is it controlling. Solvent flow rate, on the other hand, directly affects the solvent's capacity for carbon dioxide. This is really the only short-term control variable that would be effective.

Summary

The Atlantic LNG facility is very amenable to simulation using ProTreat's mass transfer rate-based approach. The simulation is truly predictive, requiring no estimates or other input beyond what is available from data sheets, P&IDs, and tower internals vendor drawings and specifications. All temperatures were matched to within better than one might expect from most thermocouple calibrations, and the simulated lean solvent loading was almost identical to measured data. The discrepancy in treated gas CO_2 content was easily explained and accounted for by a small amount of foaming suspected to be occurring in the absorber.

The Study Case LNG unit, which was designed using Raschig Super-Rings®, appears to be quite sound and stably operable, but with the proviso that the lean amine temperature must be kept between about 100 and 125°F. If there is any doubt that this can be achieved throughout the year (which may not always be the case in the region for this study case), then adequate provision must be made for operating at increased solvent flow. This requires adequate design margin in pumps, heat transfer equipment, and the regenerator itself.

The first case study (Atlantic LNG) validated the ability of the well-founded, mass transfer rate-based simulator, ProTreat®, to predict amine unit performance in LNG production with astonishing accuracy. In the second case study we have used the same model to explore the operability of a plant still on the drawing board. Without using such a simulator, it would be fairly non-obvious that there is a critical operating temperature one simply cannot exceed, and near which the absorber operation will become unstable. There may well be a lower operating temperature as well, although whether the unit will become unstable there is a moot point.

References

1. Weiland, R. H. and N. A. Hatcher, *Foundations of Failure*, Hydrocarbon Engineering, 2011.
2. Schultes, M., *Researching Rings*, Hydrocarbon Engineering, p. 57, November, 2001.
3. Schultes, M., *Raschig Super-Ring*, Trans. I. Chem. E., 81A, p. 48, 2003.

352395 "MEGA Tower" Design Considerations

Abstract:

Economic principles tell us that "Economies of Scale" drives us to larger and larger chemical plant sizes. Limits to plant sizes that were present decades ago have disappeared and the "norm" is now what we refer to as "Mega" sized plants. With this in mind, distillation tower sizes are now being designed in excess of 14 meters in diameter and towers heights are in excess of 100 meters tall. This paper deals with design parameters that need to be considered when designing such "Mega" towers. Parameters such as deflection, shop installation, support rings, bed depth and inter bed depth, beam count and sizes, vessel roundness and feed introduction will all be addressed in this paper.

Collaborations between FRI And OSU Chemical Engineering

By

James. R. Whiteley
Clint P. Aichele

School of Chemical Engineering
Oklahoma State University
Stillwater, OK 74078

Tony Cai

Fractionation Research Inc.
424 S. Squires Street, Suite 200
Stillwater, OK 74074

31 March 2014

Distillation Honors Session,
AIChE Spring Meeting, New Orleans, Louisiana

ABSTRACT

After 36 years of operation in Alhambra, CA, Fractionation Research, Inc. (FRI) relocated to the campus of Oklahoma State University (OSU) in Stillwater, OK in 1990. The potential for collaboration was a major factor that influenced OSU's successful bid to be FRI's new home. Modest amounts of collaboration were realized between OSU's School of Chemical Engineering and FRI in the 1990's and early 2000's. However, after the arrival of Mike Resetarits at FRI in 2008, a remarkable level of mutually beneficial collaboration has been realized.

This paper documents the history of collaboration between FRI and OSU with a focus on the benefits realized by both parties. Proprietary work performed by OSU for FRI is not addressed. The number of OSU faculty and students (graduate and undergraduate) who benefit from FRI/OSU collaboration is much greater than most would predict. Under the direction of Mr. Resetarits, the FRI technical and operating staff has become an integral part of the OSU Chemical Engineering education and research programs. FRI staff serves on graduate student committees, provide guest lectures in both undergraduate and graduate level courses, and routinely provide tours of the FRI unit. Exit interviews with graduating seniors always acknowledge the benefits of interactions with FRI.

The majority of the paper will highlight the wide range of topics that have involved collaboration between FRI and OSU. While all are related to the study of distillation and mass transfer, many involve applications of heat transfer, process control, safety and risk management as required implementing and studying mass transfer. These projects are performed in parallel without disruption of FRI's primary mass transfer research program. The ability to visualize and implement these parallel projects while establishing new FRI records for safety and productivity is a testament to the value added by Mr. Resetarits during his tenure as Technical Director at FRI.

DISCUSSION

Over the last few years, the collaborations between FRI and OSU have often proven to be synergistic. Many organizations have benefitted from the resultant synergies which can be classified as follows:

1. Collaborations benefitting AIChE
2. Collaborations benefitting FRI (and its 80 global members)
3. Collaborations benefitting OSU (i.e., its students, its faculty members and the organizations that hire OSU graduates)
4. Collaborations benefitting both FRI and OSU

Detailed listings are now provided regarding all of this joint work.

Collaborations Benefitting AIChE

1. Starting at the Spring 2013 AIChE meeting, FRI and OSU have contributed appreciably to the annual Young Professionals Forum. At the Spring 2014 AIChE meeting, FRI and OSU gave five of the six tutorial presentations
2. Since 2007, FRI and OSU have given at least twenty joint presentations at Spring and Annual meetings. The topics have included distillation, kettle

reboiling, steady state identification, controls strategies, spray nozzles and CO_2 absorption. A bibliography is provided at the end of this paper.

3. FRI engineers and OSU instructors have held leadership positions in the Distillation and Absorption Group of the Separation Division. Additionally, they have often served as session organizers.

Collaborations Benefitting FRI

1. FRI's present engineering staff includes three OSU graduates: Dr. Tony Cai, Mr. AY Ogundeji and Dr. Anand Vennavelli
2. Professor Jan Wagner, who was an instructor at OSU for many years, is a safety expert in several areas. For FRI, he facilitated PHA's and RMP's; he evaluated FRI's relief system; he performed a butane dispersion study
3. Professors Rob Whiteley and Ken Bell have consulted regularly regarding FRI's new research program regarding kettle reboilers. FRI's industrial size kettle reboiler now has 4 windows on it. Correlations have been developed regarding boiling pool depths and maximum vapor velocities.
4. Professor Rob Whiteley provided the control strategy for FRI's new Reflux Heater
5. Professor Khaled Gasem studied the VLE of the pentane/cyclohexane binary
6. OSU performed laboratory work regarding liquid orifice coefficients with trough configurations that included vertical and horizontal water flow
7. Mr. Anand Govindarajan correlated orifice coefficients for liquid flow through holes in distillation tray truncated downcomers.
8. An OSU ornithologist consulted regarding unexpected bird behavior before an FRI rupture disk failure
9. OSU pointed the way to the best Oklahoma boiler repair expert
10. An OSU instructor consulted for FRI regarding corrosion at the FRI Experimental Unit
11. OSU security personnel help to protect the FRI equipment. They consulted regarding improvements, following a break-in at the Experimental Unit

Collaborations Benefitting OSU

1. FRI provided monetary support regarding all of the following OSU projects:
    a. A new distillation column valve tray capacity correlation
    b. A new valve tray efficiency model
    c. A modified O'Connell correlation for distillation tray efficiencies
    d. Software for the identification of process equipment steady states
2. Regarding OSU laboratories and pilot plants, FRI has done the following:
    a. Donated spare equipment
    b. Consulted regarding instrumentation
    c. Provided training regarding instrument installation, calibration and maintenance
    d. Designed control systems and performed control system troubleshooting
    e. Provided construction management advice
    f. Provided pipefitter training
3. FRI provided OSU with an opportunity to collect high-speed pressure drop measurements and to analyze nonlinear signals. The work was aimed at flood prediction software

4. FRI engineers and technicians regularly provided tours of the FRI Experimental Unit to OSU undergraduate and graduate students. Some of those students were from the OSU chemical engineering program; some from the OSU Fire and Safety program
5. FRI engineers taught at graduate level course entitled Stagewise Operations (ChE 5633)
6. FRI engineers and technicians gave lectures at undergraduate chemical engineering classes regarding safety, distillation design, ethics, international business, etc.
7. FRI staff members participate on OSU's Industrial Advisory Committee, which includes the judging of student posters
8. FRI engineers participate on PhD thesis committees

Collaborations Benefitting Both

1. FRI and OSU jointly purchased a state-of-the-art Phase Doppler Interferometer from Artium Technologies to assist with the study of all of the following:
    a. Spray columns for CO_2 absorption (in the OSU lab)
    b. Small spray nozzle performances (in the OSU lab)
    c. Large spray nozzle performances (at the FRI Unit)
    d. Mist eliminator effectiveness (at OSU and FRI)
    e. Entrainment above operating distillation trays (at the FRI Unit)
    f. Entrainment from kettle reboilers (at the FRI Unit)
2. FRI's test columns have large diameters and data from those columns are believed to be superior to data from laboratory columns. FRI's models and correlations for structured packings were based on FRI's data and were proven, by Mr. Anil Jammula of OSU, to be superior to literature models and correlations
3. Also regarding structured packings, Mr. Jammula studied the ability of the FRI models and correlations to represent data from small-diameter laboratory test columns
4. With guidance from both FRI and OSU, Dr. Xiangpeng Li performed CFD studies on several tray and packing projects
5. FRI provided summer internship opportunities to several OSU graduate students
6. Mr. Raj Salunke, of OSU, compared valve and sieve tray efficiencies across a broad range of physical properties

CLOSING

All of the following individuals are still participating in the FRI/OSU collaborations:

For Fractionation Research Inc: Joella Redden, Jeanifer Randall, Mandy Simms, Leah Brown, Tony Cai, Simon Chambers, AY Ogundeji, Anand Vennavelli, Fred Smith, Chad Kindred, Terry Thurber, Dave Williams, Kenny Martin, Joe Rains, Allen Bowers, Brad Murphy, Kenny Grimes - and Mike Resetarits

For OSU: Rob Whiteley, Clint Aichele, Josh Ramsey, Russ Rhinehart, Jan Wagner, Ken Bell, Qingsheng Wang, Anand Govindarajan, Brett King and Yash Tamhankar

The authors also acknowledge the contributions of the previous generation of FRI/OSU collaborators.

PUBLICATIONS AND PRESENTATIONS: OSU and FRI COLLABORATIONS

Vennavelli, A. N. and Whiteley, J. R., Valve Tray Efficiency Models for Binary Distillation: A Review of the State-of-the-Art. In *AIChE Annual Meeting*, Nashville, TN, 2009; p 21b.

Vennavelli, A. N., Salunke, D. B., and Whiteley, J. R., Challenges in Fundamental Valve Tray Efficiency Modeling in the Mixed-Froth Regime. In *AIChE National meeting*, San Antonio, TX, 2010; p 109b.

Vennavelli, A. N., A Phenomenological Model for Fraction Jetting on Distillation Sieve Trays for Mutli-Regime Mass Transfer Modeling Applications. Ph.D. Thesis, Oklahoma State University, 2011.

Salunke, D. B., An O'Connell Type Correlation for Prediction of Overall Efficiency of Valve Tray Columns. M.S. Thesis, 2011.

Resetarits, M.R., Cai, T.J., Chambers, S. and Whiteley, J.R., "On Distillation Tray Entrainment", AIChE Spring Meeting, Chicago, IL, March 2011.

Vennavelli, A.N., Whiteley, J.R. and Resetarits, M.R., "Modeling Fraction Jetting on Distillation Trays", AIChE Spring Meeting, Chicago, IL, March 2011.

Resetarits, M.R., Vennavelli, A.N., Ogundeji, A.Y. and Whiteley, J.R., "Distillation Column Time-to-Steady-State" AIChE Annual Meeting, Minneapolis, MN, October 2011.

Vennavelli, A. N., Cai, T. J., and Whiteley, J. R., Reflux Temperature Control - A Case Study. In *AIChE Annual Meeting*, Minneapolis, MN, 2011; p 290a.

Jammula, A.K., Whiteley, J.R., Cai, T.J. and Resetarits, M.R., "Performance Comparison of Structured Packing HETP Correlations", AIChE Spring Meeting, Houston, TX, April 1-5, 2012.

Vennavelli, A.N., Whiteley, J.R. and Resetarits, M.R., "New Fraction Jetting Model for Distillation Sieve Tray Efficiency Prediction", I&EC Research, 51 (35), pp 11458-11462, August 24, 2012.

Jammula, A.K., Whiteley, J.R., Cai, T.J. and Resetarits, M.R., "Performance Comparison of Structured Packing Liquid Holdup and Pressure Drop Correlations", AIChE Spring Meeting, San Antonio, TX, April 28-May 2, 2013.

Vennavelli, A.N., Huang, T., Rhinehart, R.R. and Resetarits, M.R., "Distillation Column Steady State Identifier", AIChE Spring Meeting, San Antonio, TX, April 28-May 2, 2013.

Aichele, C.P., Whiteley, J.R., Tamhankar, Y. and Resetarits, M.R., "Acid Gas Absorption Fundamentals", AIChE Spring Meeting, San Antonio, TX, April 28-May 2, 2013.

Tamhankar, Y.S., Whiteley, J.R., Resetarits, M.R. and Aichele, C.P., "Spray Droplet

Characterization for Post Combustion Carbon Capture", ILASS Americas, 25th Annual Conference on Liquid Atomization and Spray Systems, Pittsburgh, PA, May 2013.

Vennavelli, A. N., Huang, T., and Rhinehart, R. R., Demonstration of the SS and TS Identifier at the FRI Unit. In *American Control Conference*, Washington, D.C., 2013

Tamhankar, Y., Whiteley, J.R., Resetarits, M.R., Cai, T.J., and Aichele, C.P., "Effect of Solvent Concentration on Spray Absorption for CO_2 Capture Applications", UTCCS2 2014, Austin, TX, January 28-30, 2014.

Vennavelli, A. N., Whiteley, J., R., and Resetarits, M. R., Predicting Valve Tray Efficiency. *Chemical Engineering Research & Design* 2014 *(Accepted, In Press).*

Aichele, C.P., Whiteley, J.R., Tamhankar, Y., Resetarits, M.R., and Cai, T.J., "Surface Area Measurements for Spray Absorption of CO_2 in Amine Solvents", AIChE Spring Meeting, New Orleans, LA, March 30-April 3, 2014.

Aichele, C.P., Tamhankar, Y., Whiteley, J.R., King, B., Resetarits, M.R., and Cai, T.J., "Characterization of Spray Nozzle Performance and De-Entrainment for Industrial Spray Applications", AIChE Spring Meeting, New Orleans, LA, March 30-April 3, 2014.

Aichele, C.P., Whiteley, J.R., Tamhankar, Y., and Resetarits, M.R., "Acid Gas Absorption Fundamentals", AIChE Spring Meeting, New Orleans, LA, March 30-April 3, 2014.

Vennavelli, A. N., Myers, L., and Govindarajan, A., Implementation of ADS on a Distillation Column. In *AIChE Spring Meeting*, New Orleans, 2014.

Tamhankar, Y., Whiteley, J.R., Resetarits, M., Aichele, C.P., "Spray Droplet Characterization Inside a Glass Column Through Dense Wall Flow," Atomization and Sprays, 24, 115-128, 2014.

352619 "Ecmd™ Trays: A Quarter Century of Excellence"

Abstract:

2014 marks the 25th year since the development and commercialization of UOP's Enhanced Capacity Multiple Downcomer (ECMD) tray. Since its initial application the ECMD tray has proven to be a mainstay in the high capacity distillation tray arena. Although primarily developed as a revamp option to maximize the capacity of bottlenecked MD or conventionally trayed columns, the advent of world scale plants today has resulted in the ECMD tray being utilized to minimize the size and cost of grassroots columns. This paper discusses the wide range of applications of the UOP ECMD tray over the last twenty five (25) years, ranging from the first application in a 2750 mm internal diameter Ethylene Plant Deethanizer revamp in 1989 to a 10400 mm internal diameter Propane Dehydrogenation (PDH) Propylene Fractionator in 2014. With over 300 towers in operation, the ECMD tray continues to be a key contributor in the UOP family of high capacity trays.

BIBLIOGRAPHY OF MR. MIKE RESETARITS

By

Tony J. Cai
Simon Chambers

Fractionation Research Inc.
424 S. Squires Street, Suite 200
Stillwater, OK 74074

31 March 2014

Paper No 28g
Topical 8: Distillation Topical Conference

New Orleans, LA

UNPUBLISHED

Abstract

After receiving BS and MS degrees in chemical engineering, Mr. Mike Resetarits' career began in 1974, at Union Carbide Corporation. While there, Mike worked on the development of the MD distillation tray and the MU liquid-liquid extraction tray. The ECMD tray was invented with Dr. Mike Lockett. Several Union Carbide businesses were acquired by UOP in 1988, at which point Mike became a UOP R&D Manager. The ECMD tray was developed and UOP acquired the Benfield hot carbonate acid gas removal process. In 1999, Mike joined Koch-Glitsch as their R&D Director and later as their Global Tray Product Manager. Mike helped to lead the development efforts associated with Ultra-Frac trays and Flexipac HC structured packing. He was on the team that blended the Koch-Glitsch and Norton technologies. Since 2007, Mike has been FRI's Technical Director, leading a team of engineers and technicians as trays and packings are being proof-tested for FRI's global membership. FRI is now a leader in heat transfer research and Yokogawa control technology.

Mike spent appreciable time in the field, running pilot plant tests, starting units, performing troubleshooting work and collecting data. He is coauthor of 17 patents and over 90 presentations and articles. This presentation describes some of Mike Resetarits' contributions to separations engineering.

Keywords: distillation absorption extraction trays packing

INTRODUCTION

Mr. Michael R. Resetarits holds a BS in chemical engineering from SUNY at Buffalo and an MS in chemical engineering from Northwestern University. He also holds an MBA from Canisius College.

Mike's work experience is summarized below:

74-88	Union Carbide	Tonawanda, NY
89-99	UOP	Tonawanda, NY
99-03	Koch-Glitsch	Dallas, TX and Wichita, KS
03-07	Consultant	Depew, NY
08-13	Fractionation Research Inc.	Stillwater, OK

At FRI, he holds the position of Technical Director.

During his 40 years of experience, Mike's primary focus area has been R&D in the separation fields of distillation, liquid-liquid extraction and acid gas absorption. He has coauthored 17 U.S. patents including a patent on the Enhanced Capacity Multiple Downcomer (ECMD) tray, which has benefitted the global chemicals processing industry appreciably.

Mike has contributed significantly to the Separations Division of the AIChE. He held the position of Director from 2008 until 2013. During that time period, Mike and his FRI colleagues presented 34 papers at AIChE meetings (Spring and Annual). Concomitantly, Mike and his colleagues chaired or co-chaired many oral sessions. In fact, Mike was Group 2A (Distillation and Absorption) Co-Chair for 3 years and Group 2A Chair for another 3 years. Mike gave 8 bibliography presentations summarizing the career achievements of other experts in the field of separations.

Mike has authored over 90 articles and presentations regarding separation technologies. This paper summarizes five of those articles, and lists some of the most important other articles and presentations.

FIVE ARTICLES

"Retraying Increases C3 Splitter Column Capacity" (1)

Enterprise Products Co. operated a propylene-propane unit in Mont Belvieu, TX. That unit employed vapor recompression to reduce energy consumption. Two side-by-side columns were employed, basically, a rectifier and a stripper. In total, there were 258 conventional trays at 610 mm (24 in) spacings. The diameter of each column was 4.3 m (14 feet).

In December of 1986, the two columns were revamped with an equal number of UOP MD trays. During that revamp many photographs were taken, especially of the old segmental tray support rings being filled in to create complete 360-degree rings. Ring levelnesses were measured and adjusted. Revamp time requirements were recorded and reported. Before the revamp, the unit would flood at a feed rate of 12,300 b/sd; after the revamp, 16,000 b/sd.

"Increase Demethanizer Efficiency and Capacity" [2]

An Asian demethanizer was originally equipped with 16 conventional 1-pass rectification trays at a diameter of 2.00 m and 41 conventional 2-pass stripping trays at a diameter of 3.00 m. All of the tray spacings were 600 mm. The rectification section was revamped with 16 UOP MD trays (a 1-for-1 changeout). The stripping section was revamped with 60 MD trays (a 3-for-2 changeout). Welding was not permitted in this particular column, and so, the new trays were supported by "surrogate rings" which were not welded to the inside tower wall. After the revamp, the tower proved capable of handling 65% more feed.

"Trays Inhibit Foaming" [3]

A DSM petrochemical plant in Geleen, The Netherlands, included a 1,3-butadiene column that contained 1-pass conventional trays at a diameter of approximately 2m. Extractive distillation was employed to effect the separation. The solvent was dimethylformamide (DMF). Those trays foamed. The foam polymerized. The column had to be cleaned out regularly, with great difficulty. Silicone oil reduced the foaming somewhat. In 1987, the conventional trays were replaced with an equal number of UOP MD trays (at spacings of 380, 570 and 610 mm). Those trays were particularly designed to inhibit foaming as follows: Large perforation diameters, very long outlet weir, reduced froth heights and operation near the spray regime. Following the revamp, the column functioned easily at 120% of the former feed rate. More importantly, the new trays did indeed inhibit foaming and fouling. After the first two years of operation, the trays were inspected and found to be completely clean. The anti-foam injection was stopped.

"Increase Aromatics Extractor Capacity with Retray" [4]

UOP Multiple Upcomer (MU) liquid-liquid extraction trays were initially developed in Union Carbide laboratories in Tarrytown and Tonawanda, NY. Pilot plant proof testing was effected at Union Carbide's BTX plant in Taft, LA. Nippon Petrochemicals operated a Sulfolane BTX unit in Kawasaki, Japan. Two extractors, with a total of 120 sealed deck trays, were employed in series. The extractor diameters were 1.5 and 1.8 m. In 1990, those trays were removed and were replaced with 114 MU trays. As a result of that revamp, the unit proved to be capable of functioning at a feed rate that was 33% higher than the pre-revamp rate. The BTX extract purity was unchanged. The BTX losses in the raffinate were greatly reduced.

"Cocurrent Flow Trays in Distillation Columns" [5]

Koch-Glitsch ULTRA-FRAC® trays employ cocurrent flow to achieve very high column capacities. With cocurrent flow, the liquid and vapor streams flow together, in the same direction, within certain sections of tray unit cells. The Koch-Glitsch trays were developed in apparatuses in Wilmington, MA and Dallas, TX. They were ultimately proof tested in a gas plant in Medford, OK. In 2001, they were employed in a 2.4/3.5 m (8.0/11.5 ft) diameter deethanizer to successfully accommodate a 50% increase in the column's internal traffic rates. That was an unusual deethanizer in that it functioned, near its critical point, at 32.1 barg (466 psig). After the first 20 utilizations of Ultra-Frac trays it was found that their efficiencies were predicted by the well-known O'Connell correlation[6].

The preceding pages summarized five magazine articles coauthored by Mike Resetarits. The reference list at the end of this paper gives several other articles and presentations (coauthored by Mike) that Mike deems to be of the most significance to the chemicals processing industry[7-24].

A very large portion of Mike's work results were deemed Confidential and Proprietary by Mike's employers. Nevertheless, the chemical processing industry still benefitted from that work.

SUMMARY

From his very first day in 1974 at Union Carbide until the present time at Fractionation Research Inc (FRI), Mr. Mike Resetarits has performed R&D work in distillation, liquid-liquid extraction and acid gas absorption. Some industrially-important products were developed. Using his work results, many new columns were designed and many old columns were revamped to achieve capacity and separation benefits. Along the way, many work results were shared with the chemical engineering public via more than 90 articles and presentations, especially AIChE presentations.

REFERENCES

1. Resetarits, M.R., Agnello, J., Lockett, M.J., and Kirkpatrick, H.L., "Retraying Increases C3 Splitter Column Capacity," Oil and Gas Journal (86), June 1988, pp. 54.
2. Resetarits, M.R. and Miller, R.J., "Increase Demethanizer Efficiency and Capacity," Hydrocarbon Processing, October 1991, pp. 79-81.
3. Resetarits, M.R., Navarre, J.L., Monkelbaan, D.R., Hangx, G.W.A. and van den Akker, R.M.A., "Trays Inhibit Foaming," Hydrocarbon Processing, March 1992, pp. 61-64.
4. Resetarits, M.R., Markowski, D.S., Teramoto, K., Syama, Y. and Maier, W.H., "Increase Aromatics Extractor Capacity with Retray," Japanese Aromatics Industry Conference, October 2, 1991.
5. Resetarits, M.R., Riter, J.B. and Weiland, R.H., "Distillation Technology: Cocurrent Flow Trays in Distillation Columns," Hydrocarbon Engineering, Vol. 7, No. 9, September 2002, pp. 85-88.
6. O'Connell, H.E., Plate Efficiency of Fractionating Columns and Absorbers, Trans. Am. Inst. Chem. Engrs., 42, 741, 1946.
7. Resetarits, M.R., Miller, R.J., Navarre, J.L., Linskeseder, M. and Reich-Rohrwig, P., "New Enhanced Capacity MD Tray Debottlenecks Deethanizer," IChemE Distillation and Absorption Conference, Birmingham, U.K., September 7, 1992.
8. Resetarits, M.R., Monkelbaan, D.R. and Miller, R.J., "Trays in Rocking Services," IChemE Distillation and Absorption Conference, Birmingham, U.K., September 7, 1992.
9. Bravo, J., Foulk, D., Imhoff, E., Resetarits, M.R., Makowski, D.S., and Kohler, R., "Sulfolane RDC Tray Revamp," AIChE Annual Meeting, Chicago, IL, November 15, 1996.
10. Nutter, D.E., Pilling, M.W., Summers, D.R., Resetarits, M.R., Urbanski, N.F. and Monkelbaan, D.R., "VGMD Trays from Nutter and UOP," IChemE Symposium, Aston Univ., Birmingham, U.K., March 25, 1998.

11. Resetarits, M.R., Urbanski, N.F., and Xu, Z.P., "Column Revamps," AIChE Spring Meeting, Chicago, IL, February 16, 1998.
12. Resetarits, M.R., "Distillation Equipment in 2020," Vision 2020: Distillation Priorities Topical Symposium, Session T1003, AIChE Spring National Meeting, Atlanta, GA, March 5-9, 2000, 138-140.
13. Angelino, S., Resetarits, M., Riter, J., Weiland, R., Alderson, H., Berkstresser, S. and Kelso, D., "Converting a Commercial Column into a Research Tower," AIChE Spring Meeting, Paper 8g, Houston, TX, April 26, 2001.
14. Resetarits, M.R. and Pappademos, N., "Factors Influencing Vapor Crossflow Channeling," AIChE Annual Meeting, Reno, NV, November 8, 2001.
15. Resetarits, M.R. and Lockett, M.J., "Distillation," Encyclopedia of Physical Science and Technology, Vol. 4, 2001.
16. Resetarits, M.R. and King, D.W., "Fifty Years of $FRI_{(SM)}$ Contributions," AIChE Annual Meeting, Philadelphia, PA, November 18, 2008.
17. Resetarits, M.R. and Ogundeji, A.Y., "On Distillation Tray Weir Loadings," AIChE Spring Meeting, Tampa, FL, April 27-29, 2009.
18. Pless, L., Carlson, R., Resetarits, M.R., and Ogundeji, A.Y., "Gamma Scanning of $FRI_{(SM)}$ Reboiler Vapor Return Lines," AIChE Spring Meeting, San Antonio, TX, March 23, 2010.
19. Resetarits, M.R., Cai, T.J., Chambers, S. and Whiteley, J.R., "On Distillation Tray Entrainment", AIChE Spring Meeting, Chicago, IL, March 2011.
20. Resetarits, M.R., Vennavelli, A.N., and Ogundeji, A.Y., "Possible Aversion of Unsafe Situations by Semi-Automatic Operation," AIChE Annual Meeting, Minneapolis, MN, October 2011.
21. Cai, T., Resetarits, M.R., and Parker, A.L., "A New Look Inside Kettle Reboilers," ChemInnovations, Houston, TX, 2011.
22. Chambers, S., Cai, T.J., Vennavelli, A.N., Ogundeji, A.Y., Resetarits, M.R. and Redden, J.C., "Effect of Tray Spacing on Tray Performance," AIChE Spring Meeting, April 3, 2012.
23. Ogundeji, A.Y., Thurber, T.C. and Resetarits, M.R., "Ruptured Rupture Disc," AIChE Spring Meeting, San Antonio, TX, April 28-May 2, 2013.
24. Aichele, C.P. Whiteley, J.R., Tamhankar, Y. and Resetarits, M.R., "Acid Gas Absorption Fundamentals," AIChE Spring Meeting, San Antonio, TX, April 28 – May 2, 2013.

PROSYS™ Maximize Your Plant.

Operator Errors and What Can be Done to Minimize

Global Congress on Process Safety

March 31, 2014

10TH GLOBAL CONGRESS ON PROCESS SAFETY

PROSYS™ Maximize Your Plant.™

10TH GLOBAL CONGRESS ON PROCESS SAFETY

Presenter

Tom Nolan

- **Graduated from Ohio University with B.S. in Chemical Engineering**
- **24 years experience in chemical process industry in a variety of roles**
- **MAIC and DFSS Six Sigma Black Belt**

PROSYS™ Maximize Your Plant™

The Cost of Errors

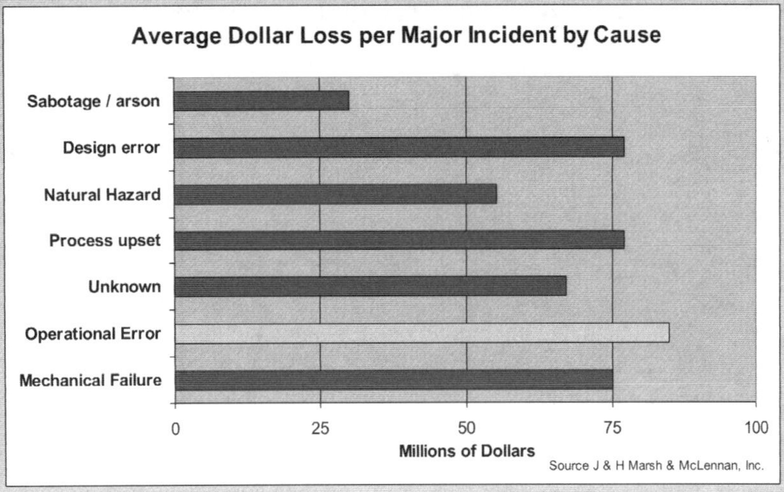

PROSYS™ Maximize Your Plant™

The Cost of Operator Errors

- **ASM estimates total loss due to operator error is $8B per year**
- **Chemical Safety Topical Committee - average of one chemical incident per day - cost of over $2 million per incident to <u>comply with requirements</u>**
- **Errors cause 42% of unscheduled shutdowns**
- **70% of process incidents occur during start-up or shutdown**

PROSYS™ Maximize Your Plant™

10th GLOBAL CONGRESS on PROCESS SAFETY

Reasons for Errors

- Lack of Skill
- Lack of Knowledge
- Carelessness
- System Design
  - Operator set up to make errors by inappropriate design – built in errors
  - Errors that are predictable are preventable by better design

PROSYS™ Maximize Your Plant™

10th GLOBAL CONGRESS on PROCESS SAFETY

Focus of this Discussion

- System Design
  - **Alarm Management**
  - **Operator Graphics**
  - Operator set up to make errors by inappropriate design – built in errors
  - Errors that are predictable are preventable by better design

Alarm Floods

In a number of industrial incidents, alarm floods were identified as a significant contributing cause to the incident...

As found by EEMUA in 1999 and CSB

Alarm Flood defined by ISA 18.2 as -
"10 or more annunciated alarms in any 10 minute period per operator"

Why do Alarm Floods Occur?

One reason is not providing dynamic alarm management

- **Alarms need to indicate abnormal situations that require operator action**
- **Processes do not operate in one state**
- **What is normal vs abnormal changes with state**
- **Alarms are typically configured for run therefore many alarms are triggered upon a Change of State: Run to Shutdown**
- **Many of the alarms are not applicable or actionable for the new state impleading the operators ability to act quickly on what is important**

PROSYS Maximize Your Plant.™

10th GLOBAL CONGRESS ON PROCESS SAFETY

What is Impacted by Alarm Floods?

- Product quality
- Operability or profitability of the process
- Loss of equipment
- Loss of containment – environmental releases
- Injury and loss of life in plant or community

PROSYS Maximize Your Plant.™

10th GLOBAL CONGRESS ON PROCESS SAFETY

What Makes Alarm Floods so Dangerous?

Can be a problem for three reasons:

- A deluge of alarms can cause critical alarms to be missed
- Floods can be a significant distraction when dealing with process upsets
- Can be an indicator of larger systemic safety issues

PROSYS Maximize Your Plant™

10th GLOBAL CONGRESS ON PROCESS SAFETY

Impacting Alarm Management Design

- **Alarm rationalization is not a process to eliminate alarms - its about quality**
- **Good rationalization will add alarms when appropriate**
- **Typical (Static) rationalization is only for run mode**
- **Dynamic rationalization considers all plant modes**

PROSYS Maximize Your Plant™

10th GLOBAL CONGRESS ON PROCESS SAFETY

Dynamic Alarm Management

- **Dynamic rationalization does everything a static rationalization does plus asks "when" for each alarm**
- **Requires dynamic software to make changes based upon operating mode of the plant**
- **Eliminates redundant alarms and lowers operator loading during transitions**
- **Only alarms what is abnormal and actionable for the given state**

PROSYS Maximize Your Plant™

10th GLOBAL CONGRESS ON PROCESS SAFETY

Actual Performance Metrics vs ISA 18.2

Type	Avg Alarm Rate/hr	Low/Hi of Avg Rate/hr	Peak Alarm Rate/hr	Low/Hi of Peak Rate/hr	% Time in Flood	Standing Alarms >24 hrs
Before Rationalization	30	Low=6.2 Hi=61	638	Low=152 Hi=2402	17.2%	9
ISA 18.2 Target Metrics	6	---	<60	---	<1%	<5
After Dynamic Alarm Management	2	Low=0.09 Hi=2.5	25	Low=13 Hi=42	0.25%	4

PROSYS Maximize Your Plant™

10th GLOBAL CONGRESS ON PROCESS SAFETY

Results After Dynamic Alarm Management

Annunciated Alarms Per Hour (Monthly) **Area: G1-G5-G6** **June 2013**

Average Counts | Peak Counts | Total Counts

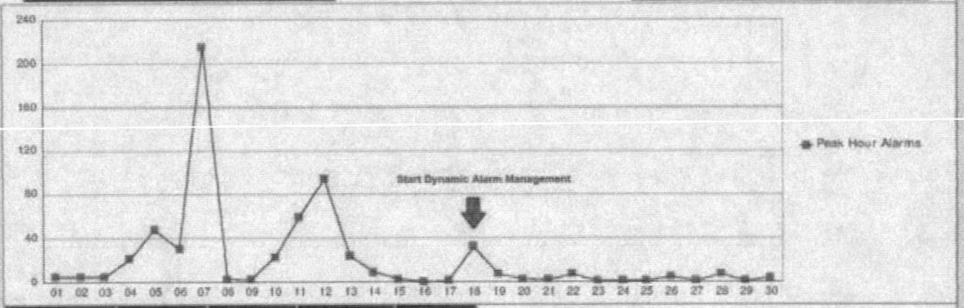

Day	Min Hour Alarms	Avg Hour Alarms	Peak Hour Alarms	Total Alarms
01-Jun-2013	0	0.3	5	6
02-Jun-2013	0	0.4	5	9
03-Jun-2013	0	0.5	5	13
04-Jun-2013	0	1.1	21	27
05-Jun-2013	0	13.3	48	319
06-Jun-2013	0	3.6	30	86
07-Jun-2013	0	11.9	215	285
08-Jun-2013	0	0.4	3	10
09-Jun-2013	0	0.3	2	7
10-Jun-2013	0	1.5	23	35
11-Jun-2013	0	4.7	59	112
12-Jun-2013	0	11.9	94	285
13-Jun-2013	0	3.8	24	90
14-Jun-2013	0	1.3	9	32
15-Jun-2013	0	0.3	3	8

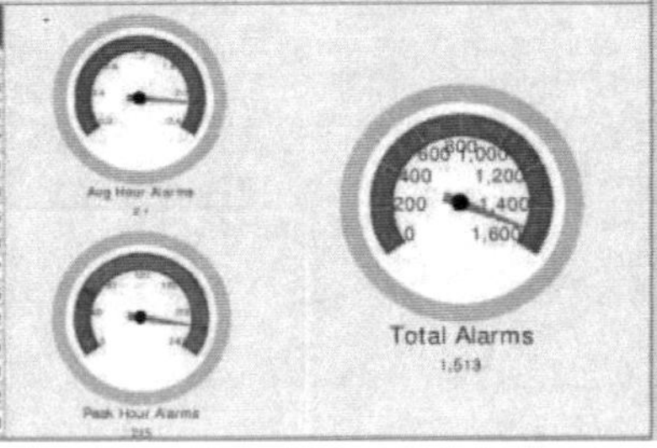

PROSYS Maximize Your Plant™

10th GLOBAL CONGRESS ON PROCESS SAFETY

Operator Graphics – The Risk of Changing Focus

- **Distraction of changing focus from process graphic to faceplate window**
- **Added workload from managing multiple open windows**
- **Increased probability of errors when changing values for un-intended tag with multiple faceplates open**
- **All of above become multiplied and more complex when process is transitioning from one state to another**

PROSYS Maximize Your Plant™

10th GLOBAL CONGRESS ON PROCESS SAFETY

Faceplate Operation

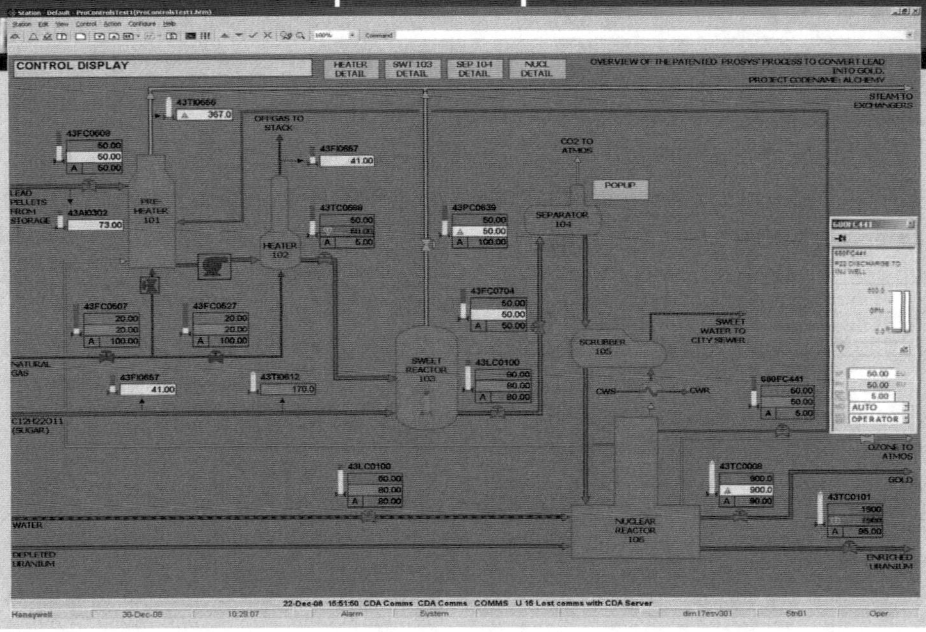

PROSYS Maximize Your Plant.™

10th GLOBAL CONGRESS ON PROCESS SAFETY

Potential for Errors

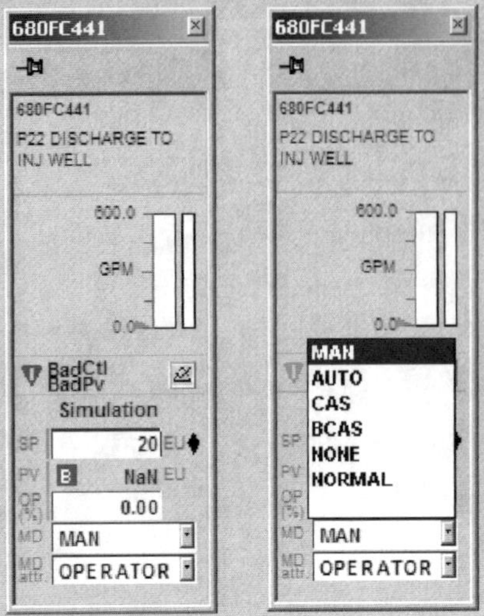

Faceplate Design -can introduce additional errors

- Mode drop down list covers SP, PV and OP values
- Operator may select wrong Mode if SP, PV and OP values are hidden from view
- Mode list offers more options than necessary

PROSYS Maximize Your Plant.™

10th GLOBAL CONGRESS ON PROCESS SAFETY

Direct Entry Operation

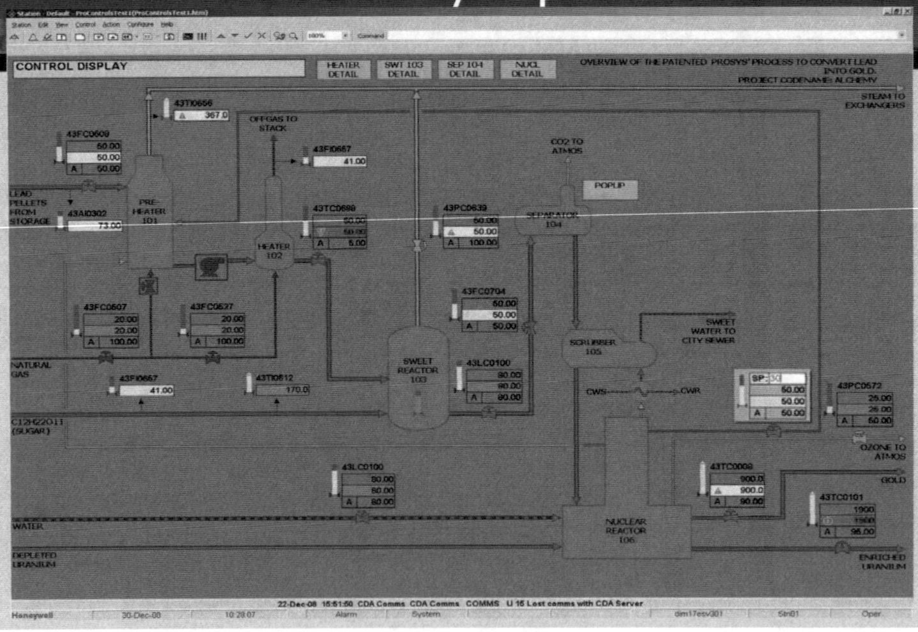

Direct Entry Fields

Values can be entered by:
- Select point
- Typing value via keypad
- Pressing enter

Key Factors:
- By maintaining focus of points after value change, it improves safety for input errors, multiple clicking functions and is more time efficient
- Easily recognizable operator enterable fields

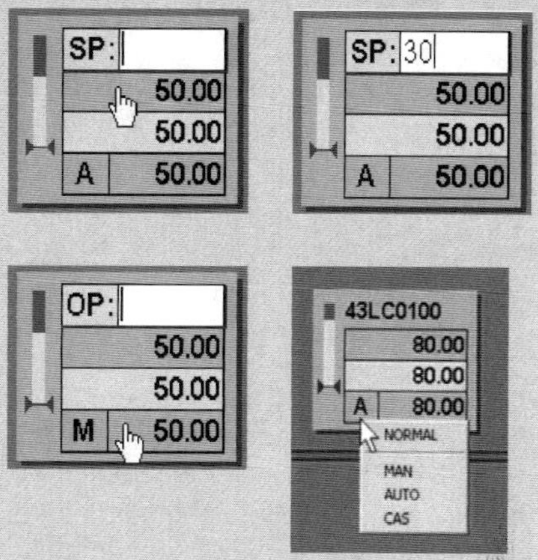

Evaluation of Methods

Comparison of Faceplate to Direct Entry
- **Keystroke Level Analysis**
- **Operator Loading Analysis**
- **Risk Analysis**

PROSYS Maximize Your Plant.™

Keystroke Level Model

Symbol	Time (s)	Description
K	0.28	Keystroke
P	1.1	Point to object
BB	0.2	Click on object
H	0.4	Home hands on keyboard or mouse
M	1.2+	Mental act or routine thinking

PROSYS Maximize Your Plant.™

Estimate Execution Time Analysis

- **Evaluates execution time by an operator comparing faceplates to direct entry for SP/OP and Mode Changes**

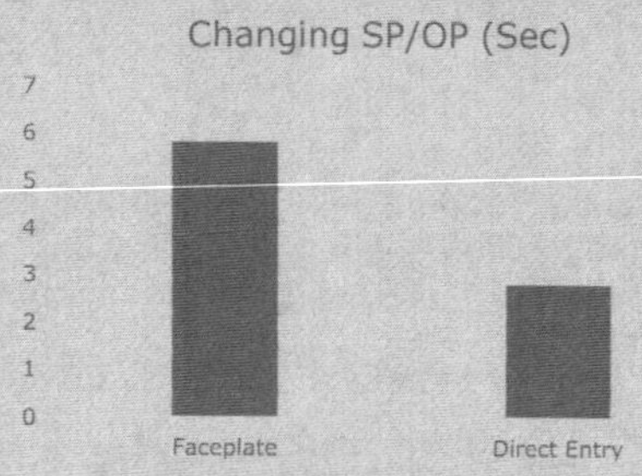

- **50 % Reduction**

PROSYS Maximize Your Plant™

10th Global Congress on Process Safety

Estimated Operator Loading Analysis

- **Evaluates Mental and Physical Load on Operator**
  - Tasks that require very little thought are removed – such as (BB), (H)

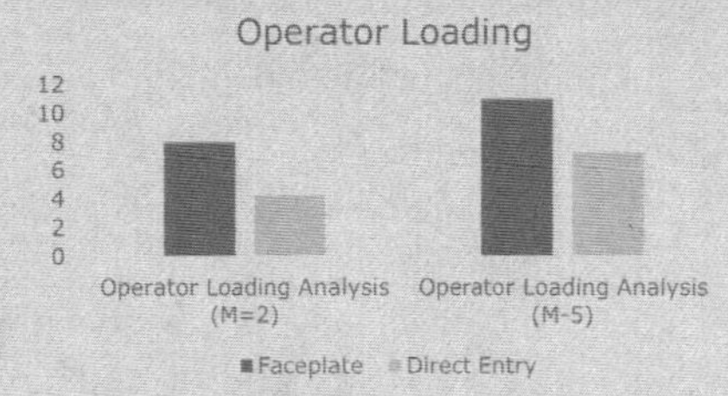

  - Mental and Physical tasks are weighted based on the amount of fatigue they cause – assigned a weight of M=2.0 to 5.0

PROSYS Maximize Your Plant™

10th Global Congress on Process Safety

Risk Analysis

- **Identifies actions where operator entry errors can occur**

PROSYS Maximize Your Plant™

10th Global Congress on Process Safety

Risk Analysis

Faceplate operation KLM for changing SP/OP	Potential Error
Move to shape (P)	Select wrong point
Click on shape (BB)	
Move to faceplate (P)	Point in faceplate is previous point
Click on SP/OP field (BB)	Change wrong parameter
Move hands to keyboard (H)	
Type in value and press enter (4K)	Mistype value and press enter
Move hands to mouse (H)	
Move to close faceplate (P)	
Click to close faceplate (BB)	
Potential error count	**4**

PROSYS Maximize Your Plant™

10th Global Congress on Process Safety

Risk Analysis

Direct Entry operation KLM for changing SP/OP	Potential Error
Move to SP/OP (P)	Select wrong parameter or point
Click on SP/OP field (BB)	
Move hands to keyboard (H)	
Type in value and press enter (4K)	Mistype value and press enter
Potential error count	**2**

PROSYS Maximize Your Plant.™

10th GLOBAL CONGRESS ON PROCESS SAFETY

Direct Entry Vs Faceplate

Summary of Benefits

- **Execution Time SP/OP Changes - 51.5% less**
- **Execution Time for Mode Changes – 50% less**
- **Operator Loading – 33 % to 47% less**
- **Risk of Errors – 50% less**

PROSYS Maximize Your Plant.™

10th GLOBAL CONGRESS ON PROCESS SAFETY

Conclusion

- **Many industrial errors are a result of operators using control systems with flawed designs**
  - Poor design and performance by alarm management distracts operators and/or occludes critical alarms
  - Operator graphics with multiple open faceplates can cause a change intended for one controller to be entered into the faceplate of another
- **The two mechanisms listed above are predictable and therefore preventable through better design available today**

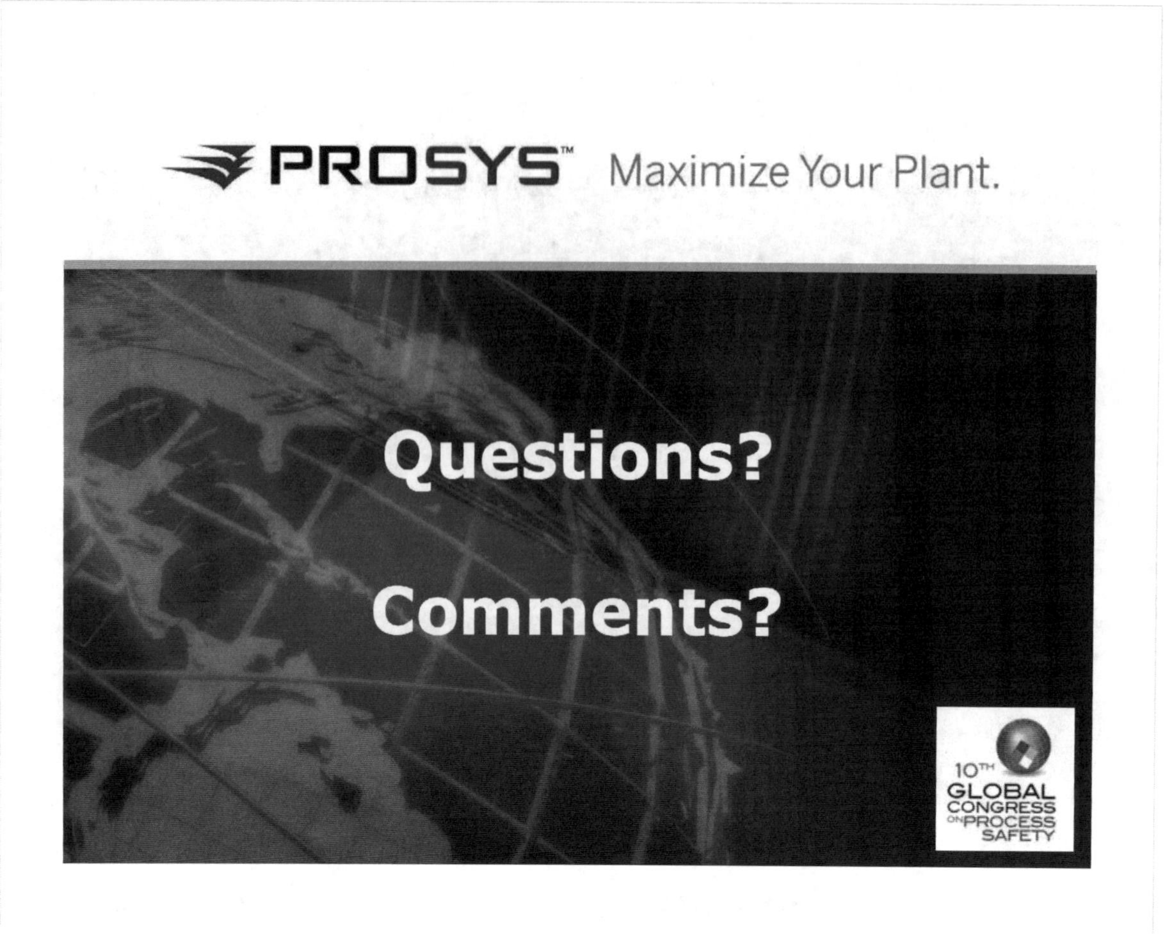
PROSYS™ Maximize Your Plant.
Questions?
Comments?
10TH GLOBAL CONGRESS ON PROCESS SAFETY

Problem Statement

visium KMS
Process Safety Management Solution

- Any organisation can improve by leveraging knowledge and lessons learned.
- Knowledge can be externally or hidden locally, in a global enterprise, where lessons learned are difficult to share beyond their initial impact.
- There is so much data, information and knowledge (opportunities) coming at your company that can make you stronger or make you look like your not paying attention.
- Not paying attention or managing the available knowledge, can sink your ship… nobody wants that!

Expert Insight Delivered 2

Osys
Part of the Rolls-Royce Group

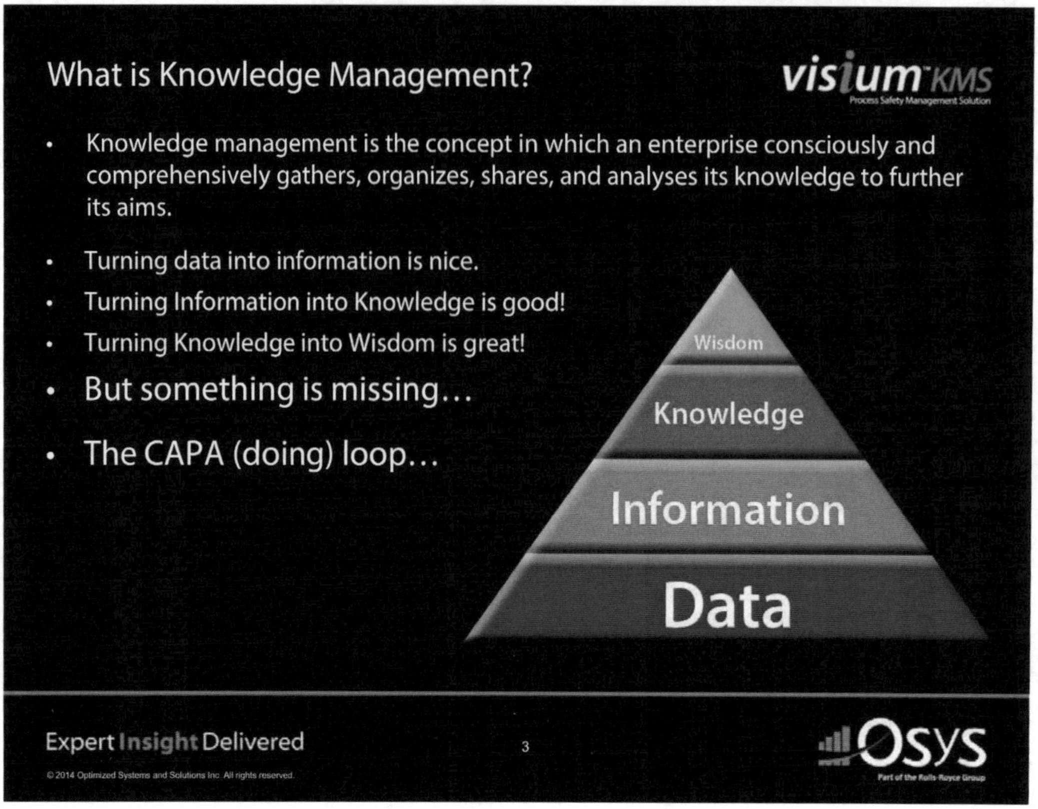
What is Knowledge Management?
visium KMS
Process Safety Management Solution
• Knowledge management is the concept in which an enterprise consciously and comprehensively gathers, organizes, shares, and analyses its knowledge to further its aims.
• Turning data into information is nice.
• Turning Information into Knowledge is good!
• Turning Knowledge into Wisdom is great!
• But something is missing…
• The CAPA (doing) loop…
Wisdom
Knowledge
Information
Data
Expert Insight Delivered
3
© 2014 Optimized Systems and Solutions Inc. All rights reserved.
Osys
Part of the Rolls-Royce Group

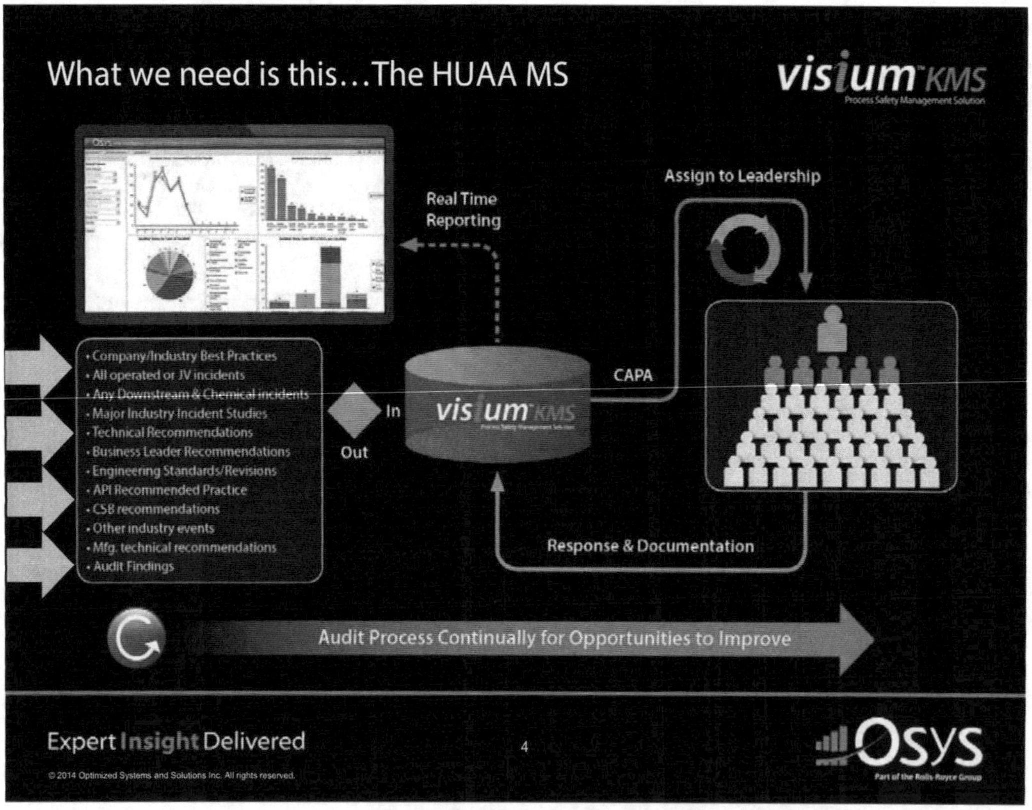
What we need is this…The HUAA MS
visium KMS
Process Safety Management Solution
Real Time Reporting
Assign to Leadership
• Company/Industry Best Practices
• All operated or JV incidents
• Any Downstream & Chemical incidents
• Major Industry Incident Studies
• Technical Recommendations
• Business Leader Recommendations
• Engineering Standards/Revisions
• API Recommended Practice
• CSB recommendations
• Other industry events
• Mfg. technical recommendations
• Audit Findings
In
Out
visium KMS
CAPA
Response & Documentation
Audit Process Continually for Opportunities to Improve
Expert Insight Delivered
4
© 2014 Optimized Systems and Solutions Inc. All rights reserved.
Osys
Part of the Rolls-Royce Group

What is the HUAA Management System…

visium KMS
Process Safety Management Solution

HUAA – Heard, Understood, Acknowledged, Actioned.

1. Opportunities identified (heard) by your experts,
2. Captured by an electronic system,
3. Expertly Reviewed (understood),
4. Accepted or rejected (acknowledged) and
5. Assigned (actioned) to Sr. Managers for resolution and tracked to closure.

Expert Insight Delivered 5

OSYS Part of the Rolls-Royce Group

HUAA – Management Systems - Detail

visium KMS
Process Safety Management Solution

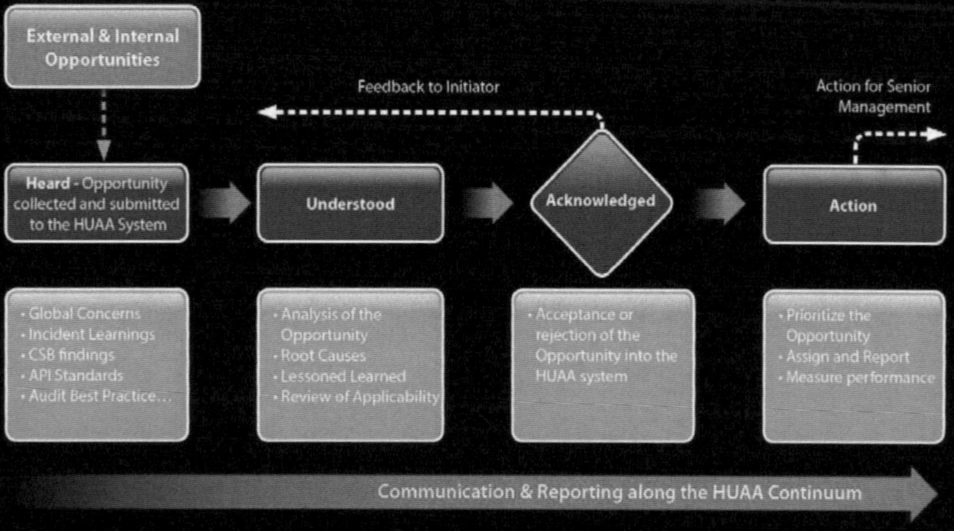

Expert Insight Delivered 6

OSYS Part of the Rolls-Royce Group

HUAA Results

- Efficient Collection - Important Opportunities are identified and collected from many sources (internal and external), and entered into the HUAA management system.
- Expert Vetting - These opportunities are vetted, parsed and resolved for inclusion or exclusion; documenting & communicating both.
- Accountable Assignment - Included opportunities are systematically assigned to Sr. Leaders and managed across the enterprise at SVP and C levels.
- Action - Improvement will be assured through the accountable actions of these leaders.
- Results - The organization will gain competitive advantage by active listening, learning from their experience, as well as the experience of others, and by taking ACTION!

Expert Insight Delivered 7

Sharing is not enough to keep your ship afloat!

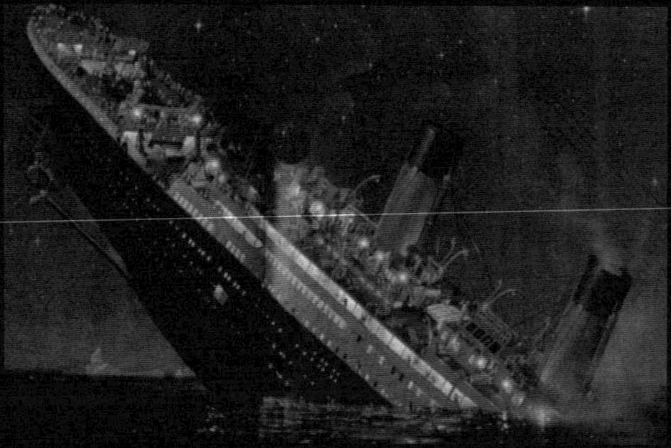

Expert Insight Delivered 8

352631 Autocad - Smart Grid – Agriculture - Architecture and the United States Government

Abstract:

The computational methods, process control, process design, process safety, public policy of the smart chip for agriculture industry, and architecture industry in the United States utilizing the computational methods of AutoCAD in the structural development stages of the smart grid for device/component/equipment and geographical development and public policy for grant funding.

352747 Production, Characterization and Catalytic Studies of Biobased Carbon Materials

Abstract:

Carbon-based nano materials were produced from wood chips, lignin and bio-char. The influence of reaction conditions on the production of nano-carbons was investigated. Carbon-based nanomaterials were characterized using scanning electron microscopy (SEM), transmission electron microscopy (TEM), high resolution transmission electron microscopy (HRTEM), and X-ray diffraction (XRD). The catalytic performance of these nanomaterials as syngas cleaning adsorbents and chemical conversion catalysts for bio-fuel processes has been evaluated.

347835 Uncertainty in Sour Gas Viscosities Estimation, What Is the Effect on Your Reservoir Inflow and Tubing Performance

Abstract:

In light of the need to find an applicable, consistent and promising correlation to accurately calculating sour gases viscosity under several reservoir conditions, this study aims to compare the existing correlations and other industry standard methods, check their accuracy and range of applicability for sour/acidic gases. Moreover, investigating the impact of uncertainties in estimating sour gas viscosities on reservoir flow performance relationship (IPR), reserves and resources estimations, tubing performance, pipeline and facilities design, the design of compressors and more importantly the need to find such correlation to be used in further advancements of reservoir simulation.

354052 Chromonic Nanocarriers for Chemotherapeutics: Size Distribution and Control Release Studies

Abstract:

Conventional chemotherapy fails to release the drug in the controlled manner; rather they release the drug instantly due to which re-administration is required as therapeutic index falls. Moreover, they target both the cancer cells and non-cancer cells which make the treatment of the cancer cells highly ineffective due to excessive toxicities. Recently, the nanoparticle mediated drug delivery for chemotherapeutics has gained much attention. The present study reports the use of a novel class of materials – *Chromonics* in cancer drug delivery. Folic acid as a chromonic material has been explored to develop nanocarrier for sustained release of chemotherapeutic drugs *in-vitro*. With the help of a biocompatible polymer as a tool, folate nano-domains were formed from self-assembled liquid crystalline folate solutions. These nano-domains were crosslinked to form a stable nanoparticle system. The effect of different factors like crosslinking agent, relative polymer and folate concentration, on the size distribution was investigated in detail. Based on the observations, we have reported that not only the synthesis but also the size of the developed nanoparticles can be controlled by varying the folate: HPMC weight ratios. The present study then explored the release of drugs from these nanocarriers. We have investigated the release of folic acid as a model drug from these nanoparticles as well as release of methotrexate and cytarabine as an anticancer drug for the period of 60 days. It was observed that release of drug was nanoparticle's size-dependent, release medium dependent, pH of the release medium dependent, crosslinking agent and its concentration dependent. The significant part of the current study is the synthesis of chromonic nanoparticles, how the size can be controlled, how controlled release can be obtained as well as how the release profiles can be modified depending on the dosage forms. Therefore, we can say that folate nanoparticles can be a potent nanocarrier for the sustained release of cancer drugs.

Double Containment Piping Systems: A Fail-Safe, Leak-Free Solution

IPEX
Committed to Excellence

Darin Johnson, P.Eng.
Applications Engineer
IPEX USA LLC

AIChE – 2014 Spring Meeting
New Orleans, LA

Monday, March 31, 2014

Why Double Containment Piping Systems?

IPEX
Committed to Excellence

- December 1988 – law change within the ***Resource, Conversation, and Recovery Act*** in the United States regarding underground piping.
- U.S. Environmental Protection Agency (EPA) requirement: Code of Federal Regulations (CFR), Title 40, Part 280.
  - Petroleum Products
  - Hazardous Substances (over 3200, CFR Title 40 Part 302.4)
  - Hazardous Wastes

Why Double Containment Piping Systems?

- Buried tank - all buried piping connected to it must be double contained.

- Above ground tank – associated buried piping consists of 10% or more of the tank volume, it also must be double contained.

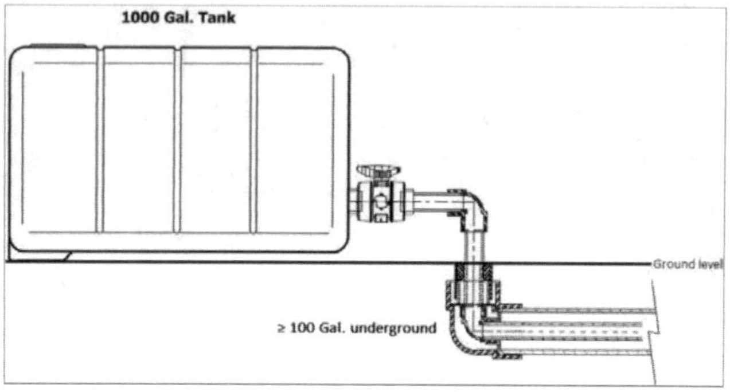

Why Double Containment Piping Systems?

- If the piping is less than 10% of the volume of the aboveground tank, and you choose not to double contain the buried piping, ***responsibility for any clean-up or remediation resulting from a release will still fall on the Client.***

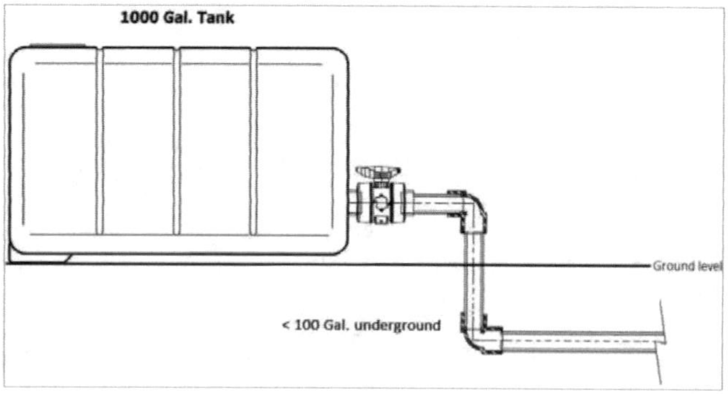

Why Double Containment Piping Systems?

- **Above Ground Piping Systems**
  - Different jurisdictions have to ability to require chemicals be double contained.

Nomenclature of Double Containment Piping

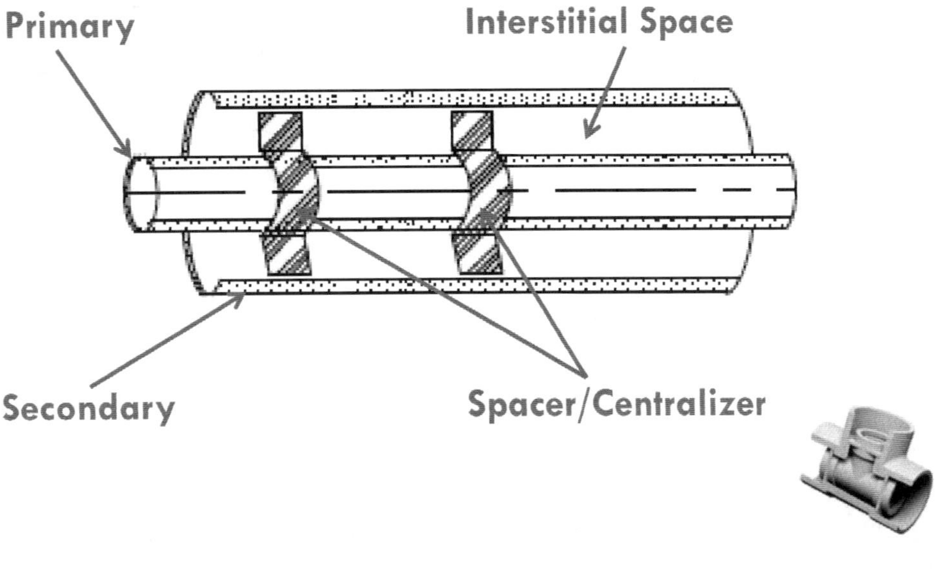

Double Containment Piping Systems: Material Combinations

SYSTEM Carrier	Containment	Carrier SIZE inches
PVC 80	PVC 80	1/2 - 12
PVC 80	PVC 40	1/2 - 12
PVC 40	PVC 40	1/2 - 12
PVC DWV	PVC DWV	1-1/2 - 12
PVC 80	CS SCH 10	1/2 - 12
CPVC 80	CPVC 80	1/2 - 8
CPVC 80	PVC 80	1/2 - 12
CPVC 80	PVC 40	1/2 - 12
CPVC 80	CS SCH 10	1/2 - 12
NAT PP 80	PVC 80	1/2 - 2
NAT PP 80	PVC 40	1/2 - 2
RED PVDF 80	CPVC 80	1/2 - 2
RED PVDF 80	PVC 80	1/2 - 2
NAT PVDF 80	PVC 80	1/2 - 2
ENFUSION PP	PVC DWV	1/2 - 6
S40 A53B CS BW	PVC 80	1/2 - 8
S40 A106B CS BW	PVC 80	1/2 - 8

SYSTEM Carrier	Containment	Carrier SIZE inches
S40 A53 BLK SW	FRP	1/2 - 2
S40 A53 BLK SW	PVC 80	1/2 - 2
S80 A53B CS SW	PVC 80	1/2 - 2
S10 T304L SS BW	PVC 80	1/2 - 8
S10 T304L SS BW	PVC 40	1/2 - 8
S10 T0304L SS BW	CPVC 80	1/2 - 8
S10 T304L SS BW	FRP	1/2 - 12
S10 T304L SS BW	S10 T304L	1/2 - 12
S40 T316L SS BW	PVC 80	1/2 - 8
S40 T316L SS BW	CPVC 80	1/2 - 8
S40 T316L SS BW	FRP	1/2 - 12
S40 T304L SS SW	PVC 80	1/2 - 2
S40 T304L SS SW	PVC 40	1/2 - 2
S40 T304L SS SW	CPVC 80	1/2 - 2
S40 T304L SS SW	FRP	1/2 - 2
S10 T316L SS SW	PVC 80	1/2 - 2
S10 T316L SS SW	CPVC 80	1/2 - 2

Double Containment Piping Systems: Design Considerations

- **Material Selection**
  - Primary Pipe
    - Chemical Compatibility
    - Pressure Capabilities
    - Thermal Properties
  - Secondary Pipe
    - Chemical Compatibility
    - Loading (Static & Live) for buried applications
    - Hanging requirements for above ground
    - Leak detection requirements
    - UV requirements

IPEX
Committed to Excellence

Double Containment Piping Systems: Design Considerations

- **Thermal Expansion & Contraction**
  - Two basic thermal relationships may be present:
    - When only one of the two systems experiences thermal expansion/contraction.
    - When both systems experience different magnitudes of thermal expansion/contraction.
  - The Floating System
  - The Restrained System

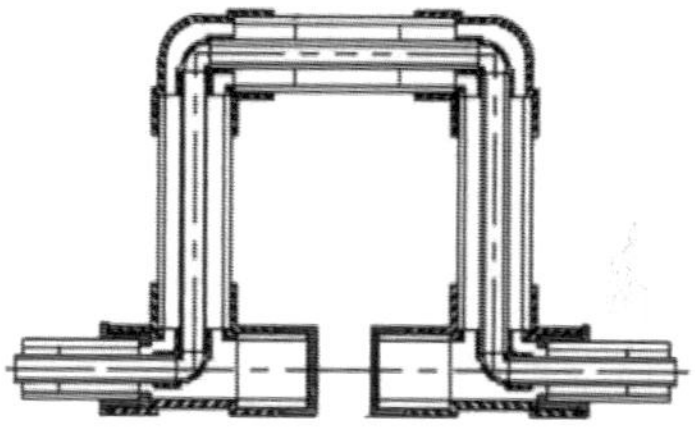

IPEX
Committed to Excellence

Double Containment Piping Systems: Leak Detection Requirements

- EPA requirement: (CFR), Title 40, Part 280
  - Pressurized systems require electronic leak detection that must be checked once per year.
  - Drainage systems (gravity flow) require visual leak detection with monthly manual inspection for a media leak.

Double Containment Piping Systems: Leak Detection Requirements

- The following forms of leak detection are acceptable as per regulations

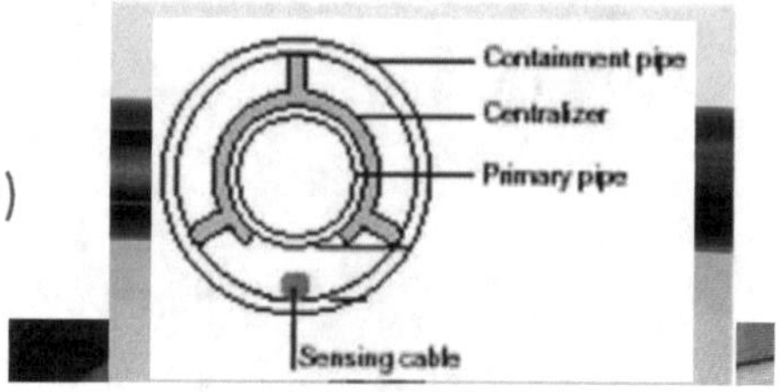

  - Visual (drainage applications only)
  - Electronic low point probes
  - Continuous cable

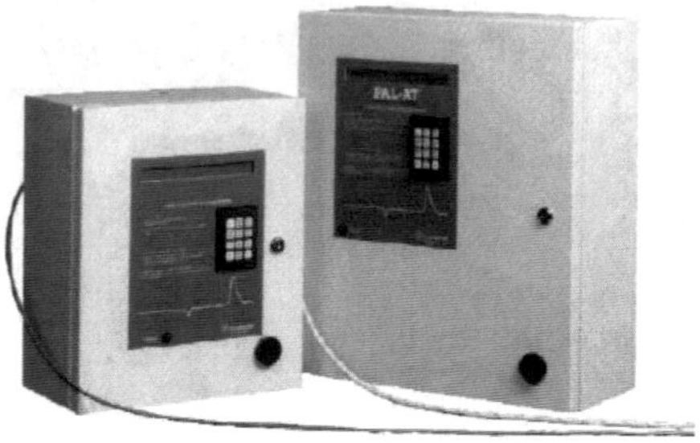

Double Containment Piping Systems: Applications

- Underground Applications
- WTP & WWTP
- Oil & Gas
- Hospitals
- Museums
- Food and Beverage
- Pharmaceuticals
- Laboratories
- Data Centers

Questions?

Thank you!

DTP Process

"Now Ready to License"

JGC Corporation
Mitsuo Morita

2014

JGC CORPORATION

MITSUBISHI CHEMICAL

What is DTP Process ?

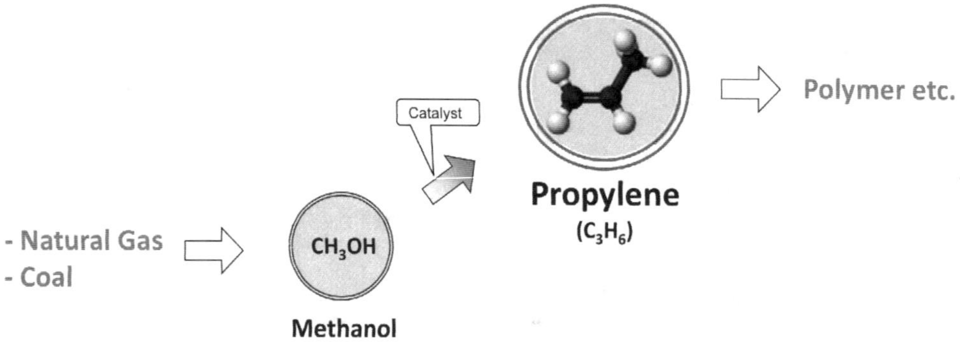

-On-Purpose Propylene Production from Methanol

JGC CORPORATION

- 2 -

MITSUBISHI CHEMICAL

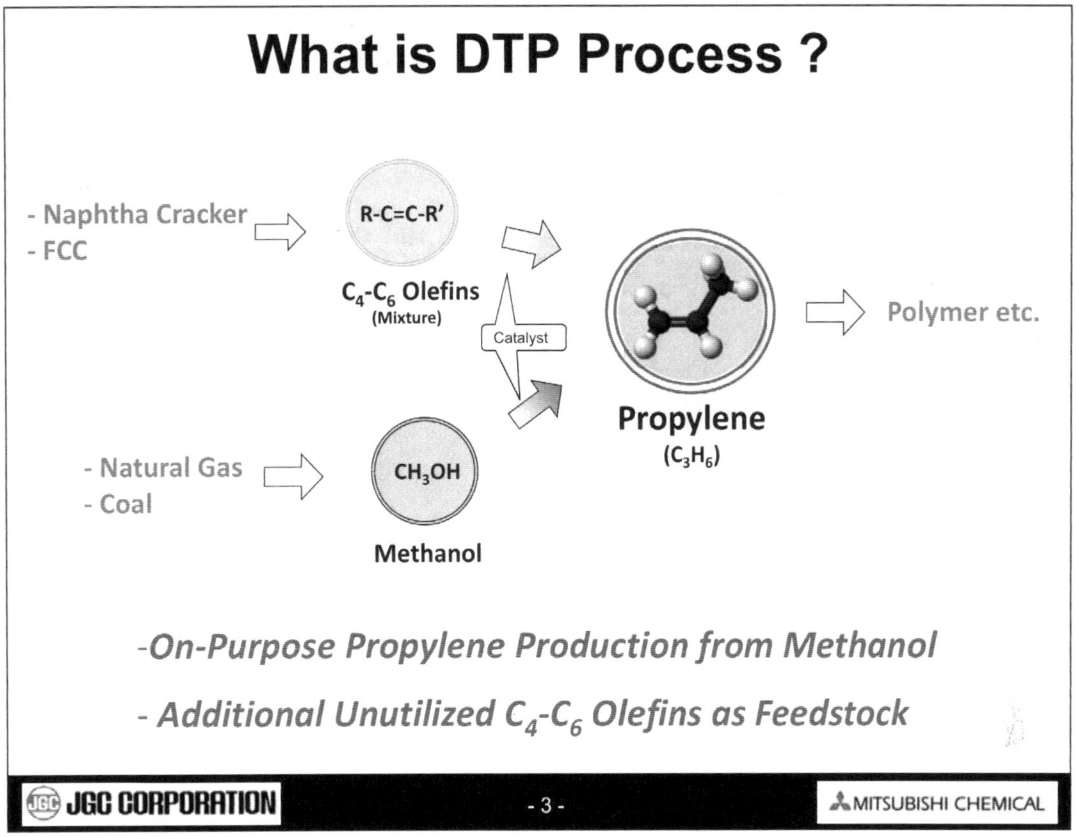
What is DTP Process ?
- Naphtha Cracker
- FCC
R-C=C-R'
C4-C6 Olefins
(Mixture)
Catalyst
Propylene
(C3H6)
Polymer etc.
- Natural Gas
- Coal
CH3OH
Methanol
-On-Purpose Propylene Production from Methanol
- Additional Unutilized C4-C6 Olefins as Feedstock
JGC CORPORATION
- 3 -
MITSUBISHI CHEMICAL

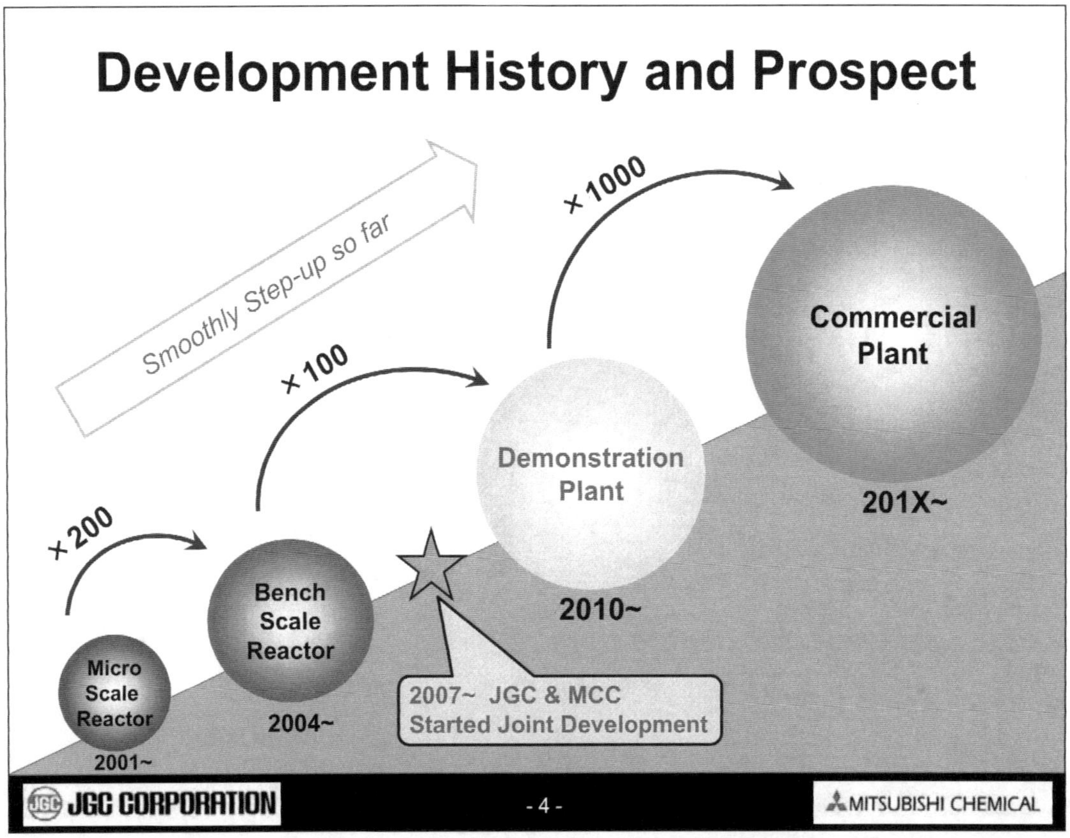
Development History and Prospect
Smoothly Step-up so far
×1000
×100
×200
Commercial Plant
201X~
Demonstration Plant
2010~
Bench Scale Reactor
2004~
Micro Scale Reactor
2001~
2007~ JGC & MCC
Started Joint Development
JGC CORPORATION
- 4 -
MITSUBISHI CHEMICAL

Demonstration Plant Overview

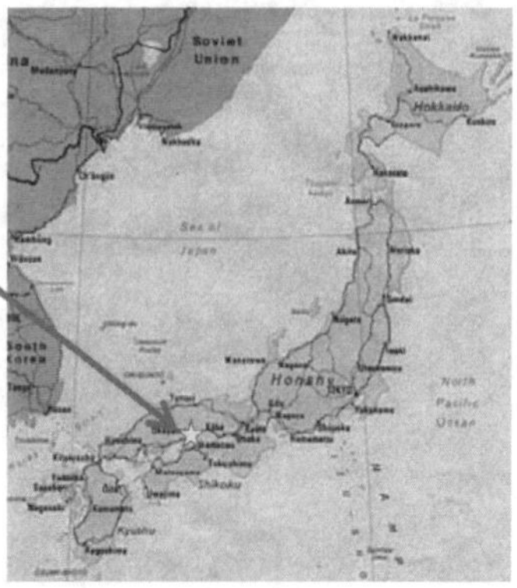

- **Located in Mitsubishi Chemical Mizushima Plant**
- **Successfully Start-up in August 2010**
- **Continuous Operation for Two Years**

Process Flow (Standalone Case)

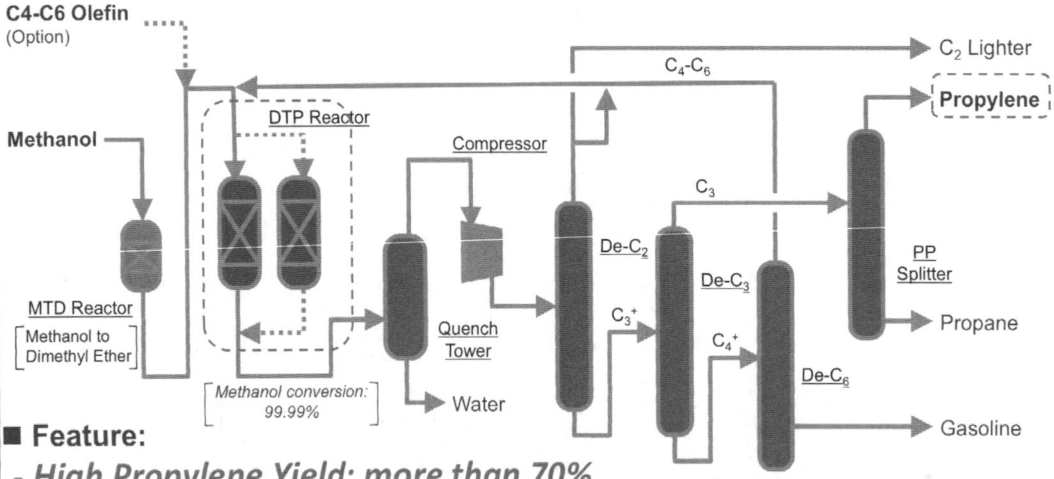

■ **Feature:**

- ***High Propylene Yield: more than 70%***
- ***Applied Existing Proven Technologies***

Process Flow (Integration Case)

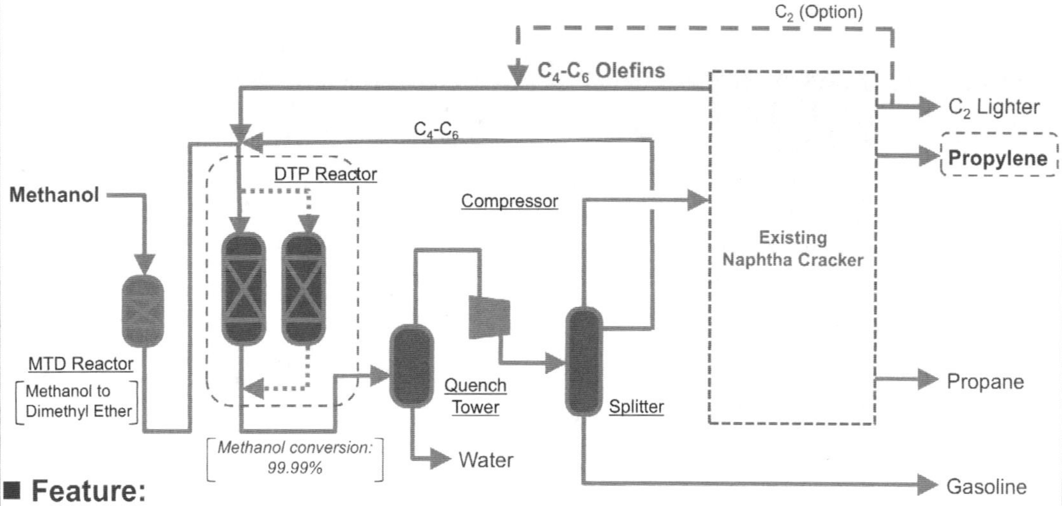

■ **Feature:**

- *High Propylene Yield: more than 70%*
- *Applied Existing Proven Technologies*
- *Low Investment Cost in case of Integration with Naphtha Cracker*

Application

(1) Naphtha Cracker Integration Case (Propylene Capacity: 20KTA – 200KTA)

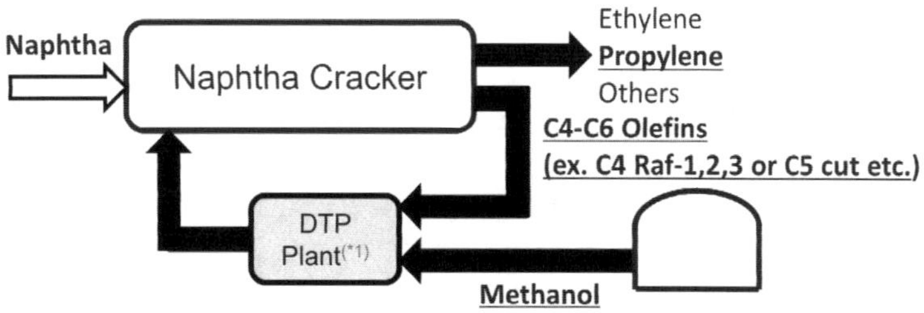

(*1) Utilizing facilities in Naphtha Cracker.

(2) Standalone Case (Propylene Capacity: 250KTA – 500KTA)

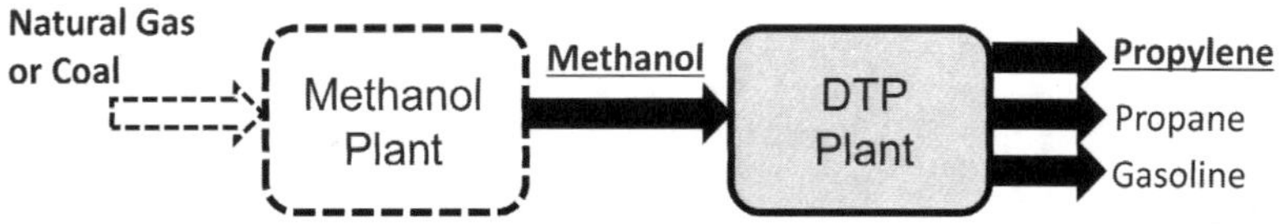

- *Two Cases available depending on Client requirements !!*

DTP is…

- *High Yield Propylene Production Process*
- *Flexible Feed Methanol to Olefin Process*
- *Process that increase P/E Ratio for Cracker*

Thank you very much !!

Contact:
Mitsuo Morita
Market Development Dept.
Global Marketing Division
2-3-1, Minato Mirai, Nishi-ku,
Yokohama 220-6001, Japan

Phone: 81-45-682-8371
FAX: 81-45-682-8183
E-Mail: morita.mitsuo@jgc.co.jp

JGC CORPORATION - 9 - MITSUBISHI CHEMICAL

352533 Self-Cleaning "Bernoulli" Type Filters Used in Onshore and Offshore Applications

Abstract:

Georg Schuenemann GmbH (SAB-BREMEN) has developed a new Filter Type F-480, dubbed the "Compact Giant" for applications in onshore water intake and cooling water systems, sea water desalination plants, and in offshore projects, such as FPSO's, Drillships, and Production Platforms.

This paper will describe the essential design features of the filters, based on the "Bernoulli principle" which result in a fully automated self-cleaning filter system, and choice of materials in a sea water service with high corrosion rates.

A case study will be presented in which stringent project requirements such as substantial flow rates, adaptibility to space limitations and pipe connections had to be observed. Reliability of the system, extended run times without the need for frequent replacement of filter elements, as well as ease of monitoring and operator involvement have been achieved.

352653 Effects of Support on Sulfur Tolerance and Regeneration of Pt Catalysts Measured By Ethylene Hydrogenation and EXAFS

Abstract:

We investigated the effect of support on sulfur tolerance and regenerability under reducing environments by measuring the rate of ethylene hydrogenation, combined with hydrogen chemisorption, and extended x-ray absorption fine structure (EXAFS). Alumina and titania (P25) supports were used to prepare Pt catalysts at 1 wt.%. The catalysts were tested after treatment in H_2S/H_2 at 250 °C followed by regeneration treatments in H_2 at 250 °C, 350 °C and 450 °C. The sulfided catalysts exhibited a 20-27 times decrease in the rate of ethylene hydrogenation and a 4-6 times drop in the Pt surface area. Regenerations up to 450 °C were unable to remove all the sulfur, as shown by the presence of Pt-S bonds by EXAFS. However, a partial recovery of the hydrogenation rate per mole of Pt was observed on sulfided Pt/Al_2O_3 after reduction at 450 °C, while the induction of strong metal support interactions (SMSI) at reduction temperature above 350 °C was observed on Pt/P25, regardless of the presence of sulfur. For Pt/P25, the reversal of the SMSI state together with sulfur removal by mild oxidation suggests that sequential reduction/oxidation treatments may be more effective in restoring the S-free state of TiO_2-supported catalysts.

352799 Improved Operational Efficiency and Reliability through Insulation Materials Selection

Abstract:

This paper will explore the implications of insulation materials selection on the energy efficiency and reliability for Ethylene production units. Considerations will include insulation properties, structural support, installation methods, installation and maintenance costs.

BOILER MACT COMPLIANCE: MERCURY IN FUEL GAS

PATRICK LAINE, WES YONGE, & VIC VICKERY
PEI, HOUSTON, TX 77043

Patrick Laine, Ph.D.
Chief Scientist
Houston, TX
850-866-6686
plaine@pei-tx.com

Wes Yonge, PE
Chief Engineer
Houston, TX
713-806-8396
wyonge@pei-tx.com

James 'Vic' Vickery
Chief Technology Officer
Houston, TX
281-536-0899
vvickery@pei-tx.com

1 Introduction & Objectives

For most ethylene producers the primary objective of this sampling and analysis effort will be to verify and "establish a mercury fuel specification through which gas-fired boilers that use a fuel other than natural gas or refinery gas may be considered Gas 1 unit" as identified in 40 CFR Part 63.7521.(f).

Measurement and Monitoring Requirements: Decision Rules

❑ If Hg concentration ≤ 50% of Gas 1 specification* then no further testing required

❑ If Hg concentration > 50% and ≤ 75% of the specification* then semi-annual sampling required

❑ If Hg concentration > 75% of the specification* then monthly sampling required

*Specification is **40 μg/Sm³**

The goal will be to execute a sampling and analysis campaign that will produce compliance quality data. The key to producing and validating compliance quality data is a well-designed and executed QA/QC protocol. At a minimum, a performance based QA/QC protocol is recommended that includes the following components:

- Sorbent Trap Breakthrough Bed:
  - ❑ Sorbent traps consisting of a minimum of two sorbent beds in series so that the trailing bed can be analyzed to verify that no breakthrough occurred from the leading bed.
- Duplicate Samples:
  - ❑ The sampling system should be configured with two trains to accommodate the collection of simultaneous duplicate sorbent trap samples, in parallel, providing a means of verifying precision in the sample collection process.
- NIST (National Institute of Standards and Technology) Traceable Spike:
  - ❑ Sorbent traps with a third bed (downstream) that is spiked with a NIST traceable spiked mass of mercury serves as a matrix spike and provides a means of verifying that components in the matrix (the fuel gas) are not affecting the trapping efficiency of the sorbent trap.

2 Methods

Sorbent Trap Mercury Sampling Methods Comparison

Method	Description	Method Type: S = Sampling A = Analysis	Evaluation	Sample Collection Phase QA/QC
ISO 6978	Hg in Natural Gas by Thermal Desorption (CVAFS)	S & A	Method calls for low sample volumes can lead to non representative results.	No
EPA Method 30B	Determination of Gaseous Hg Emissions from Coal Fired Combustion Sources using Carbon Sorbent Traps	S	Requires modification (Modified M30B)	Yes
ASTM D6350	Hg S&A in Natural Gas (AFS)	S & A	Method calls for short term samples (i.e., low sample volume) which leads to poor accuracy.	No
ASTM D5954	Hg S&A in Natural Gas (AAS)	S & A	AAS is less sensitive than AFS.	No
EPA Method 1631	Measurement of Hg in Water	A	No	Yes

3 EPA Method 30B QA/QC Performance Criteria

QA/QC Test or Specification	Acceptance Criteria	Frequency
Analyzer calibration	Correlation Coefficient: 0.999	Daily
Transfer efficiency of analyzer	%Transfer Efficiency: 90%–110%	Once per sorbent trap
Initial calibration blank	< 0.5 ng	Once before each calibration
Sorbent Trap Section B breakthrough	≤ 10% of Section A Hg mass, for Hg concentrations > 1μg/dscm ≤ 20% of Section A Hg mass, for Hg concentrations ≤ 1μg/dscm	At each sample point
Spike Recovery Test	Average recovery between 75% and 125 % for Hg_0	Once per sample point
Paired Sorbent Traps Agreement	≤ 10% RD for Hg concentration > 1μg/dscm ≤ 20% RD or 0.2 μg/dscm absolute difference for Hg concentrations ≤ 1μg/dscm	At each sample point

4 Equipment

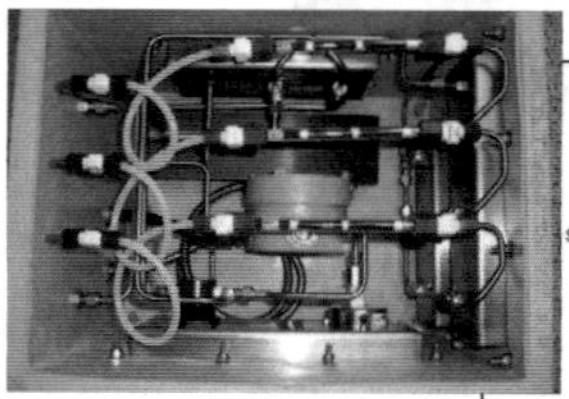

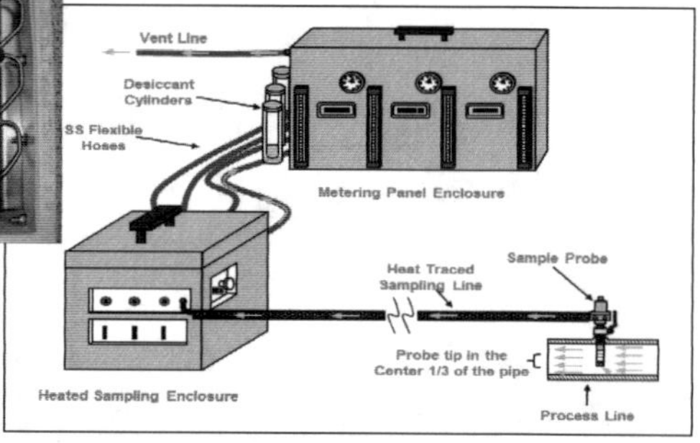

Flow

	Primary Bed	Breakthrough Bed	Spiked Bed
	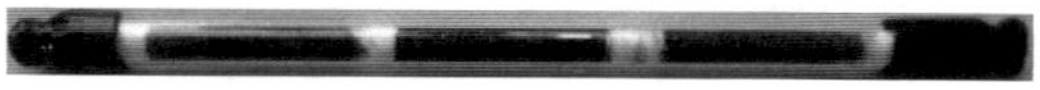		
QA/QC Acceptance Criteria:	99% of Total Collected Hg	≤ 10% of Primary Mass	Recovery between 75% and 125%

Gold Trap:

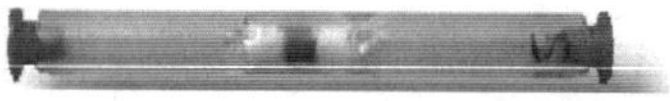

5 Conclusions/PEI Recomendations

❑ There are five methods accepted by the Boiler MACT rule. PEI has extensive experience and expertise employing **all five** of these methods.

❑ Operators can achieve compliance without submitting a Fuel Analysis Plan by using ASTM D6350, however, we have noticed variable results with this method.

❑ The **PEI recommendation** for obtaining the most accurate compliance quality data is to use Modified EPA Method 30B in conjunction with EPA 1631 where traps should be analyzed in a certified ISO 17025 laboratory. The recommended pathway includes a rigorous, performance based QA/QC protocol that validates accuracy and precision of all sample collection procedures.

351376 Young Professional Simulation Tutorial

Abstract:

18C00 Young Professional Simulation Tutorial

Authors: Naomi Hua and Mike Donahue, Invensys

Process simulation is a complex skill to master. Even with years of experience, a typical user is constantly facing new problems that need to be addressed. There are many issues that can lead to unexpected and unnecessary errors with the use of process simulation tools. Regardless of what simulation program you are using, these issues are common from one program to another. In this presentation, targeted for the young professional, we will discuss the "Most Common Pitfalls in Process Simulation". For each topic, we will discuss various techniques and methods these troublesome problems and prevent their occurrence.

Topics to be discussed include:

1. Erroneous data entry

2. Simulation defaults

3. Bad plant data

4. Thermodynamic method selection

5. Simulation tolerances and convergence

6. Recycle identification and handling

7. Model complexity

8. Difficult or impossible specifications

9. Assay pseudo-component development

10. Assay property blending

1. Erroneous data entry is the most common mistake made when building a process simulation model. This presentation will cover methods that can help ensure proper data entry in your process simulation model, such as checking the data you entered, particularly the units of measure, using "alternate forms" of data input such as copy/paste and data import, and reviewing all warning and error messages.

2. Defaults are methods, values, and conditions that are automatically selected for you by the program when you begin a new simulation model. These defaults will affect the results calculated by your model and every simulation program has them. Most defaults within simulation programs fall under categories such as convergence criteria, thermodynamic methods, tolerance and units of measure. The problem with defaults is unintended results from not fully understanding the effects a particular default has on the model. Take the time to understand the defaults of the program and how these defaults will affect the results. If you do not agree with the defaults selected by the program, most simulation programs will allow the user to change the default.

3. Plant data changes over time. The feed information and process conditions of the unit operations may be different. Not only does plant data change over time, but measurement information is often very inaccurate. This could be due to faulty or poorly calibrated measurement equipment, leaks in lines, sensor failures, and other issues. Because of these changes and inaccuracies over time, you always have to validate the information that is entered in your model to ensure that you are obtaining the proper results. To do so, you can utilize programs that can help with data validation and reconciliation like SimSci's ROMeo online optimization program. With it, you can connect your model to your data historian to constantly import current data measurements into your model and then reconcile these measurements to determine when problems arise within your plant and where those problems are occurring. You will also need to use your engineering judgment to determine if the data is valid. More details will be provided during the presentation.

4. The proper selection of thermodynamic methods for your simulation is the most important decision that you make when modeling a process. The method you select could have a drastic effect upon your results. Often, any number of thermodynamic methods is suitable for a given application. It is up to the user to determine the best method to use based upon knowledge, personal experience, and best judgment. Some important factors that need to be considered are the components being used and the operating conditions.

Oftentimes, the selected thermodynamic method still needs to have additional inputs such as binary interaction parameters to simulate the process accurately.

5. Many of the calculations in process simulation are iterative. When these calculations iterate, the program compares the results calculated from one iteration to the results calculated in the next iteration in order to determine

whether it is converged to a solution or needs to continue calculating. This method of comparing results is where tolerances come into play and simulation programs use tolerances to determine when a solution is reached. One of the major mistakes when modeling a process is not understanding or using the correct tolerance values. A looser tolerance may make the file easier and quicker to converge, but it also reduces the accuracy of the model. A tighter tolerance may be more accurate, but might make it impossible for the model to solve or increase the convergence time exponentially. A common practice is that when you have reached a converged solution with your model, tighten the tolerances within the model and solve it again. If you notice that your answers have changed significantly, you need to continue tightening the tolerance values until your results change little from one iteration to another. You will then know that you have reached an accurate solution for your model.

6. Recycles in process simulation occur when you return to a previously calculated unit operation to recalculate the unit due to a change made elsewhere in the model. They are very common and can often be difficult to converge. Some suggestions on troubleshooting recycle convergence include initially solving your model without the recycles to see how it behaves, providing initial estimates for recycle streams, using tighter tolerances on unit operations within recycle loops than the total recycle tolerances themselves, and ensuring that all components have a way in and out of a recycle by performing mass balances and using purge or makeup streams where necessary.

7. The general rule of thumb when modeling is to start simple and add complexity as you go. Each time you add a few streams or unit operations converge and save the model before moving on. Make sure you justify any additions that you make to the model. You can most likely model your system without including every unit from your PFD in the simulation. When modeling, think about the goal of your simulation and not about literally translating every real-world piece of equipment into the simulation model. Make the judgment call on what is most important and what the purpose is of every unit you add to your simulation. For example, if your real-life process shows a heat exchanger, a pump and a splitter, but all that is necessary is that the temperature, pressure, and separation are achieved properly, consider using a flash drum to accomplish all three.

8. Specifications within process simulation are everywhere and every stream and unit operation you place in your model contains specifications. In its simplest form, a specification is a constraint that you place on the model that the model needs to meet in order to successfully converge. For streams, these are usually temperatures, pressures, flowrates and compositions. For unit operations, they range from process conditions, like temperatures and pressures, to product conditions, like an overhead composition in a column. There are some pitfalls, however, that you can run into with specifications such as azeotropes in columns, high purity specifications, and the possibility of multiple solutions.

9. Petroleum assays are sets of distillation data that are used to characterize all of the components that reside in a given crude sample. When you enter distillation data into a process simulator, the program converts it into components that it can use when modeling your process. The components that are generated are called pseudo-components, and each pseudo-component represents all of the unknown components that would be within a given boiling point range in the crude. Some of the things you will want to ensure when properly developing your assay components are did you use the correct distillation curve basis, did you enter the data correctly, did you use enough cuts and are you using the correct characterization methods .

10. When you have more than 1 assay in a simulation model, you need to be concerned with assay blending and how it will affect the results of your file. Assay blending is the mixing of 2 or more groups of components (i.e. ¨C sets of cuts/pseudo-components) in order to calculate averaged component properties for cuts that are in the same boiling point range as one another. Once these new averaged component properties are determined, the program will use these averages from this point on in the simulation model any time it needs to access component

properties for a given pseudo-component. All component properties are blended, including normal boiling points, specific gravity, molecular weight and refinery inspection properties, and usually the blended component properties are calculated using a weighted average

These topics are based on experience and feedback from end-users to technical support representatives for process simulators. They often represent the most commonly asked questions or areas that most users find to be problematic. While the specific examples highlighted in each topic come from SimSci software, the intent of this tutorial presentation is to be broad-based and is particularly applicable to young professionals seeking to develop their experience with simulation.

Polymeric Sulfonic Acid Composites for Fuel Cell Applications: Effects of Casting Solvent and Sonication Time on Membrane Performance

Jimoh K. Adewole[1], *Abdullah S. Sultan*[1,*], *Amir Al-Ahmed*[2], *and S. M. Javid Zaidi*[3]

[1]*Center for Petroleum & Minerals, King Fahd University of Petroleum & Minerals, Dhahran 31261, Saudi Arabia*

[2]*Center of Research Excellence in Renewable Energy, King Fahd University of Petroleum & Minerals, Dhahran 31261, Saudi Arabia*

[3]*School of Chemical Engineering, The University of Queensland, Australia*

Abstract

Polymer blends comprising sulfonated poly (ether ether ketone) (SPEEK) and 3wt% (2-acrylamido-2-methyl-1-propanesulfonic acid) (PAMPS) were prepared by solution casting method. Two solvents (N,N-Dimethylacetamide (DMAc) and N-Methyl-2-pyrrolidone (NMP)) were uesd separately. Residual solvents in the membrane samples after drying were quantified using thermo gravimetric analysis. Effects of periods of sonication (0 to 90 minutes) and types of solvent on the membrane performance were evaluated interms of morphology and water uptake. An attempt was made to correlate membrane performance with the time of sonication in order to select the most suitable solvent and membrane casting conditions for subsequent membrane preparations. For both solvents, the water uptake was observed to first decrease at 10 minutes of sonication and then increased as the sonication time was increased. The minimum water uptake recorded for DMAc samples was 19.23% at 20 minutes sonication time while that of NMP sample was 24.24% at 10 minutes. However membrane morphology was found to deteriorate with increase in sonication time. No correlation was obtained between the membrane properties and the sonication time. In general, membrane performance was better for membranes samples using DMAc than NMP solvent. Therefore, DMAc was selected as the casting solvent for membrane preparation on which further studies will be carried out.

Keywords: morphology, water uptake, residual solvent, thermogravimetric analysis

Introduction

Water soluble polymers have received academic and industrial interests due to their potential applications in fuel cells and other electrochemical devices (Ye et al. 2012). One of the water soluble polymers that have attracted publication interest is poly (2-acrylamido-2-methyl-1-propanesulfonic acid) (PAMPS) (Florjańczyk et al. 2003). Generally, some of the major performance criteria for fuel cell materials include proton conductivity, water and methanol uptake, good mechanical properties, chemical stability, thermal stability and methanol permeability (Gil et al. 2004; Swier et al. 2005; Ye et al. 2012). PAMPS was reported to posses higher proton conductivity than partially hydrated Nafion. In addition, it has been found to be more tolerant to water fluctuation in high temperature operations. Some of the major drawbacks in the use of PAMPS are its poor mechanical properties, thermal stability, water and methanol uptake and solubility in many solvents (Florjańczyk et al. 2003). One of the approaches that have been proposed to solve these problems is development of polymer composites.

This research paper investigates the synergistic effect of combining water soluble PAMPS with sulfonated poly (ether ether ketone) (SPEEK) which is non - water soluble polymer to prepare composite membranes for application in fuel cells. Specifically, preliminary results on the performnance of the composite were presented. Performance were evaluated based on the composite morphology and water uptake. Effects of sonication time and types of casting solvent on membrane properties were also evaluated.

Material and Method

Material

Sulfonated poly (ether ether ketone) (SPEEK) purcahsed from FuMA-Tech, Germany, and poly (2-acrylamido-2-methyl-1-propanesulfonic acid) (PAMPS) purchased from Aldrich Chemical Co., are the two major polymers used in this research. The SPEEK has a DS of 0.533 and solubility of 99% in DMAc. SPEEK is derived by sulfonation of Poly (ether ether ketone) (PEEK). PEEK is a commercially successful polymer due to its excellent thermal stability and solvent resistance (Mohr et al. 1991). For sometimes, further application of PEEK have been stalled due this solubility issues. By sulfonation, PEEK can be made soluble in most commercial solvents. Importantly, the solubility of SPEEK also depends on its degree of sulfonation (DS). Since, this work involves the use of two polymers, a solvent which will disolve both the PAMPS and SPEEK must be selected. Literature review showed that PAMPS dissolves in N,N-Dimethylformamide (DMF), N,N-Dimethylacetamide (DMAc), propylene carbonate (PC), and water (Florjańczyk et al. 2003). Initial solubility studies that was carried out in our laboratory was done for N,N-Dimethylacetamide (DMAc) and N-Methyl-2-pyrrolidone (NMP). The choice of NMP was due its unique solvent properties and its structtural distiction from DMF and DMAc (Figure 1). Moreover, no literature was found where NMP was used as casting solvent for PAMPS. Thus, DMAc and NMP were used as solvents in this work. All materials were used as received from the manufacturers.

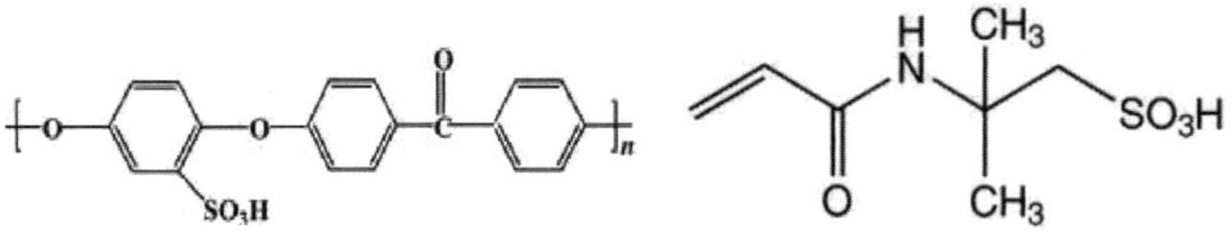

Figure 1 Structure of SPEEK and PAMPS

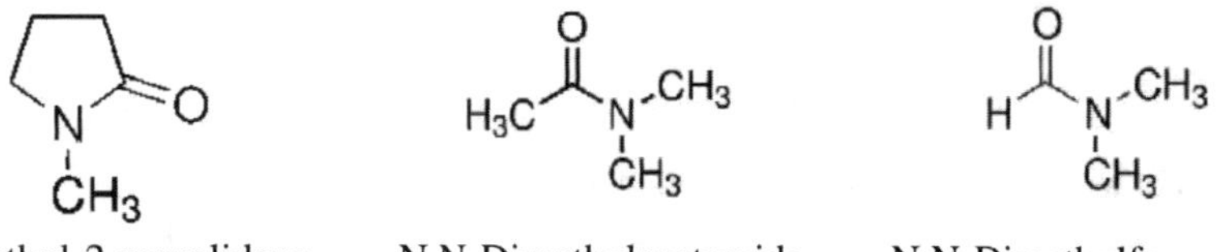

N-Methyl-2-pyrrolidone N,N-Dimethylacetamide N,N-Dimethylformamide

Figure 2 Structure of Solvents

Experimental Procedure

Membrane Preparation

SPEEK was first dissolved in solvents (DMAc and NMP) and stirred for 2hours at 70°C. PAMPS was then added and the blend was stirred for 24hours. The solution was sonicated for various periods of time ranging from 10 to 90minutes using an Ultrasonic bath. In untralsonic bath, the final temperature achieved for a given time of sonication depends on the tempearutre of the laboratory (summer or winter). It is therefoe genrally advisable that maximum temperature of the bath is determined and maintains (as the so - called equilibrium tempearture), when operating continuously under ambient conditions (Santos et al. 2009). During the sonication, equilibrium temperature was maitained by simply filling the bath with water that was preheated to sample temperature.

Membrane Characterization

Water uptake and morphological tests were performed. Water uptake was carried out by first vacuum drying all membrane samples at 100°C until constant weights were obtained. Dried samples were then immersed in deionized water for 24h at room temperature (25°C). Samples were taken out of the water after 24h, wipped to remove surface water and then weighed immediately. Water uptake was calculated using Eq. 1.

$$water - uptake = \frac{W_{wet} - W_{dry}}{W_{dry}} \qquad \text{Eq. 1}$$

Surface morphology of the membrane was done using scanning electron microscopy (SEM). Thermogravimetric analysis (TGA) was done from room temperature to 350°C using TA Q600 analyzer at heating rate of 20°C/min under nitrogen atmosphere with flow rate of 20mL/min using ~ 8 -10mg membranes samples.

Results and Discussion

Membrane Morphology by Scanning Electron Microspcopy (SEM)

Results of surface morphological analysis performed on membrane were shown in Figure 3 and 4. The overall pictures of this figure depicts a dense membrane with some surface defects observed for most of the sonicated samples. For example, Figure 3A showed some kind of layered arrangement with no visible pores at the surface. Also, the PAMPS seems to be well distributed throughout the SPEEK. The micrograph in Figure 3B can be observed to be rough and possesses some uneven ramdom distribution of PAMPS agglomerates indicating that the distribution of PAMPS have been disturbed by the 10minutes sonication. An increase in sonication time to 20 minutes was observed to improve the distribution of PAMPS. This was however accompanied by some cracks believed to be due to the slight increase in the viscousity of the blend solution. The vicousity increase was brought about by the excape of some fraction of the solvent during sonication. As the sonication time was increased, PAMPS was observed to become more evenly distributed and the cracks become more visible and larger in size (Figure 3D to F). SEM images of samples at sonication time of 10 and 30 minutes were reproduced in higher magnification of 10μm (Figure 4) in order to further revealed the effect of sonication on the morphology of the somposites. Similar morphological behaviour has been reported in literature, though for inorganic/organic clusters of nanoclay/epoxy composites membranes (Lam et al. 2005).

Water Uptake

Water uptake have been reported to be affected by the degree of sulfonation, membrane pretreatment, hydration, ambient relative humidity and water temperature (Xing et al. 2004; Wu et al. 2006). Figure 5 showed the effect of sonication time and solvent type on water uptake that was measured at room temperature (25°C). For both solvents, water uptake is same at zero sonication time. There was however a further decrease in the water uptake for NMP at 20 minutes sonication time before it beging to increase. Thus, the best sonication time for both solvent lied between 10 and 20 minutes with respect to water uptake. Interestingly, this result correspond to that reported by Lam (2005) on the effect of ultrasonication on microhardness of composites.

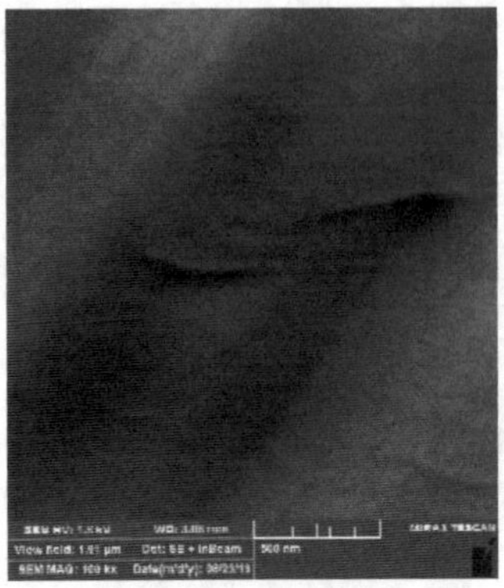

A) No sonication

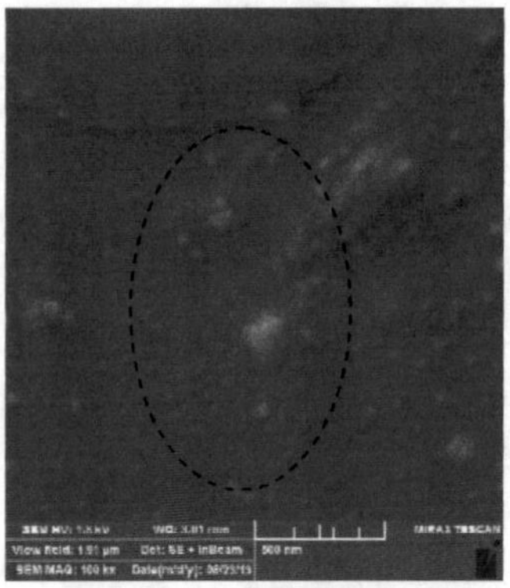

B) 10 minutes sonication

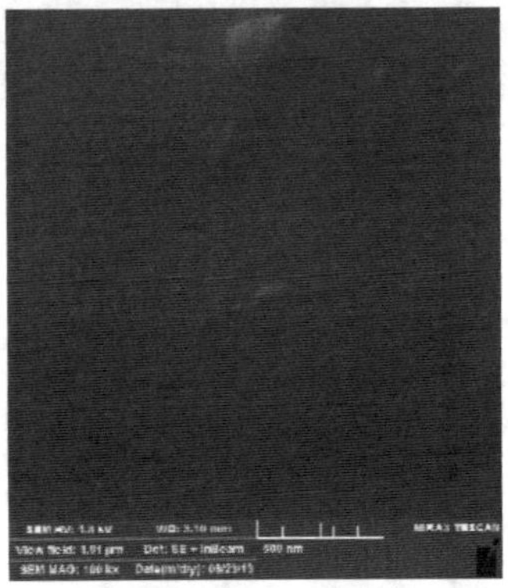

C) 20 minutes sonication

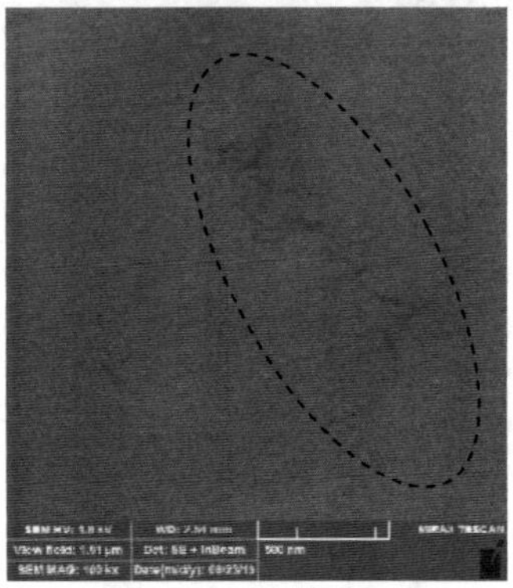

D) 30 minutes sonication

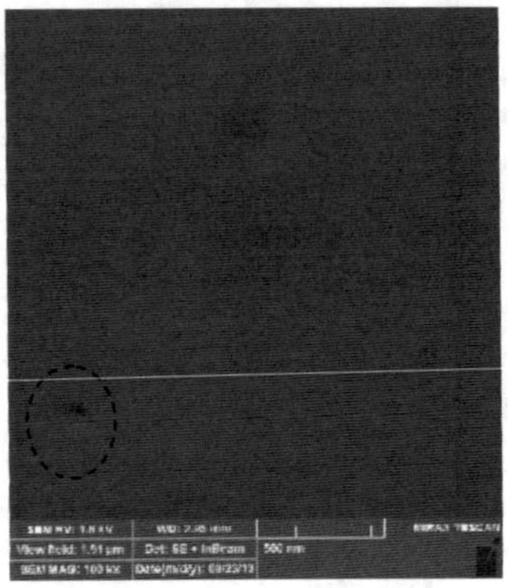

E) 60 minutes sonication

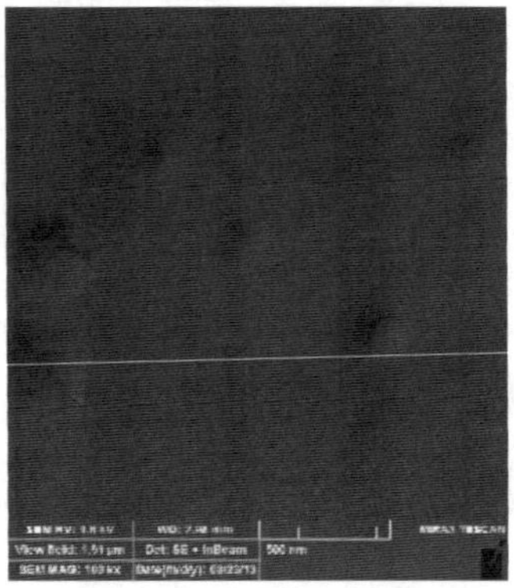

F) 90 minutes sonication

Figure 3 SEM micrograph of SPEEK-PAMS composite with various times of sonication with magnification of 500nm

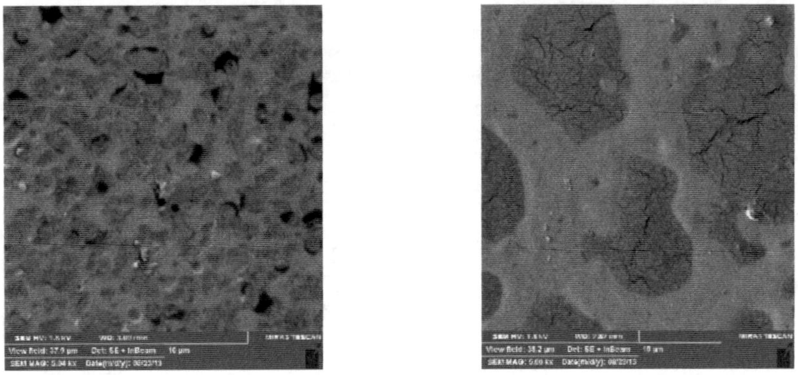

A) 10 minutes sonication B) 30 minutes sonication

Figure 4 SEM micrograph of SPEEK-PAMS composite with various times of sonication with magnification of 10μm

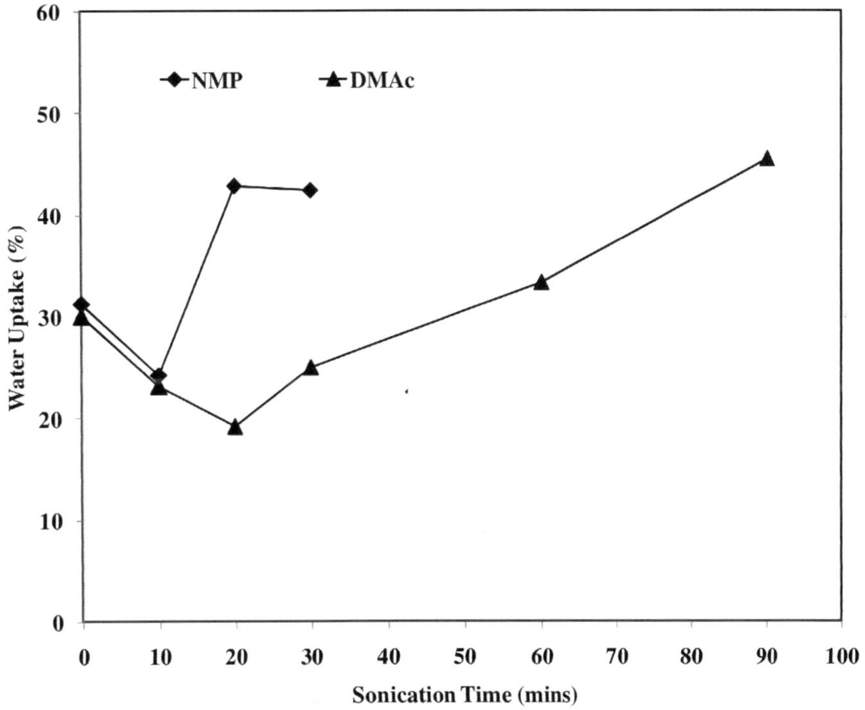

Figure 5 Water uptake of SPEEK-PAMS (3wt%) composite with variuos sonication times

Thermogravimetric analysis (TGA)

The difference in membrane performance resulting from the use different casting solvents have been attributed to the presence of residual solvents in the membrane samples after evaporation (Joly et al.

1999). TGA was thus used to quantify the residual solvent and evaluate its effect on the membrane performance as has been previously demonstrated in lieterature (Recio et al. 2007; Iqbal et al. 2008). Figure 6 shows the TGA curve for samples prepared from NMP and DMAc with different post vacuum heat treatment conditions. From this figure, weight loss can be attributed to evaporation of variety of substances, including solvents, from the samples (Rich et al. 1997). Table 2 contains the solvent and its boiling points, and weigth loss at the boiling point range that were extracted from TGA results. For sample heat treated for 24hrs, the overall percentage weight loss up to 204°C was about 8wt% of membrane sample for NMP while it is about 6.3wt% for DMAc. These will definitely include compnenents such as water, solvents and other unidentified substances. To quantify residual solvents, weight loss was estimated between 192 -214°C for SPEEK-PAMPS-NMP and between 154 and 176 for SPEEK-PAMPS-DMAc. The results obtained showed that the amount of the residual solvent in NMP is larger than that of DMAc by 1.25wt% (Table 1). To further investigate if this difference is responsible for higher value of water uptake for the NMP sample, membrane samples made from both solvents were further heat treated under vacuum for 72hrs. Residual solvent after this treatment was slightly higher in DMAc than NMP. Despite this, the value of water uptake of the NMP membrane is still higher than that of DMAc by approximately the same value as before the heat treatment (1.71wt%).

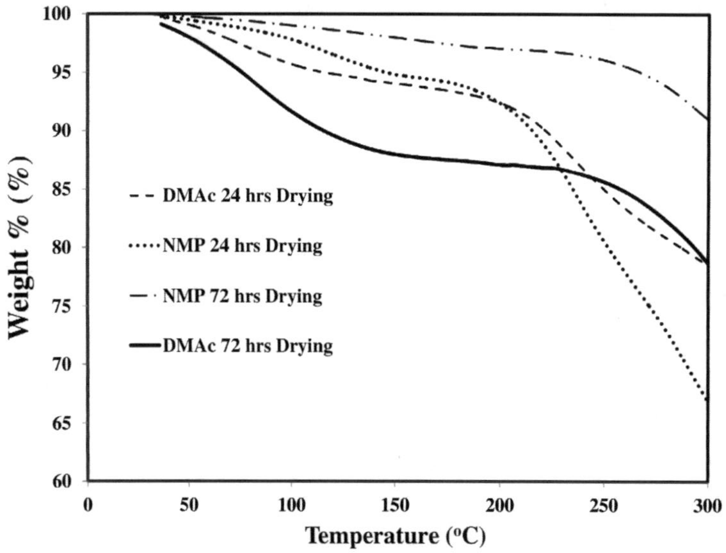

Figure 6 TGA of SPEEK-PAMPS-NMP and SPEEK-PAMPS-DMAc

Table 1 Residual solvent in SPEEK-PAMPS composites membrane samples

Membrane Sample	Solvent	B. Pt. (°C)*	Residual Solvent (wt%)	Water Uptake
DMAc 24 hrs Drying	DMAc	164.5 – 166	0.57	30.00
NMP 24 hrs Drying	NMP	202 – 204	2.85	31.25
DMAc 72 hrs Drying	DMAc	164.5 – 166	0.367	21.21
NMP 72 hrs Drying	NMP	202 – 204	0.217	22.92

*Boiling Points of Solvents

On a general note, results from Figire 1 – 5 showed that the best perfomance can be obtained for variety of sonication time dependin the property of preference. Also, performance of DMAc solvent was observed to be better than that of NMP on the basis of water uptake. For this reason, DMAc was selected as the casting solvent for membrane preparation on which further studies will be carried out. Water uptake of DMAc membrane, prepared without sonication, was comapred with those reported in previuos literature. Table 2 illustrates the comparison between SPEEK-PAMPS and other relevant membranes previuosly reported in literature. In addition, this table serve to provide insight into the possible potentials of the composites. From this table, it was observed that the water uptake for SPEEK-PAMPS is within the range of values of other membranes on the table. Its water uptake is less than that of Nafion 112 which is a commercial membrane. The differences observed between the water uptake values of SPEEK – PAMPS and other membranes from the table can be explained based on three important factors - SO_3H group, the degree of sulfonation and experimental temperature (Gil et al. 2004; Shan et al. 2006; Wu et al. 2006). Water uptake of SPEEK – PAMPS is higher due to the presence of SO_3H group in the PAMPS molecules (Figure 2). SO_3H group have been reported to enhance membrane acidity and hydrophilic properties (Shan et al. 2006). Infact, the higher the SO_3H group the larger the water uptake. The degree of sulfonation has been reported to affect the water uptake of membranes. The higher the degree of sulfonation, the larger the value of water uptake (Gil et al. 2004). Thus, water uptake of SPEEK – PAMPS is lower than SPEEK -3 due their different degree of sulfonation, 53.3 and 72% respectively and the experimental temperature 24 and 60°C, respectively. The experimental temperature of SPEEK – 2 is close to that of the composite. However, the results cannot be compared in this case since the DS of SPEEK – 2 was not known. Above explanation is also true for the values of water uptake for SPEEK – 1.

Table 2 Comparison of water uptake values obtained with those from Literature (Xing et al. 2005)

Membrane	Water Uptake (wt%)	Experiment Temperature (°C)	**D S (%)	Reference
Nafion 112	22.85	N/A	-	(Xing et al. 2005)
SPEEK – 1*	16.00	30	45	(Ren et al. 2005)
SPEEK – 2*	44.90	25	N/A	(Zhong et al. 2008)
SPEEK – 3*	85.00	60	72	(Wu et al. 2006)
SPEEK– 4*	15.00	N/A	N/A	(Handayani and Dewi 2011)
SPEEK/PAMPS (3wt%)	21.21			This work

*The numbering have been included for explanation purposes.

**DS means Degree of Sulfonation

Conclusion and Recommendations

Blend of SPEEK and PAMPS was prepared by solvent casting method and its performance was evaluated with respect to its morphology and water uptake. Effect of sonication time (10 – 90minutes) and solvent types (DMAc and NMP) were also investigated. Performance of membrane prepared using DMAc was observed to be better than that of NMP. There was no correlation between the sonication time and membrane properties, but it was observed that the sonication deteriorate the membrane morphology. Membrane prepared without sonication shows better surface morphology, whereas the best water uptake was observed between 10 and 20 minutes. Thus, other properties need to be investigated in order decide the best sonication time. Therefore, further investigation such as thermal stability, methanol permeability, ion exchange capacity and gas (O_2 and H_2) permeability of the membrane are recommended for a complete evaluation of the membrane performance for direct methanol fuel cell (DMFC) application. All these investigations are in progress in our laboratory.

Acknowledgement

Authors would like to thank the National Science, Technology and Innovation Program (NSTIP) of KACST for funding this research under project number 10-ENE1374-04. Also, authors extend their appreciation to Center of Petroleum & Minearls, Center of Excellence in Renewable Energy, Mr. Dan Jerry D.Cortes of the Nanotechnology Laboratory, Chemical Engineering Department, King Fahd University of Petroeleum & Minerals, Saudi Arabis for their supports.

REFERENCE

Florjańczyk, Z., et al. (2003). "Proton conducting electrolytes based on poly(2-acrylamido-2-methyl-1-propanesulfonic acid)." Electrochimica Acta **48**(14–16): 2201-2206.

Gil, M., et al. (2004). "Direct synthesis of sulfonated aromatic poly(ether ether ketone) proton exchange membranes for fuel cell applications." Journal of Membrane Science **234**(1–2): 75-81.

Handayani, S. and E. L. Dewi (2011). "Influence of Silica/Sulfonated Polyether-Ether Ketone as Polymer Electrolyte Membrane for Hydrogen Fueled Proton Exchange Membrane Fuel Cells." Internat. J. of Sci. and Eng. **2**(2): 26-30.

Iqbal, M., et al. (2008). "Solvent effect on morphology and CO2/CH4 separation performance of asymmetric polycarbonate membranes." J. Membr. Sci. **318**(1–2): 167-175.

Joly, C., et al. (1999). "Residual solvent effect on the permeation properties of fluorinated polyimide films." Separation and Purification Technology **16**(1): 47-54.

Lam, C.-k., et al. (2005). "Effect of ultrasound sonication in nanoclay clusters of nanoclay/epoxy composites." Materials Letters **59**(11): 1369-1372.

Mohr, J. M., et al. (1991). "Gas transport properties of a series of poly(ether ketone) polymers." Polymer **32, 13**: 2387-2394

Recio, R., et al. (2007). "Gas separation of 6FDA–6FpDA membranes: Effect of the solvent on polymer surfaces and permselectivity." J. Membr. Sci. **293**(1–2): 22-28.

Ren, S., et al. (2005). "Surface modification of sulfonated poly(ether ether ketone) membranes using Nafion solution for direct methanol fuel cells." Journal of Membrane Science **247**(1–2): 59-63.

Rich, D. C., et al. (1997). "Effect of cure conditions on Probimide 32 polyamide-imide." Journal of Applied Polymer Science **63**(9): 1113-1126.

Santos, H. M., et al. (2009). The Power of Ultrasound. Ultrasound in Chemistry: Analytical Applications. J.-L. Capelo-Martinez. Weinheim, Wiley-VCH.

Shan, J., et al. (2006). "Sulfonated polyether ether ketone (PEEK-WC)/phosphotungstic acid composite: Preparation and characterization of the fuel cell membranes." Pure Appl. Chem **78**(9): 1781 - 1791.

Swier, S., et al. (2005). "Sulfonated poly(ether ketone ketone) ionomers as proton exchange membranes." Polymer Engineering & Science **45**(8): 1081-1091.

Wu, H.-L., et al. (2006). "Preparation and characterization of poly(ether sulfone)/sulfonated poly(ether ether ketone) blend membranes." European Polymer Journal **42**(7): 1688-1695.

Xing, D. M., et al. (2005). "Characterization of Sulfonated Poly(Ether Ether Ketone)/Polytetrafluoroethylene Composite Membranes for Fuel Cell Applications." Fuel Cells **5**(3): 406-411.

Xing, P., et al. (2004). "Synthesis and characterization of sulfonated poly(ether ether ketone) for proton exchange membranes." Journal of Membrane Science **229**(1–2): 95-106.

Ye, Y.-S., et al. (2012). "Water Soluble Polymers as Proton Exchange Membranes for Fuel Cells." Polymers **4**(2): 913-963.

Zhong, S., et al. (2008). "Modification of sulfonated poly(ether ether ketone) proton exchange membrane for reducing methanol crossover." Journal of Power Sources **180**(1): 23-28.

Fixed Plant Facility Siting – A Mega Project Lifecycle Framework

M.F. Haider
mech_haider@yahoo.com
mhaider@suncor.com

Prepared for Presentation at
American Institute of Chemical Engineers
2014 Spring Meeting
10th Global Congress on Process Safety
New Orleans, LA
March 30 – April 2, 2014

Fixed Plant Facility Siting – A Mega Project Lifecycle Framework

M.F. Haider
mhaider@suncor.com
mech_haider@yahoo.com

Keywords: Facility Siting, Project Management, Mega Project, Consequence Analysis (CA), Quantitative Risk Assessment (QRA), Adaptive, Change driven, Agile methodology

Abstract

Facility Siting based on tools such as Consequence Analysis (CA) and Quantitative Risk Assessment (QRA) is often times successfully applied for Brownfield operating unit projects, projects with relatively smaller footprints and for projects lasting relatively short time-periods. Challenge remains to apply and realize the full potential and value of Facility Siting for Mega-Projects. This is mostly due to project complexity and length – this paper will introduce a Management Framework to successfully apply CA and QRA as part of Facility Siting for Mega-Projects. The paper will also outline key strategies to ensure successful (i.e. value adding) implementation of facility siting, suggest a list of activities, associated sequence and pitfalls to avoid. Project Management team (e.g. Project Sponsor, Project Engineers, Managers) of both owner/operator and contractor/vendor may find this paper more useful as it provides a broad overview (i.e. when to initiate, how to plan, execute and close such assignment to maximize value) including real life examples of facility siting process over the project lifecycle.

1. Introduction

The driving force for process safety has been primarily based on catastrophic events [1]. With an increasing number of incidents, the process industry and governments started to take measures such as Facility Siting initiatives with hope to minimize loss of life and property. In the 1970s and 1980s, some of the world's most tragic industrial accidents took place. The Flixborough explosion in 1974 was reinforced by that of the Seveso tragedy in 1976 [2]. The Macondo blowout in the Gulf of Mexico on April 20, 2010, killed 11 employees and led to an uncontrolled oil spill lasting 87 days. This blowout was the most significant offshore incident in the US to-date. Finally, the Fukushima Daiichi nuclear plant incident in Japan in March 2011 drew the attention of the global process and power industries and encouraged them to incorporate natural disaster risks in a hazard analysis study [3]. As part of inherently safer design, many companies now-a-days apply CA and QRA as part of facility siting for their new projects. Due to the scarcity of information at the beginning, the studies in many cases start later in the project lifecycle for them to be effective and value adding. This paper will focus more on how to manage such initiatives for new Mega-Projects over the technical intricacies of the CA/QRA processes. The intention of the paper is to enhance the Body of Knowledge publicly available for management team in as much non technical languaga as possible to make their greenfiled Mega-Project initatives a success i.e. safer for people, environment and property.

So, what is different for industrial Mega-Projects?

- Industrial Megaprojects, as the name implies, are very large. To provide a simple and simply applied definition, we are defining a megaproject as any project with a total capital cost of more than $1 billion (U.S. dollars) as measured on January 1, 2003 [4].
- Most often engineering and construction activities run in parallel and not in sequence
- Multiple sub-projects under the 'Mega-Project umbrella' run in parallel
- Complex – design is almost 'never' complete in early stages – goes through many iterations
- Progressive elaboration is evident
- Scarcity of critical workforce with mega-project experience
- Scarcity of information – requires project team to make assumptions for CA/QRA
- Timeline: 2-5 years

2. Key Strategies

2.1 Facility Siting Charter

It is not only important or good to do but critical that the stakeholders need to identify, understand and agree on the purpose and value to conduct facility siting for their projects. The Author's personal experience over the years has been in order to maximize value of Facility Siting the project team has to start 'early' to develop project Charter and start with the end in mind. As can be seen from the following generalized Figure 1 that as the company starts to appraise opportunities, it is

Figure 1: Stages and Milestones to apply CA/QRA in a Mega-Project

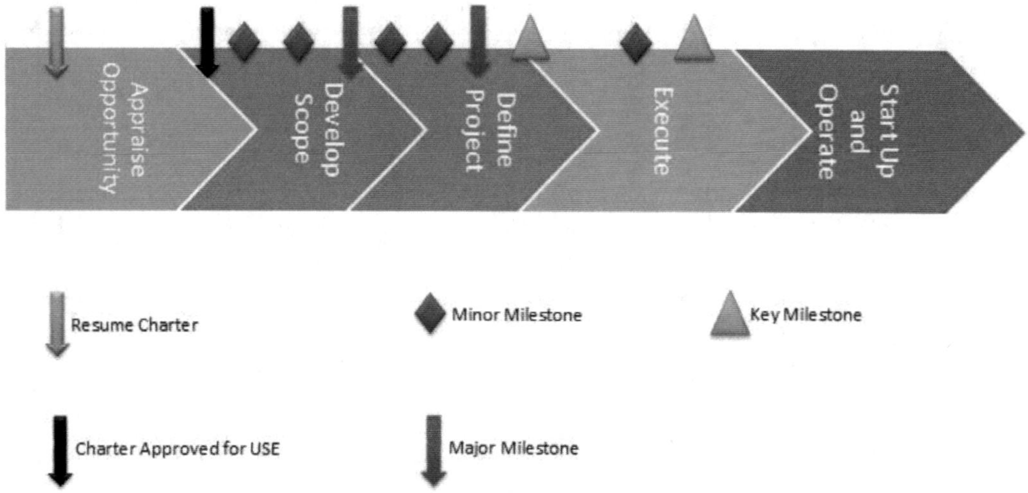

prudent to resume developing the Facility Siting Charter by bringing in the stakeholders together. Figure 1 shows a generalized project lifecycle for a typical Petro-Chemical Mega-Project. The

Figure also shows that the Key Deliverable is developed through a successive series of repeated cycles over the project life-cycle. The Milestones will be defined later in the paper. Part of starting with the end in mind is to clearly outlining ultimate goal/s, e.g. vision is to design a safer facility and utilize Mitigated QRA Report to formally document and communicate that facility is within the Risk Threshold or justification for accepting risk profile as is by Senior Management. The Vision in the Charter also provides opportunity to strike a balance between business goals and climbing the Risk Mountain – at each stage of increasing safety (decreasing risk). As part of striking a balance, the Project Engineer has to Identify options up-front if and/or when it becomes apparent that aggregate Residual Risk will not be within the Risk Threshold or acceptable so that team can put more effort in developing justification for either accepting or rejecting the risk profile by Management

Figure 2: Strike a balance between Business goal and Climbing the *expensive* Risk Mountain

2.2 Management Strategy

In order to successfully manage Facility Siting, CA and QRA for a Mega-Project, the stakeholders have to align on a management strategy. There are two primary management approaches for the purpose of managing such work:

1. Predictive or Plan driven
2. Adaptive or Change driven

The Author's past experience has been if the organization undertook very 'similar' projects in the near past (<5 years), critical experienced workforce is retained, Lessons Learned is readily available, and most importantly, the past project is still considered to be a success, it is advisable to use Plan driven approach. On the other hand, if the organization is introducing new technology, has limited past experience with similar project, involves relatively high number of stakeholders, capital cost > $1B, project life spans >2 years, it is recommended to utilize Change driven approach to maximize value. It is to be noted that it is not essential to meet all the suggested criteria mentioned above to pick one management strategy over the other.

2.2.1 Benefits of Adaptive or Change Driven Approach

In Adaptive/Change Driven Planning, project team intentionally repeats one or more project activities as the project team's understanding of the product increases ……….Also known as …agile methods …..intended to respond to high levels of change and ongoing stakeholder involvement [5]. The heart of Facility Siting involves making timely but informed decision. Some of the characteristics of this framework:

1. To add small incremental improvements that will deliver value to stakeholders
2. Milestones develop the product (e.g. Mitigated QRA Report) through a series of repeated cycles, while increments successively add to the functionality of the product
3. For each Milestone, a deliverable (read 'model') or set of deliverables will be completed. Future iterations may enhance those deliverables or create new ones. Each iteration incrementally builds the deliverables until the exit criteria for the phase are met, allowing the project team to incorporate feedback.

The biggest challenge with Mega-Project is scarcity of key information needed as inputs to develop scenarios and models. Some of the critical benefits, a project may gain by applying Change Driven approach:

1. The approach acts as Leading Indicator on Process Safety for Plant Layout - allows for multiple opportunities to make critical decisions on plant layout over the project life-cycle capitalizing on what is known at a given time. This lets the project team to use "what they know", to ensure "what they don't know" falls into the right place
2. Opportunity to make decisions over the project lifecycle not only works to save life but also helps with striking a balance between business needs and climbing the risk mountain
3. Aides design-development with progressive elaboration while producing final CA/QRA deliverable for the project

2.3 Execution Strategy

The main idea of Change Driven process starts with decomposing the known preliminary layout and process information into a set of requirements, work to be performed and their priority sequence, sometimes referred to as a *Product Backlog.*

It is advisable that the owner/operator forms a multi-use agreement (contract) with the chosen vendor. This way the plant areas under the Mega-Project umbrella can get their specific work done while the CA/QRA Project Team provides overall oversight and guidance.

At the beginning of an iteration, the team will work to determine how many of the highest priority items on the backlog list can be delivered within the next iteration. The time-length of each iteration should be less than 4 weeks. As shown in Figure 1, during "Develop Scope" and "Define Project", the length of each iteration could be even less than 2 weeks. This iterations (otherwise known as "Minor Milestone" as shown on Figure 1) will develop scenarios, assumptions and preliminary CA/QRA models. Assumptions are required due to lack of available quality information in the early stages of a project.

At the end of each iteration, the Minor Milestone/deliverable should be ready for review by the customers/stakeholders. This does not mean that the customer is required to accept delivery, just that the product should not include unfinished and/or incomplete features.

The sponsor and customer representatives should be continuously engaged with the other project stakeholders to provide feedback to Contractor on deliverables as they are created and to ensure that the *Product Backlog* reflects Sponsor/stakeholders' current needs. Below, you will find two examples using two different management strategies. For confidentiality purpose, the specific name/location is kept generic and few items were modified to suit.

2.3.1 Project X Facility Siting Using Change Driven Approach

A CA/QRA team was given the assignment to support design development during "Develop Scope" phase (Figure 1). The Sponsors already selected the site for the project. But the fixed plants were yet to be located on the site. The Vision for the initiative was for the QRA result to come under the organization Risk Threshold. Multiple EPC's were given to engineer different plant areas. The CA/QRA team selected a vendor to support them with defining the *Product Backlog* based on the preliminary Process Design Basis, intended product the facility is expected to sell, the feedstock necessary etc. The CA/QRA team had prioritized the Product Backlog based on hazard level of feedstock, preliminary PHA schedule and project schedule. This also involved obtaining feedback from different Process Engineering and Operations teams to collect necessary inputs to put together the assumptions and scenarios to be modeled within the 1st iteration.

Out of 25+ scenarios, only 7 perceived to be the most hazardous were modeled for CA that gave ample input (e.g. explosion, toxic and fire contours) for the CA/QRA team to work with the stakeholders to recommend location for specific process units on the given site. Note that, CA/QRA team integrated the potential consequence to understand local incident and global consequence events. At this time QRA was not performed. This was used as input for minor modification of preliminary plant layout. It is to be noted that the site plan was still preliminary after the first iteration but assisted different EPC accountable for their specific area to recognize where to locate the specific plants i.e. finalized key plans for the overall site plan.

With further design-development, more refined information became available from each EPC. This new information led to 130+ scenarios and was used for further CA/QRA iterations that created either new model and/or embellished the past ones. Each EPC was accountable for facility siting within their boundary while the CA/QRA team provided technical oversight and integration support for the overall project. The Sponsor/stakeholders were engaged on a regular basis to ensure that the Product Backlog reflects future modeling requirements to properly site facilities e.g. Fire Water Pump house, Tank Farms, site Medical Centre etc. Early engagement of CA/QRA team and higher management willingness to incorporate their feedback with the progressive elaboration of the project led to QRA result meeting the organization risk threshold and ultimately safer facility. Changes suggested based on CA/QRA work in earlier stage (i.e. Develop Scope, Define Project) were easier to incorporate into the design.

2.3.2 Project Y Facility Siting Using Plan Driven Approach

The sponsors for project Y selected the site for the project and decided to hold on to not perform CA/QRA work with the progressive elaboration of the project activities. This was mostly decided based on the fact that not all information was available to properly conduct the 'studies'. Fundamentally, the sponsors looked CA/QRA as deliverables requiring a more 'complete' plan. Once the Develop Scope (as referred in Figure 1) is almost complete, the sponsors decided to engage CA/QRA team and a vendor. By this time, the project had significantly high number of stakeholders. Many of the stakeholders were not very familiar with CA/QRA and the value it could bring to them. There was not much time to schedule and train the stakeholders as well to bring them up to speed. As a result, the stakeholders were unable to align on a Vision. The Vendor was asked to provide explosion, toxic and fire contour maps for specific scenarios and locations that were almost 'fixed' on the site plan at that point in time. The resultant contour maps showed significant consequence on nearby facility with over 700+ people. It was also noted that the plant area consuming the most hazardous substance would impact the nearby facilities significantly from consequence perspective. It should be noted that the result was only based on a selected few scenarios. By the time the project reached the "Execution" phase, more scenarios were added as part of the scope. But it was already too late to incorporate most of the recommended changes resulting from CA/QRA work. The project went through several area by area risk reduction workshops only to accept (by the company executives per organization risk matrix) the aggregate residual risk level which was also found to be not within the organization risk threshold.

2.3.3 Importance of Starting Early

Figure 3: Impact of variable based on project time [5]

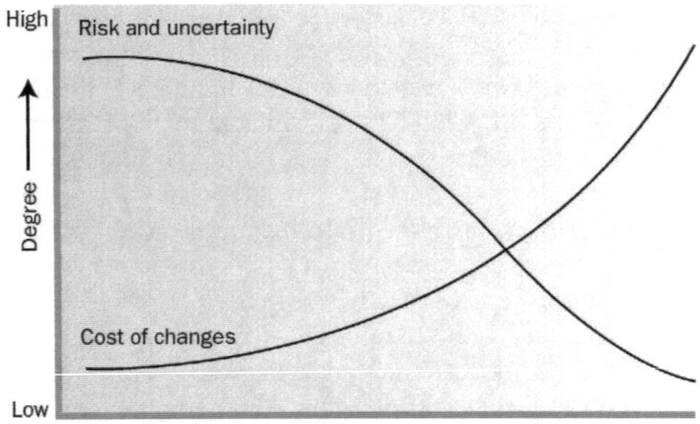

The above two examples highlight the importance of starting CA/QRA work early enough to realize and maximize the value CA/QRA brings to a Mega-Project. The high-level view provides common frame of reference for comparing projects for CA/QRA initiative, even if the projects were dissimilar. Change driven methodology allows for applying the known information at early stage in a way that supports laying out plant facilities to reduce risk and thus maximize potential value. While these characteristics remain present to some extent in almost all project life cycles, they are not always present to the same degree [5]. Change Driven framework, in particular, keeps stakeholder influences higher and the costs of changes lower especially in the early stages of the life cycle than in predictive/plan life cycles. The Figure 3 highlights a well-known fact of project management that even though risk and uncertainty remains high in the early stages of a project, cost of incorporating changes remains much lower. This translates into a key opportunity for project stakeholders to align on laying out the key areas on a site plan intelligently to reduce risk and maximize value by applying tools such as CA and QRA.

2.4 Key Decision Points

Facility Siting, from a management point of view, is all about bringing the stakeholders closer and find the 'right' spot for the facilities. Finding the 'right' spots requires making timely decisions. The challenge has always been to make timely decisions as the stakeholders more often find themselves forced to make those decisions based on information that are frequently of poor quality and at times uncertain, especially in the early stages of a project. At times, due to the scarcity of information, stakeholders rely on their past experience to make those decisions on plant layout. As a result, depending on the individual experience, two different teams for a project area may come up with two very different plant layout. CA/QRA work, if started early enough, will also depend on input that most often requires embellishment through assumptions to create needed scenarios and models. Even with this limitation, CA/QRA provides a key 'input' in allocating real estate for key facilities on a plant site at early stage of a project lifecycle. With each iteration and deliverable 'Change driven' methodology allows for multiple opportunities to make critical decisions on plant layout over the project life-cycle. It allows the stakeholders to decide when to 'sharpen the pencil' (e.g. PHAST to Computational Fluid Dynamics model), relocate facility or accept the consequence and/or risk, 'pull the plug' on further modeling/refinement and start developing justification, if required, to recommend management to accept/reject the aggregate risk profile.

2.5 Alignment with Process Hazard Analysis (PHA) Planning

CA/QRA results are a useful input to PHA (e.g. HAZOP, LOPA) activities. This is particularly true as projects move forward after establishing the plant layout on a high level and starts to conduct "Define Project" (Figure 1). PHA activities may also result in recommendations (i.e. new scenarios) that goes into the *Product Backlog* and requires project team to develop new models or refine old ones. Mega-projects have multiple sub-projects running in parallel most of the time, having an Integrated Schedule for all the reviews is a good strategy so that the right stakeholders are booked for the sessions. The integrated PHA schedule also supports prioritization of *Product Backlog* for CA/QRA activities over the lifecycle of the project.

2.6 Integration Strategy

It is critical for a Mega-Project involving multiple plant areas/silos to understand the BIG picture so that site risk can be dealt with on an aggregate basis. There is strong possibility of one area to impact other plants in the vicinity. It is recommended to review integration aspect and aggregated results on a regular basis. The following Figure 4 shows that the origin of an incident may reside in one particular plant area, while it may impact other areas in the vicinity.

Figure 4: Consequence of an incident in one area impacting multiple areas

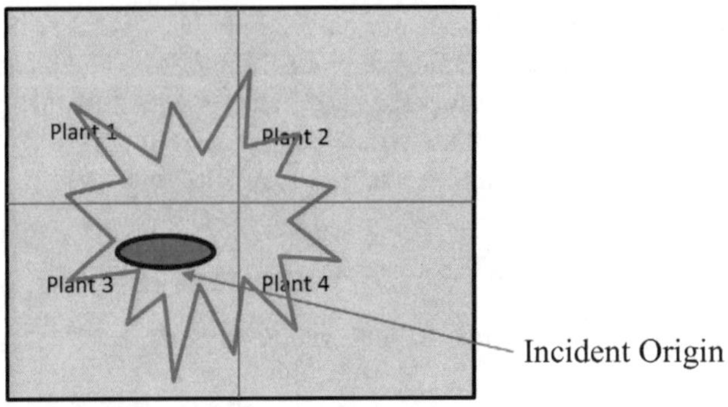

Recommended Strategy:

- Setup Strong Center - Task Force Single Point of Contact (SPOC)
- Collaborate – SPOC will promote Collaboration among Plants/Areas

3. List of Activities (suggested)

3.1 Overall Sequence

The CA/QRA work typically should start with the approved Charter as shown on Figure 1. A project Task Force i.e. SPOC is formed comprising of Project Engineer, Operations rep, Process rep and CA/QRA Technical Expert. The SPOC is ultimately accountable to produce the CA/QRA deliverable studies but also acts to support the design development and site plan finalization efforts in the early stages of the project.

The team collects 'input' such as Preliminary site Layout and design information, works with area process and operations team to identify scenarios, develop assumptions for known-unknowns and unknown-unknowns, develop modeling 'Product backlog', prioritize backlog based on PHA schedule and produce modeling output (i.e. deliverable) for stakeholder use. The stakeholder usually has the option of accepting or rejecting consequence. If the result of an iteration is acceptable within area causing the incident – communicate to outside stakeholders to ensure there is no cross boundary impact. If the result of the iteration is NOT acceptable to outside of the area – use the result as input to PHA for the areas to mitigate. Collaboration recommended between origin area and impacted area/s to identify appropriate mitigations. There is also a possibility that the consequence resulting from the iteration is not acceptable to the area causing the incident, in that case the area will initiate a PHA to mitigate the consequence. As part of developing mitigations, consider the following:

- Develop QRA if appropriate
- Perform Building Assessments for Critical Buildings
- Apply Cost-Benefit analysis to optimize risks/opportunities

- Determine Safe Haven locations and their characteristics

If Mitigation is not possible from Business perspective – perform As Low As Reasonably Practicable (ALARP) justification for Senior Management to formally accept the posed risk
- To apply mitigations on a site wide basis, review Milestone deliverables by taking on a BIG picture perspective
- Future iterations may enhance those deliverables or create new ones as new information is available. Each iteration incrementally builds the deliverables to produce the Mitigated CA/QRA report to support Facility Siting for the stage

The overall effort is expected to produce a maximum of two single reports for the overall site:

1. Mitigated version of QRA Report
2. and Consolidated CA Report

It is recommended to develop as-built, pre-start up CA/QRA reports prior to start-up. Table 1 and 2 outlines typical activities performed by SPOC and plant areas respectively once key process areas are established/finalized on Site Plan. Please note that the activities are project dependent and may warrant modifications to suit specific situation.

Table 1. SPOC Sequence of Activities to conduct Facility Siting

Initiative	Key Steps
Setup Facility Siting Team	Sponsor to support to setup a team (i.e. Project Engineer, Technical Expert, Operation, Process representative) to act as SPOC
Integration	Provide functional (technical and management) guidance to Areas
	Assess cross boundary impact/opportunity
	Communicate incremental outputs after each Area PHA/Risk Reduction
IFA versions (prior to "Define Project" stage)	Consolidated CA Report
	Mitigated and Integrated QRA Report
IFD CA/QRA (prior to "Execution" stage)	Re-validate IFA CA/QRA post PHA and vendor PHA
As-built versions	Post-oil version

3.2 Typical Area Activities

Table 2. Area (Typical) Sequence of Activities to conduct Facility Siting

Initiative	Key Steps
Plant 1 (Typical)	Perform Modeling and Calculations
	Conduct Risk Reduction Workshop (if appropriate) – Apply mitigations
	Incremental output for use based on reduced risk level
	Assess Buildings/facilities
	Identify Safe Haven/shelter Locations (if required)
	Estimate Area Project Scope impact
	Perform area scope updates/changes e.g. passive fire protection requirement
	Carry out QRA (if appropriate)
	Update Product Backlog
	Develop/update contract to support area specific activities

4. Pitfalls to Avoid

The following observations were noted by the Author during his time to manage Facility Siting scope:

1. Facility Siting does not replace good and common sense engineering practice e.g. locating a plant that uses high volume of hazardous material will significantly increase risks to other plants/areas (e.g. camp) within the vicinity
2. Evaluate effectiveness of 'Cost Reduction' philosophies as opposed to added risk e.g. different modularization techniques may decrease the plant footprint by reducing pipe lengths but makes the plant more compact and likely to increase severity of explosion consequence
3. Avoid developing models with hope to realize risk reduction
4. Sharpening the pencil for models most likely may not reduce risk
5. Avoid jumping into CFD or other similar methods without evaluating cost-benefit for undertaking such effort
6. Utilization of QRA results is much more controversial than the methodology. Watson [6] has suggested that QRA should be considered as an argument, rather than a declaration of truth. In this view – focus on accuracy over precision
7. CA/QRA Task Force should be part of Management of Change (MOC) process
8. Evaluate carefully before locating facilities e.g. tank farms close to main entrance/egress as part of Facility Siting

9. Keep the focus on VISION for the overall effort to avoid scope creep
10. Note the difference between Facility Siting as a Deliverable and tool for design-development activities

5. Conclusion

This paper presented Adaptive/Change driven framework to manage initiatives like facility siting/CA/QRA as part of Mega-Projects. Change driven framework does not mean "lack of planning or no planning". As demonstrated in the paper, this methodology starts with a vision and then develops the ultimate product both iteratively and incrementally when it is possible to define small incremental improvements that will deliver value to stakeholders. This allows making key decisions on laying out the facilities in early stages of a project rather than waiting for design to develop to a significant extent. Often times, project team wait too long for design to mature enough before applying CA/QRA to validate consequence impact and/or risk level. Most often, especially for Mega Projects, this leads to delays in executing the CA/QRA work properly at the right time to influence site layout. This delay in execution leads to scenarios when scope changes cannot be incorporated anymore. This translates into value erosion as project was unable to reap the full benefit of Facility Siting. The management framework recommended in this paper i.e. Change driven/adaptive framework allows for the stakeholders to start early in the project lifecycle to apply CA/QRA in laying out the plant facilities. The key is to start EARLY but START with the END in mind. Alignment among stakeholder on a VISION up-front is also important as this is expected to save life while striking a balance between business needs and climbing the Risk Mountain. Vision also allows avoiding scope creep while applying Adaptive/Change driven approach. It is highly recommended that stakeholders focus more on accuracy over precision. In the early stages of project lifecycle, quality information is scarce. Change driven/Adaptive framework gives you as the project stakeholder the opportunity to use "what you know" intelligently to ensure "what you don't know" falls into the right place.

6. References

[1] M. Mannan, A. Chowdhury and O. Reyes-Valdes, "A portrait of process safety: From its start to present day," College Station, 2012.

[2] M. Mannan, Lees' Loss Prevention in the Process Industries, Volumes 1–3 (3rd Edition), , 2005, Elsevier, 2005.

[3] K. McAndrews, "Consequences of Macondo: A Summary of Recently Proposed and Enacted Changes to US Offshore Drilling Safety and Environmental Regulation," in *Society of Petroleum Engineers, Americas E&P Health, Safety, Security and Environmental Conference*, 2012.

[4] E. W. Merrow, Industrial Megaprojects Concepts, Strategies and Practices for Success, New Jersey.: John Wiley & Sons, Inc., 2011.

[5] A Guide to the Project Management Body of Knowledge (PMBOK® Guide) – Fifth Edition, Newtown Square: Project Management Institute, Inc., 2013.

[6] S. Watson, "The Meaning of Probability in Probabilistic Risk Assessment," *Reliability Engineering and System Safety,* vol. 45, pp. 261-269, 1994.

348480 Global Energy and Transportation and Mobil Oil

Abstract:

Biofuels, Computational Methods, Engineering Economics, Polymers, Thermodynamics. Polymers-plastic, Thermodynamics- oil, Engineering Economics – the money it takes to finish the method of Global Engineering Research, Computational Methods – numerical input of analytical data, Biofuels – the transport fluids that utilized in more than one method from industrial usage to everyday living which enters the atmosphere. Exxon has completed a $600 million dollar deal for production of algae biofuel, amongst Mobil Oil has performed major research Global Energy and the effects of oil. Reading a most recent article on Mobil website.

348748 Characterization of Iron Phthalocyanine As the Active Material for Lithium Batteries

Abstract:

Results on the determination of the lithium-ion diffusion coefficient in the iron phthalocyanine (FePc) active material as a function of intercalation level, using a modified electrochemical impedance spectroscopy technique, will be presented. Also, developed thermodynamic functions such as the Gibbs free energy, entropy, and enthalpy changes for the lithium insertion reaction with FePc, derived from cell open-circuit voltage measurements as a function of intercalation level, are presented. Also presented are estimates of the maximum amount of lithium that can be intercalated into the FePc based on constant volume and temperature conditions.

349086 Design of a Free-Fall Reactor for Fast Pyrolysis of Waste Plastics

Abstract:

Thermal pyrolysis has earned significant reputation in the recovery of valuable chemicals, especially petrochemical based fuels from various materials, including biomass and plastics. Of many available technologies, *fast* pyrolysis was preferred in his study owing to its minimal heat and mass transfer limitations and better control on process conditions. The focus of this study is towards efficient conversion of waste plastics into gaseous products consisting primarily of alkanes, alkenes and aromatics. Lab scale experiments were performed using a micro reactor to determine favorable conditions, including reaction temperature and particle size that would result in maximum yield of ethylene. Using these results, a simple industrial size radiatively heated free-fall reactor was designed for the purpose of pyrolyzing waste plastics. The goal is to design a *fast* pyrolysis reactor that improves upon inefficiencies in traditional reactors while resulting in high yield of ethylene.

349732 Biosorptive Dehydration of Ethanol/Water Azeotropes Using Compound Starch-Based Adsorbent

Abstract:

Abstract: Recently, the interest in the production of bioethanol has greatly increased as a result of the increasing fossil fuel prices and associated environmental issues. Bioethanol is made from either grains or biomass and the broth contains 6 to 12 percent ethanol. In addition, water removal represents a serious problem as ethanol form azeotrope with water at atmospheric pressure. Distillation to water-free ethanol consumes 50 to 80 percent of energy used in the fermentation ethanol manufacturing process. Various techniques have been developed to break the azeotropic point, such as salting-out method, azeotropic distillation, extractive distillation, reactive distillation, adsorptive distillation and pervaporation. Among the methods, the adsorptive option is the most attractive, especially the biosorption, which has distinct advantages over conventional methods as it is environmentally friendly, highly selective, easily available, easy to operate, cost effective and reusable in repeated cycles in the treatment of aqueous organics.

In this work, a specially formulated compound starch-based adsorbent (CSA), which consists of corn, sweet potato and foaming agent, was developed. As an exploratory study, this research was focused on the investigation of CSA for biosorptive dehydration of aqueous ethanol and optimization of conditions for its maximum adsorption under various conditions of vapor feed flow rate, bed temperature and bed height through an orthogonal design method. Furthermore, to investigate the adsorption mechanism, dynamic saturated adsorbance of CSA was studied and characterization of the adsorbent before and after adsorption, as well as after regeneration, was performed.

In the first step, the net retention time and separation factor of ethanol and water were obtained using inverse gas chromatography (IGC), the results indicated that the net retention time of water on CSA is significantly longer than that of ethanol which attested the feasibility of this adsorptive separation purpose and low temperature could be propitious to the process which was consistent with the traditional theory of adsorption.

Then, through a batch adsorption experiment, the optimum adsorption condition was determined under vapor feed flow rate, bed temperature and bed height by orthogonal design method, which indicated that the optimization condition for ethanol was at vapor feed flow rate of 1.1g/min, bed temperature of 83 and bed height of 25cm. The packing density was 445kg/m^3. The residence time is determined by vapor feed flow rate and bed height. The residence time is more influenced than bed temperature. The longer the residence time, the more the output is. But it cannot be ignored that the adsorption selectivity may decreases. Low bed temperature is benefited to the adsorption process which agrees with the result of IGC. Five cycles experiments of adsorption-regeneration were carried out at the optimum adsorption and regeneration condition, the results showed that there is practically no change in the regeneration efficiency of CSA after the cyclic process adsorption–regeneration. It is due to the properly chosen regeneration conditions assuring complete desorption of the water adsorbed without changes in the CSA structure. This unchanged adsorption capacity after a number of cycles demonstrates reusability of the adsorbent.

To study the adsorption mechanism, experiment was conducted to study the dynamic saturated adsorbance of CSA. The water adsorption capacity of CSA has been found to be similar to some molecular sieves, which is about 0.15g/g. Besides, field emission scanning electron microscopy (FESEM) and mercury porosimetry (MP) were applied to investigate the change of surface morphology and microstructure of CSA before and after the adsorption, as well as after regeneration, as shown in Fig.1 and Fig.2. The FESEM images showed that compared with the particles before the adsorption, they became plump and the void spaces became smaller after adsorption, which indicated that the CSA was saturated with water. However, the CSA particles after regeneration became wizened like the nature particles, which demonstrated that almost all the water had been desorbed from CSA. The pore size distribution (PSD) indicated that compared with the PSD curve before adsorption, the peak at 6000nm disappeared and the peak at 17000nm became remarkably smaller after adsorption, which manifested that CSA is mainly composed of macropores which play a major role in the adsorption. However, the peaks at 6000 and 17000nm reappeared after regeneration, and the curve looked exactly the same with that before adsorption, which further demonstrated that CSA after regeneration can be totally regenerated and recovered to the original structure.

These results indicated that the starch-based adsorbent can be considered as an effective and low-cost bio-sorbent to dehydrate alcohol.

349781 Purification of 2-Amino-1-Phenylethanol Enantiomers By a New Technique Combining Distillation and Crystallization

Abstract:

Stripping crystallization (SC) is a new separation technique which combines distillation and crystallization. Basically, SC is operated at a triple-point condition, in which the liquid is simultaneously vaporized and crystallized due to the three-phase equilibrium. In essence, the process is continued until the liquid phase is completely eliminated and only the pure solid crystals remain in the feed. Thus, crystal washing is not required since only pure solid crystals remain in the feed and no impurities are adhered on the crystal surfaces at the end of SC. In the previous research, SC has been successfully applied to separate p-xylene from the mixed xylenes by Shiau et al. (2005, 2008).

Phenylethanolamine, also known as 2-amino-1-phenylethanol (2-APE), is an important basic skeleton in many adrenergic drugs and is often used as ingredients of cold medicines and anorectics. As the biological activity and pharmacokinetics of the enantiomers of a chiral drug can be different, the objective of this research is to apply SC to purify R-2-APE from the chiral 2-APE mixture. The experiments show that no crystallization occurs during SC without seeding due to subcooling of the liquid enantiomers. On the other hand, seeding can eliminate subcooling during SC, leading to the R-2-APE crystals gradually formed preferentially on the seeds. Thus, seeding with magnetic stirring can be successfully employed to purify R-2-APE from the 2-APE enantiomers.

1. Shiau, L. D., Wen, C. C. and Lin, B. S. Separation and purification of p-xylene from the mixture of m-xylene and p-xylene by distillative freezing, *Ind. Eng. Chem. Res*, **44(7)**, 2258-2265, 2005.

2. Shiau, L. D., Wen, C. C. and Lin, B. S. Separation of p-xylene from the multicomponent xylene system by stripping

350841 Fabrication of Low Cost Insulating Material from Kaolin Clay for Construction Purposes

Abstract:

A new low cost, thermal insulator, construction mineral polymeric material was synthesized from local Saudi Arabia resources using Inorganic Mineral Polymerization technique (MiP). In this technique, the mineral polymer is produced from the following components: clay mineral rich in koalinite, sand, sodium hydroxide solution, and water. Inorganic polymerization experiments are conducted by mixing known amounts of clay minerals and sand in a variable speed stainless-steel mixer for some period of times until a homogenous solid mixture is obtained. Sodium hydroxide solution is added to the mixture and the polymerization reaction takes place at different mixing speeds and times. The produced mixture is divided into a number of samples of known weights. Each sample is moulded in a stainless-steel cylinder at a medium pressure. Then the moulded samples are cured in an oven for 24 hours. The curing temperature was 80°C. The mineral polymerization conditions such as the ratio of clay to sand filler, particle size of the clay, particle size of the sand, concentration and amount of sodium hydroxide, speed and time of mixing were optimized to obtain stable and hard mineral polymer. Compressive strength for different specimens of fabricated mineral polymers was tested under dry and wet conditions. Values of compressive strength with values ranging from 11.5 MPa to 27.9 MPa were obtained.

351141 Thermodynamic Study of Binary PAH (Anthracene + Phenanthrene) Solid Mixtures

Abstract:

Polycyclic aromatic hydrocarbons (PAHs) are common components of many materials, such as petroleum and various types of tars. They are generally present in mixtures, occurring both naturally and as byproducts of fuel processing operations. It is important to understand the thermodynamic properties of such mixtures in order to better understand and predict their behavior (i.e., fate and transport) in the environment and in industrial operations. The Knudsen effusion technique was combined with gas chromatography and differential scanning calorimetry to measure the vapor pressure, composition, and phase behavior of binary anthracene and phenanthrene mixtures at near ambient temperatures. Mixtures of anthracene and phenanthrene exhibited non-ideal mixture behavior: In mixtures with less than 80% phenanthrene, this more volatile component preferentially sublimed and the vapor pressures decreased as the mixtures became enriched in anthracene. The vapor pressure data do not follow Raoult's law (that is, ideal mixture) predictions. Mixtures with high phenanthrene content (> 80%) showed azeotrope like behavior, in that anthracene and phenanthrene sublimed at nearly constant vapor pressure and molecular ratio. These results confirm that the behavior of few-component PAH mixtures does not follow ideal mixture models. Hence, as tars and petroleum (and potentially other organic compound-containing mixtures) weather to few component mixtures over time, complicated, non-ideal phase behavior might be observed.

351160 How Confident Are You That Your Major Accident Risks Are Under Control?

Abstract:

Major accidents in the process industries including those at Texas City , Buncefield UK and in the Gulf of Mexico have highlighted the critical role played by senior managers and executives. Effective leadership is essential if companies are to develop a positive safety culture that remains constantly vigilant towards the risk of a major process safety incident. In the UK the HSE has made leadership a key priority for the high hazard industries and there is a cross-industry initiative to develop and deliver training focussed towards board members responsible for process safety performance.

This paper explores the leadership failings that contributed to recent major accidents, and the cost to both the company reputation and the individuals involved. Key principles are discussed that can help improve leadership in the major hazard industries. These include; attitudes and decisions of senior manager affecting the investment strategy and the safety culture of the organisation, reinforcing the importance of safety by personal example, thorough understanding of major accident hazards and key risk control systems, investigating process safety incident and near misses to find the underlying causes, developing world class safety management systems, and identifying weaknesses in these systems using targeted performance indicators.

The author bases this paper on the development of a training standard for 'Process Safety Leadership' involving a cross-industry 'expert panel'. Training of senior executives in compliance with this standard has been carried out from late 2011 and feedback from these courses and the key requirements for effective leadership will be discussed.

351236 Sil Determination: Shortcomings with the Use of LOPA

Abstract:

Proposed Topic Area: Managing Risks

For a number of years, industry has been trying to apply the principles outlined in IEC 61508. In the process sector, this has resulted in a focus on the requirements of IEC 61511 (2003) and ANSI/ISA 84.00.01 (2004). These are risk based standards. They involve the identification of potential hazardous events, the causes that could lead to each event and the necessary risk reduction for each event. Where the risk reduction involves an instrumented function (e.g. a trip or an interlock), the required performance of the instrumented function can described in terms of the average probability of failure on demand. In the standards, the performance for such instrumented functions is assigned to one of four Safety Integrity Levels. The assessment of what performance level is needed in respect to a specific hazardous event is referred to as SIL Determination. In the field of SIL Determination, IEC 61511 provides examples of a number of different approaches – these are shown in IEC 61511 Part 3. These include Event Trees, Risk Graphs, Safety Matrices, Fault Tree Analysis and Layer of Protection Analysis (LOPA).

In the first few years, following the publication of IEC 61508 Edition 1 in 2000, industry generally embraced the use of Risk Graphs as the method of choice. However, over the subsequent few years, many companies using this method found that it was (a) giving higher safety integrity level targets than really needed and (b) was less flexible than other methods and consequently more difficult to use in many real situations. This has resulted in a general move to other methods and in particular to the use of Layer of Protection Analysis. Indeed, LOPA may now be the predominant "Method of Choice" for SIL Determination. As a method, LOPA gives the impression of simplicity and appears quite straightforward. It almost invites anyone with a minimum of understanding of the method to be a practitioner.

However, there are a number of shortcomings in the use of the LOPA method that have become evident through its use across the globe.

This paper will review the application of LOPA and provide insights into what may be seen as the proper use of the method and its core principles. It will also highlight various shortcomings in use that have been found during expert reviews carried out of a number of existing studies. The paper will be of benefit both to those who are new to the LOPA method, but also of value to those who have been using the method for some time.

TRU-SCAN® and TRU-CAT™ Scanning of a Structured Packing Operating at Deep Vacuum

By

Simon Chambers
Fractionation Research Inc., Stillwater OK

Lowell Pless
Tracerco, Pasadena TX

Ron Carlson
Tracerco, Pasadena TX

Paper 61a

Presented at the Topical 8, Distillation Topical Conference,
AIChE Spring Meeting, New Orleans, Louisiana, April, 2014

***Simon Chambers*[1], *Lowell Pless*[2] *and Ron Carlson*[2]**
[1] *Fractionation Research, Inc., Stillwater, Oklahoma, USA*
[2] *Tracerco, Pasadena, Texas, USA*

Abstract
Pioneering gamma scanning was conducted on a high capacity structured packing operating under deep vacuum during proprietary testing at FRI The packing and multistage trough liquid distributor were tested with the para-/ortho- xylene system at 98.6 mbara in FRI's 1.22 m diameter Low Pressure Column. From top to bottom, the column was baseline scanned when dry and followed by scanning at 24, 30, 67, 100 and 105 % of the packing's Maximum Useful Capacity. Distributor liquid levels derived from the scan data compared very well against liquid level data from FRI bubblers. CAT-scans were performed at one column elevation near the top. At this elevation, 81 data points through the bed of packing were recorded using 9 scanning source positions with 9 different detector positions for each source position around the column circumference. Processing of the CAT-scan data using a proprietary mathematical technique provided significant insights regarding liquid holdup and liquid flow patterns in columns containing structured packings.

Keywords: packing, gamma scan, maximum useful capacity, bubblers

1. Introduction

Gamma scanning employs small radioactive devices to measure directly vapour-liquid flow behavior on trays and packings during specific operating conditions. The Tracerco TRU-SCAN® and TRU-CAT™ scanning techniques[1, 2] provide valuable information that is seldom available from conventional scans. With packed columns, Tracerco employ the PackView™[1] method to dry scan the packing which is set equal to a 'zero' density. With this baseline, additional densities are the result of the liquid volume fraction or liquid withheld inside the packing. Using this pioneering approach in a Fractionation Research Inc. (FRI) packed column test was of particular interest to Tracerco because of the wide range of operating conditions employed with a hydrocarbon system.

On receiving permission from Julius Montz, potentially unique first-of-a-kind gamma-scans were performed on the Montz-Pak high capacity structured packing during proprietary device testing in the FRI Low Pressure (LP) column. Tracerco's primary focus was operations at deep vacuum (98.6 mbara).

2. Results and discussion

2.1 Column Set-up and Operations

Figures 1 and **2** provides elevation drawings and photographs describing the FRI Low Pressure (LP) Column as it was configured for the Montz-Pak high capacity structured packing tests. The column diameter was 1.22 m and the bed height was 2.66 m. The packing specific surface area was 350 m^2/m^3. A Montz Type-R multi-stage trough and drip tube liquid distributor was employed.

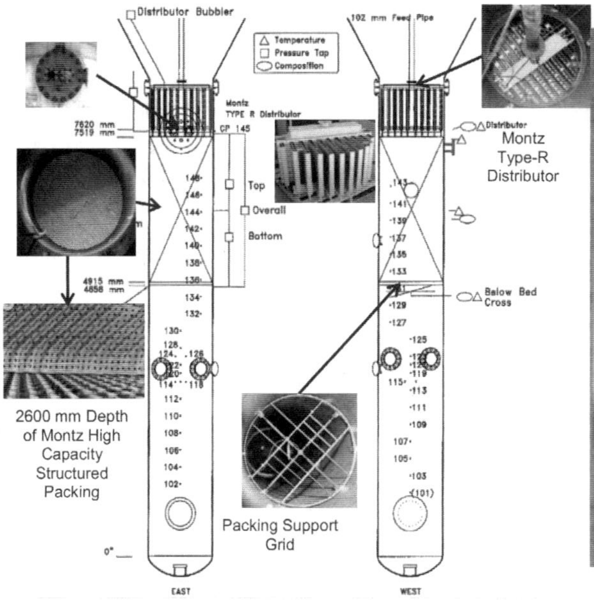

Figure 1. FRI Low Pressure (LP) Column Set-up For the High Capacity Structured Packing Tests

Figure 2. Photograph of the FRI Low Pressure (LP) Column

With this rare and unique gamma scan work, the only system employed was the para-/ortho- xylene system at 98.6 mbara. For multiple gamma scans at different flow rates, all of the 'wet' runs were at total reflux (TR) conditions. The FRI measured HETP's were obtained from the Fenske equation using liquid sample compositions. The HETP's were employed in a graphical analysis to determine the Maximum Useful Capacity (MUC) of the structured packing. The HETP's for the five 'wet' conditions are provided in **Figure 3**. **Table 1** shows all of the wet runs are expressed as a percentage of MUC.

Scanline	Scan 1	Scan 2	Scan 3	Scan 4	Scan 5	Scan 6
% MUC	Dry	24	30	67	100	105
Data Curve	**Blue**	**Orange**	**Black**	**Green**	**Red**	**Brown**

Table 1. Montz-Pak High Capacity Structured Packing MUC, P/O-Xylene 98.6 mbar

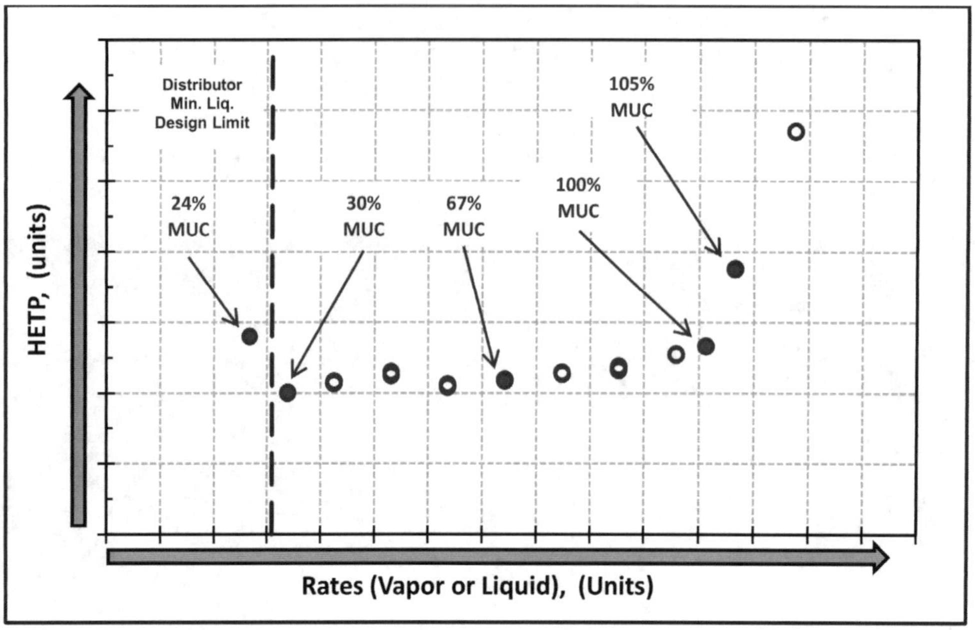

Figure 3. **Montz-Pak High Capacity Structured Packing Efficiency (HETP)**
P/O-Xylene at 98.6 mbara System

2.2 *TRU-SCAN® Method*

The LP Column was gamma scanned vertically from near the top of the installed liquid distributor through the entire bed of packing. A gamma scan generates a density profile of the column, which can be used to identify the integrity of internals and vapor-liquid behavior while a column is operating. **Figure 4** is a representation of the scanline orientation used to obtain the *TRU-SCAN®* data.

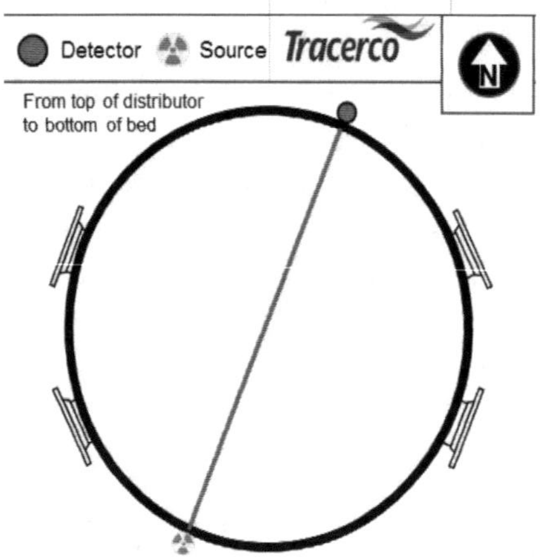

Figure 4. TRU-SCAN® Scanline Orientation in LP Column

2.3 *TRU-SCAN®* *Results*

Figure 5 is a *Tru-Tower™ Tru-View™* view of the scan results through the LP column, with reference column elevation on the y-axis and radiation intensity detection (counts) on the x-axis. In addition, **Figure 5** includes a "clear vapor" reference line, where the radiation intensity corresponds to the lowest vapor density observed. The "clear vapor" reference is presumed liquid free and is used as the reference for calculating relative densities of the test system.

Tracerco

	% MUC	Data Curve
Scan 1	Dry	Blue
Scan 2	24	Orange
Scan 3	30	Black
Scan 4	67	Green
Scan 5	100	Red
Scan 6	105	Brown

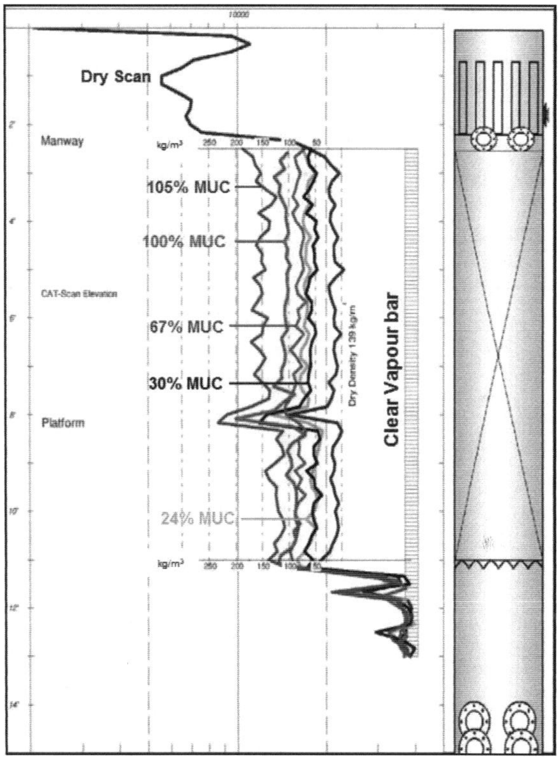

Figure 5. Tru-Tower™ Tru-View™ Complete Vertical Scan at Five 'Wet' Conditions

Superimposed on **Figure 5** is a density retention scale showing average calculated density (in kg/m³) of the liquid holdup in the packed bed for each test run. The calculated average dry bed density was 137 kg/m³ compared to the reported manufacturer's dry packing density of 139 kg/m³. The total overall density through the packed bed is the sum of dry packing density and retention density of the liquid hold-up.

The gamma scan profiles show how that at 30% MUC, there was slightly less liquid hold-up in the bottom third of the packed height compared to the top two thirds. At 24% MUC, liquid accumulation was surprisingly higher in the upper bed section compared to 30% MUC. These data are highly suspect. At the Tracerco's request, FRI purposely operated the LP column at this condition such that the liquid rate was below the liquid distributor minimum design limit. Flow instability was highly possible at these conditions.

The scan profiles at 67% MUC displayed uniform (top to bottom) liquid retention through the entire packing height. As the LP column approached 100% MUC and up to 105% MUC, the scan profiles show a disproportionate liquid build-up in the top and middle zones of the packed bed.

2.4 *TRU-CAT™ Method*

TRU-CAT™ is a series of gamma scans covering a full horizontal plane at a given column elevation. The Montz-Pak was *TRU-CAT™* scanned near the top of the packed bed, approximately 230 mm into the bed. **Figure 6** is a representation of the CAT scan orientation used on the Low Pressure Column.

The TRU-CAT™ scan consisted of 81 scan chords or discrete path lengths. Each chord represents data points that collectively were regressed to provide a cross-sectional relative density profile of the LP Column fluids. The data regression used a fourth-order least-squares mean-value algorithm to construct an image of relative density, expressed in kilograms per cubic meter, versus the cross sectional position (3-D surface plot).

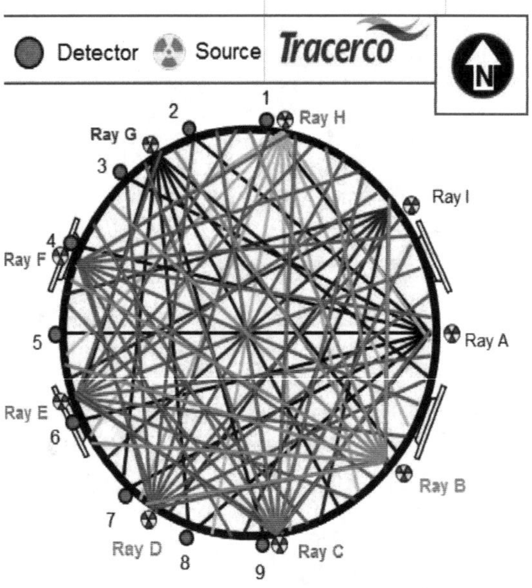

Figure 6. Representation of TRU-CAT™ Orientation on LP Column

2.5 *TRU-CAT™ Results*

The TRU-CAT™ scans in **Figure 7** revealed no gross liquid mal-distribution at rates below 100% MUC. The narrow band of green-to-yellow around the column perimeter for the 30% MUC scan showed preferential liquid flow on the column wall (higher densities). At higher liquid rates, the green-to-yellow band at the wall expanded corresponding to an increased liquid migration toward the column wall.

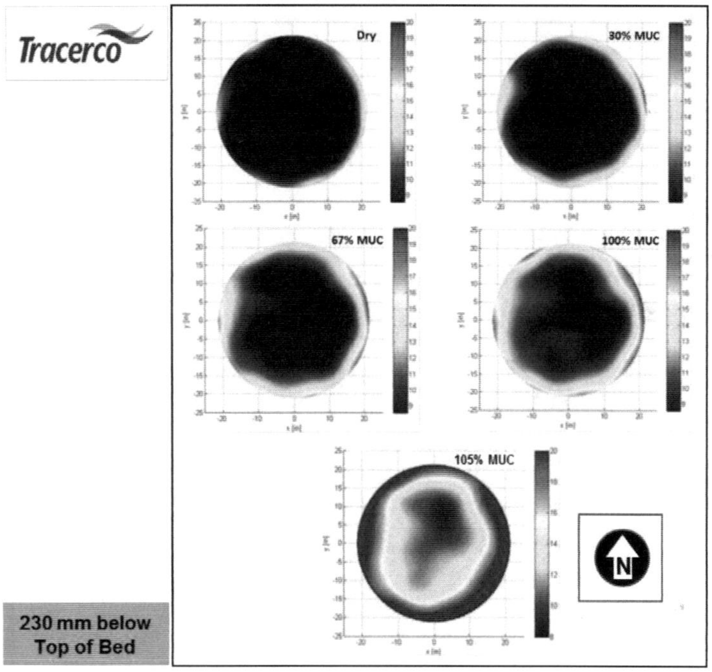

Figure 7. TRU-CAT™ 3-D Surface Plot of Relative Density Profiles Near Top of The Bed

2.6 *Liquid Distributor Levels: Gamma Scan Data versus FRI Bubbler*

The gamma scan distributor liquid levels are a composite level detected through the distributor arm channels that the scan radiation beam passed through. The measurement method is shown below:

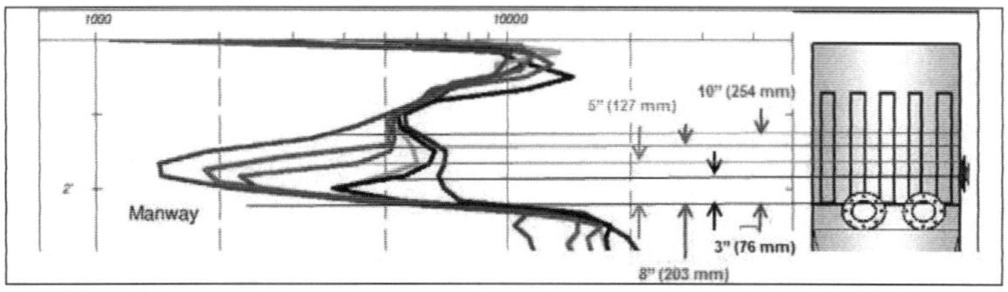

For all five wet total reflux conditions, the gamma scan measured liquid heads in the distributor arm channels agreed very well with hydrostatic liquid heads measured with the FRI bubbler. The results are summarized in **Table 2**.

Percent of Maximum Useful Capacity	Data Curve	Average Bed Density, kg/m^3	TRU-SCAN® Distributor Liquid Level, mm	FRI Bubbler Distributor Liquid Level, mm
Dry	**Blue**	137	Dry	Dry
24	Orange	178	76 ± 25	80
30	**Black**	179	76 ± 25	100
67	Green	207	127 ± 25	157
100	Red	226	203 ± 25	198
105	Brown	264	254 ± 25	217

Table 2. Distributor Liquid Head Measurements: TRU-SCAN® vs. Bubbler

3. Conclusions

Generally:

- CAT-scanning sometimes provides valuable information that is seldom available from conventional scans
- Before column commissioning, dry gamma scans can aid appreciably in the interpretations of wet scans for a column in operation.
- Top-to-bottom scans facilitate troubleshooting efforts e.g. liquid maldistribution.

Regarding the scanning and CAT scanning of the High Capacity Structured Packing:

- Approaching the Maximum Useful Capacity (MUC), there was build-up of liquid in the *top* and *middle* of the packing height.
- At rates slightly above the MUC, there was a build-up of liquid throughout the bed, but appreciably higher liquid buildup in the top and middle of the bed. Additionally there was appreciable liquid migration toward the column wall.
- The gamma scan determined liquid heads in the distributor corroborated with the FRI bubbler (hydrostatic head) readings extremely well.

Abbreviations

HETP Height Equivalent to a Theoretical Plate
LP Low Pressure
MUC Maximum Useful Capacity

Acknowledgements

The authors wish to thank Mr. Thomas Rietfort of Julius Montz GmbH, for permitting the gamma scanning of their structured packing during proprietary testing at FRI.

References

1. Pless, L., "Tower Scanning As a Quantitative Tool – Pack View™", Topical 8, Distillation Topical Conference, AIChE Spring Meeting, San Antonio, TX, April 28-May 2, 2013.
2. Chambers, S., Pless, L., Carlson, R. and Schultes, M., "Gamma Scan and CAT-Scan Data From a Distillation Column at Various Loadings", AIChE Spring Meeting, San Antonio, TX, April 28-May 2, 2013.

351754 Performance Characteristics of an Intermediate Area High Performance Structured Packing

Abstract:

Results are presented of total reflux distillation experiments carried out with Montz-pak B1-350MN in a 1.21 m internal diameter column at Fractionation Research Inc. using two tests systems at two operating pressures, i.e. cyclohexane/n-heptane, at 0.31 and 1.62 bar, and paraxylene/orthoxylene at 0.1 and 1 bar, respectively. In all cases an impressive packing efficiency was achieved, ranging from five stages- at low to four stages per unit bed height at high vapor loads (F-factor). Comparison with data obtained earlier with B1-250MN under similar conditions indicates a relatively higher maximum efficient capacity of B1-350MN, while the relative efficiency and pressure drop correspond with the ratio of installed specific geometric areas. Owing to its rather low pressure drop per stage B1-350MN appears to be a versatile packing, considered presently as first choice for design of dividing wall columns. The Delft model captures observed packing geometry related effects quite well qualitatively. Quantitatively, the efficiency and pressure drop predictions are on the safe side, and only one adjustable parameter is needed to ensure perfect match.

Absorption Hydraulic and Mass-Transfer Characteristics of Raschig Super-Pak 250

L. Valenz, F.J.Rejl, J. Haidl, V. Linek
Institute of Chemical Technology, Prague, Czech Republic

1. Introduction

Mass-transfer coefficients in the liquid and in the gas phase, k_L, k_G, and effective interfacial area, a, are the basic rate-based mass-transfer characteristics of the packings. They are used together with their hydraulic properties for the design of the absorptions columns and strippers as well as for comparison of the packings. As the models predicting mass-transfer and hydraulic properties of the packings are still not sufficiently evolved with respect to the prediction of the exact packing shape influence on these properties, they have to be measured for each new packing.

In this contribution above mentioned data have been measured for Raschig Super-Pak 250, RSP 250, belonging to the class of the high capacity packing which provide significantly larger capacity without sacrificing of the mass-transfer properties. The packing differs substantially from the classical corrugated sheet packings as it forms sinusoidal, 50% open channels from the single metal sheet.

Absorption mass-transport properties in terms of the mass-transfer coefficients and effective interfacial area of the RSP 250 packing has been measured by Wang et al. [4] at SRP using column of 0.42m i.d. Hydraulics as well as *HETP* of the RSP 250 measured on 0.42m i.d. SRP column has been published by Chambers and Schultes [5].

This contribution presents hydraulic and mass-transfer data measured on the 0.29m i.d. column measured using proven standard systems [1] and has been correlated using Billet-Schultes model [2], thus providing set of the constants for this packing, which has not been published yet.

2. Experimental part

The experiments have been carried out in the column with a 0.29 m inner diameter. The packed bed consisted of four elements of the packing rotated mutually by 90° with a total bed height of 0.93 m. Liquid distribution over the cross-section of the column was carried out by the pipe type distributor with drip-point density of 630 dp/m^2. Gas entered the column through a large drum where the flow was calmed and preceded into the column through the grid which secured its uniform distribution.

Volumetric mass-transfer coefficient in the liquid phase, k_La, has been measured by oxygen striping from the water to the stream of the nitrogen. The system exhibits mass-transfer resistance solely in the liquid phase. k_La has been evaluated from the oxygen concentrations in the liquid phase under the assumption of the plug flow of the liquid phase out of the relation (1). The oxygen concentration has been measured by optical oxygen probes.

$$k_L a = \frac{u_L}{H} \cdot \frac{c_{O_2}^{in}}{c_{O_2}^{out}} \tag{1}$$

Volumetric mass-transfer coefficient in the gas phase, k_Ga, has been measured by sulfur dioxide absorption from the air into the aqueous solution of the sodium hydroxide. Due to the instantaneous reaction at the interface the system exhibits mass-transfer resistance solely in the gas phase. k_Ga has been evaluated from the sulfur dioxide concentrations in the gas phase under the assumption of the plug flow of the gas out of the relation (2). The sulfur dioxide concentration has been measured by infrared analyzer.

$$k_G a = \frac{u_G}{H} \cdot \frac{c_{SO_2}^{in}}{c_{SO_2}^{out}} \tag{2}$$

Effective interfacial area has been measured by absorption of the diluted CO_2 from the air into the aqueous sodium hydroxide solution, which is the system, where the gas-phase relative mass-transfer resistance is low and the rapid reaction of CO_2 with OH^- consumes quantitatively CO_2 within the liquid film. In such case, the mass-transfer rate is determined by the reaction-transport phenomena in the liquid film and is determined solely by the physical quantities of the system. Effective interfacial area can be calculated by combination of the relation for the CO_2 local mass-transfer rate and its balance along the height of the packing

$$a_{R_G=0} = \frac{(K_{OG}a)_{CO_2}}{He\sqrt{D_{L,CO_2}k_{OH}c_{OH,ave}}} = \frac{1}{He\sqrt{D_{L,CO_2}k_{OH}c_{OH,ave}}} \cdot \frac{u_G}{H} \cdot \ln\frac{c_{CO_2}^{IN}}{c_{CO_2}^{OUT}} \quad (3)$$

Value of the reaction term $He\sqrt{D_{L,CO_2}k_{OH}c_{OH,ave}} = 2.24 \cdot 10^{-3}$ m/s (20 °C and 1 M NaOH solution) was determined by still level experiments (Linek et al. [6]) as this procedure circumvents the errors induced by application of different physical data of solubility and diffusivity by the authors.
Pressure drop of the packing has been measured with the air-water system. The pressure drop related to 1 m of the packing has been evaluated out of the relation (4).

$$\frac{dp}{dH} = \frac{p^{in} - p^{out}}{H} \quad (4)$$

3. Results and discussion

The liquid phase volumetric mass-transfer coefficient of the RSP 250 packing follows approximately power law relation to the liquid phase velocity with power of 0.67. In this respect the behavior is very similar to that of the Mellapak 250Y, M250Y. The absolute k_La values of the packings are quite similar up to the B~40m/h, at higher liquid loads the RSP 250 k_La are 10-15% larger (see Figure 1). In comparison with the data provided by Wang [4] we have found them to be ~40% higher. This comparison involves recalculation of the k_L data between air/toluene/water and nitrogen/oxygen/water system and is therefore indirect.
The data has been correlated using Billet-Schultes correlation [2] with the relative mean deviation of 16% and resulting C_L value is 1.446 (see Table 1).

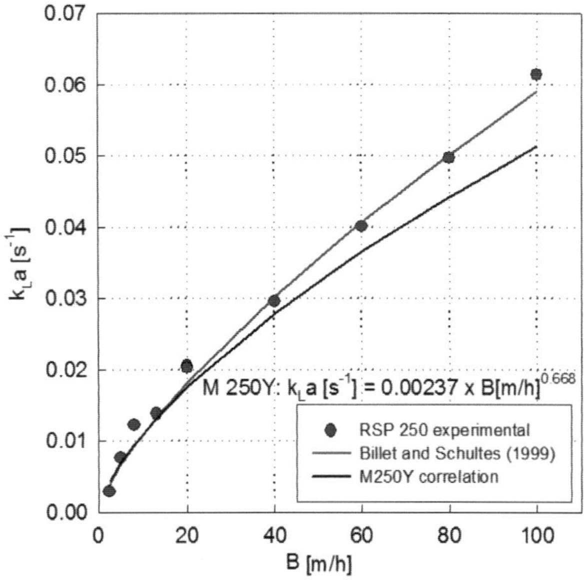

Figure 1: k_La of the RSP 250 measured with the

nitrogen/oxygen/water system

The volumetric mass-transfer coefficient in the gas phase, k_Ga, of the RSP 250 has been measured under various gas and liquid loads (see Figure 2). Resulting k_Ga are 39-81% higher than that of the M250Y which is caused by both, higher effective interfacial area as well as higher mass-transfer coefficient k_G under similar phase flows. The obtained data are ~80% higher than those obtained by Wang [4], possibly as a result of different experimental arrangement differing significantly in the packed bed height and corresponding SO_2 concentration range.

The data has been correlated using Billet-Schultes correlation [2] with relative mean deviation of 26% and resulting C_V value is 1.023 (see Table 1). The correlation predicts more significant effect of the liquid flow on the k_Ga values than was found experimentally.

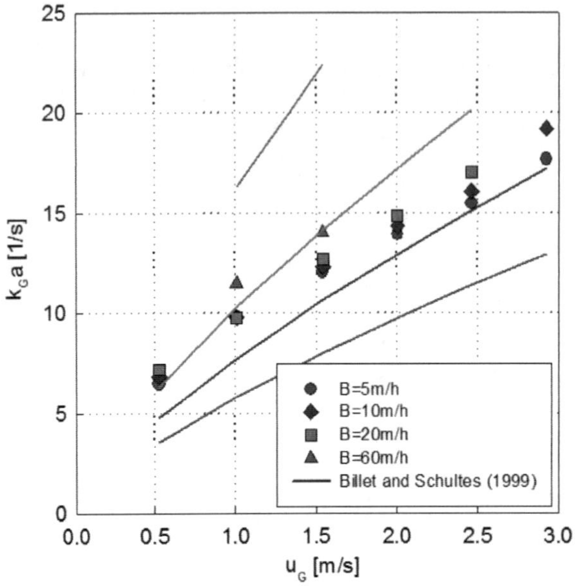

Figure 2: k_Ga of the RSP 250 measured with the air/sulfur dioxide/sodium hydroxide aq. sol.

The effective interfacial area of the RSP 250 packing evaluated from our data is in close agreement with that found by Wang [4]. The effective interfacial area is larger than RSP 250 geometrical area for the liquid loads above B~10m/h (see Figure 3). This is in contrast with what was found for M250Y. The effective interfacial area of the RSP 250 is 20-50% higher than that of M250Y.

The pressure drop of the RSP 250 has been measured under various gas and liquid loads (see Figure 4). The obtained data correspond well with the ones published by Chambers and Schultes [5] up to the loading point achieved at our column. The capacity is about 20% lower in comparison with data from Chambers and Schultes [5] possibly as a result of different experimental arrangement differing in the column inner diameter and packed bed height. The capacity of the packing is comparatively high in comparison with M250Y packing measured in the same column.

The pressure drop values has been correlated using Billet-Schultes correlation [2] with relative mean deviation of 16% and resulting hydraulic parameters, C_S, C_{Fl}, $C_{P,0}$, are summarized in Table 1.

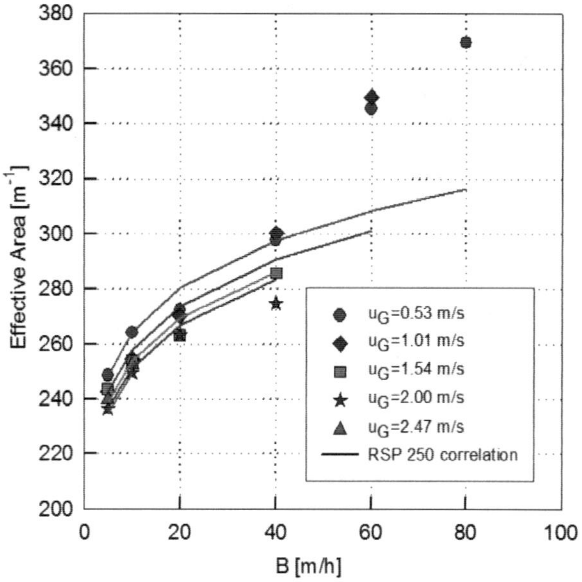

Figure 3: effective interfacial area, *a*, of the RSP 250 measured with the air/CO_2/sodium hydroxide aq. sol.

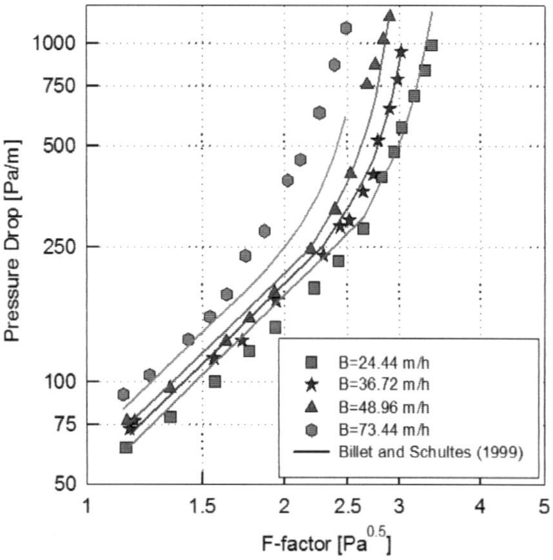

Figure 4: pressure drop of the RSP 250 measured with the air/water system

Table 1: Hydraulic and mass transfer parameters of Billet-Schultes model [2]

parameter	C_S	C_{Fl}	$C_{P,0}$	C_L	C_V
value	3.89	2.67	0.260	1.446	1.023
mean relative deviation	16.4 %			15.8 %	25.7 %

4. Conclusions

Basic mass-transfer characteristics k_La, k_Ga, a and pressure drop of the metal sheet Raschig SuperPak 250 has been measured under absorption conditions in the i.d. 0.295m column using standard systems.

There was found, that the liquid phase volumetric mass-transfer coefficient, k_La, of the RSP 250 is very close to that of Mellapak 250Y up to the B~40m/h, at higher liquid loads it is 10-15% larger.

The interfacial area of the packing is found to be larger than its geometrical area for the liquid loads above B~10m/h which is in contrast with what was found for M250Y. This behavior suggests that there exists some phenomenon forming additional mass-transfer area, possibly generation of the droplets. The effective interfacial area of the RSP 250 is 20-50% higher than that of M250Y.

The volumetric mass-transfer coefficient in the gas phase, k_Ga, of the RSP 250 is 39-81% higher than that of the M250Y which is caused by both, higher effective interfacial area as well as higher mass-transfer coefficient k_G.

The capacity of the packing is comparatively high in comparison with M250Y packing. The hydraulic and mass transfer parameters of the Billet-Schultes model [2] for the metal RSP 250 structured packing have been evaluated on the basis of the obtained absorption data.

Acknowledgements

Support of Grant Agency of the Czech Republic through the project No. 13-01251S is gratefully acknowledged.

References

[1] Rejl, F.J., Linek, V., Moucha, T., Valenz, L., 2009. Methods standardization in the measurement of mass-transfer characteristics in packed absorption columns. Chemical Engineering Research and Design 87, 695-704.

[2] Billet, R., Schultes, M., 1999. Prediction of mass transfer columns with dumped and arranged packings. Updates summary of the calculation method of Billet and Schultes. Trans IChemE 77 (Part A), 498–504.

[3] Rejl, F.J., Valenz, L., Haidl, J., Kordač, M., Moucha, T. On the modeling of gas-phase mass-transfer in metal sheet structured packings. To be published in the Chemical Engineering Research and Design.

[4] Wang, C., Perry, M., Rochelle, G., T., Seibert, A., F., 2012. Packing characterization: Mass transfer properties. Energy Procedia 23, 23-32.

[5] Chambers, C., Schultes, M. 2007. Reaching new performance levels with surface enhanced Raschig Super-Pak structured packings. AIChE Spring meeting, April 22-26 2007, Houston.

[6] Linek, V., Sinkule, J., Brekke, K., 1995. A critical evaluation of the use of absorption mass transfer data for the design of packed distillation columns, Chem. Eng. Res. Des. 73 (Part A), 398–405.

Gas-Phase Mass-Transfer Characteristics of Mellapak 250Y Structured Packing under Distillation Conditions

František Rejl, Lukáš Valenz, Jan Haidl

Institute of Chemical Technology, Prague
Technická 5, 166 28 Prague 6, Czech Republic
frantisek.rejl@vscht.cz

Introduction:

Characterization of the packing efficiency as well as the column design using HETP value is a proven method, which, however, brings several known problems based on the weak relation between real physical phenomena in the column and their description by stage-wise concept. This is overcome by rate-based approach, for which, the parameters (volumetric mass-transfer coefficients k_La, k_Ga) must be known.

Measurement of the volumetric mass-transfer coefficients under absorption conditions is generally possible as there exist systems with the mass-transfer resistance limited to the single phase as well as the systems in which the mass-transfer rate is controlled by the interfacial area. For the distillation the situation is far more complicated, as the mass-transfer resistance is usually distributed between the both phases and separation is not straightforward. Currently there are only limited attempts to determine effective interfacial area during distillation process.

Table 1:

M250Y		C_6/C_7		M/E	
u_G [m/s]		0.75	1.18	0.86	1.79
B [m/h]		12.8	20	5.66	11.8
k_La [s^{-1}]	EXP	0.032	0.04	0.012	0.021
	RBF	0.031	0.04	0.02	0.03
	BS	0.025	0.034	0.016	0.027
	DELFT	0.079	0.094	0.055	0.075
k_Ga [s^{-1}]	EXP	2.9	5.07	4.07	9.3
	RBF	2.78	4.80	2.84	7.05
	BS	2.94	5.65	3.61	7.26
	DELFT	1.82	2.77	2.28	4.29
a [m^{-1}]	RBF	181	217	135	180
	BS	83	112	68	92
	DELFT	240	245	215	231

This situation is reflected by the insecurity of the predictions of the mass-transfer coefficients and active interfacial area values provided by three respected models (RBF [1], BS [2], DELFT [3]). While predictions of the volumetric mass-transfer coefficients for C_6/C_7 and for the methanol/ethanol system are similar for RBF and BS models, the DELFT model predicts much larger mass-transfer resistance for the gas phase (see Table 1). As has been told, which of these predictions is closer to the truth can't be easily told, as the *HETP* values calculated on the basis of these data don't differ much and roughly correspond with the experimental ones. On the other hand, the predictions of the fundamental mass-transfer parameters (k_L, k_G, a) by individual models differ in multiples. This is a direct consequence of no reliable method for the effective area determination under distillation conditions. While this outcome could be (hardly) satisfactory for the column designers, the unclear fundamental of the distillation process on the packing is rather alarming.

The distribution of the mass-transfer resistance between the phases considerably affects the shape of the binary concentration profiles along the column packing (see Figure 1) and this fact enables evaluation of the k_La, k_Ga as optimized parameters of the distillation rate-based model, which produces concentration profiles matching the experimental ones. This so called "profile method" was used to analyze distillation data measured on several distillation systems (methanol/ethanol, ethanol/propanol, methanol/propanol, cyclohexane/n-heptane) and two types of the metal structured packing (Sulzer Mellapak 250Y and Mellapak 452Y).

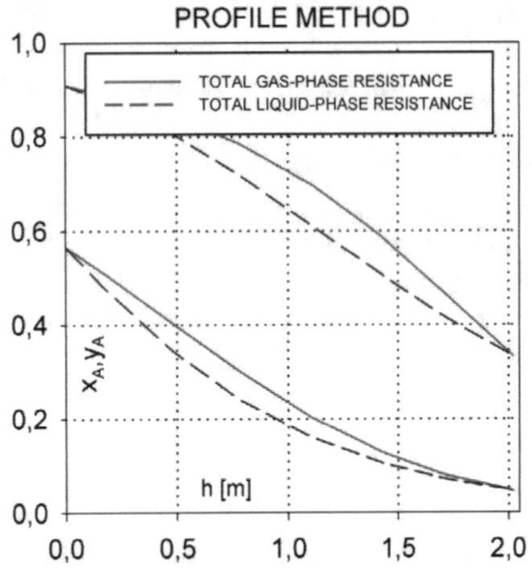

Figure 1: molar fraction profiles in the column under total reflux

This contribution presents volumetric mass-transfer coefficients in the gas phase, k_Va, measured under distillation conditions and attempts to correlate them in the dimensionless form on the basis of the variation of the following parameters a) physical properties of the system through involvement of several types of the system b) flow of the phases through change of the reboiler power, c) characteristic dimension through change of the packing.

Speaking in advance, it seems that the values of k_La and k_Ga under distillation conditions are rather well predicted by the RBF and BS models. Under just slight simplifications with minor impact on the results the key equations of these models describing the gas-phase mass-transfer are

$$Sh_G = C_{BS}(packing) \cdot Re_G^{3/4} \cdot Sc_G^{1/3} \quad \text{BS} \quad (1)$$

$$Sh_G = C_{RBF}(packing) \cdot Re_G^{0.8} \cdot Sc_G^{1/3} \quad \text{RBF} \quad (2)$$

These equations differ in the values of the multiplication constant and also in the value of the Reynolds criteria power, but consistently predict form of the system-related term as "$Sc_G^{1/3}$".

In contrast to the distillation the values of the mass-transfer coefficients k_G can be deduced under the absorption conditions. For several Mellapack packings [4] the k_Ga data have been measured using the air/SO_2/lye and interfacial area, a, using the air/CO_2/lye system and the resulting k_G data have been correlated in the form

$$Sh_G = 0.409 \cdot Re_G^{0.622} \cdot Re_L^{0.0592} \quad (3)$$

Now let us simply assume that the system-related term "$Sc_G^{1/3}$" works well and enables transfer of the gas-phase mass-transfer data between absorption and distillation. Under this assumption the absorption derived correlation (3) can be generalized into the form

$$Sh_G = 0.439 \cdot Re_G^{0.622} \cdot Re_L^{0.0592} \cdot Sc_G^{1/3} \quad (4)$$

Effective mass-transfer area under the distillation conditions can be subsequently estimated from the values of the k_Ga determined using "profile method" and k_G values computed on the basis of the correlation (4).

$$a = \frac{k_Ga}{k_G} \quad (5)$$

Experimental:

The experiments were performed in the stainless steel atmospheric distillation column with an interior diameter of 0.15 m, packed with 10 pcs of Mellapak 250Y or Mellapak 452Y packing, to the height of 2.1 m. The characteristics of the packing are described in Table 2. An electrically heated reboiler, equipped with an electric power regulator (up to 60 kW) was used to maintain a constant rate of boiling. The heat-loss, measured by independent experiments, was less than 3% of the reboiler duty.

The column was operated under total reflux, which was maintained by PI regulation of the liquid level in the reflux head. The reflux flow was measured with a precise turbine flow-meter. Reflux flow was distributed on the packing using a pipe-type distributor with 25 openings (1 472 holes/m^2). Vapor from the reboiler was introduced into the column by a chimney distributor.

The column was equipped with seven openings for placing thermistors and sampling devices for the withdrawal of liquid and vapor samples out of the packing along the column. The holes for the sampling devices were drilled into the structured packing with a water jet. The column also enables the withdrawal of samples of the liquid reflux and of the vapor from the bottom of the column.

Four distillation systems were used: methanol/ethanol (M/E), ethanol/propanol (E/P), and methanol/propanol (M/P) and cyclohexane/n-heptane (C_6/C_7). About 45 liters of the distillation mixture was introduced into the reboiler and the electric power input to the reboiler was adjusted to the desired level. The column was operated for a sufficient lengths of time to achieve a steady state, which lasted about one hour. Liquid and vapor samples (2.5 ml) were withdrawn by a means of glass syringes from all the sampling devices simultaneously over a period of approx. 6 minutes. Analysis of the samples was carried out using a tempered digital refractometer, the dependence of the mixture refracting indexes on their concentrations does not manifest any extremes and the accuracy of the molar fraction determination is better than 0.001.

These experiments were repeated four times for every power input with four different concentrations of the mixture in the reboiler. The concentrations were chosen in order to encompass the entire concentration range of the measured concentration profiles in the column. The result of these experiments is twelve to sixteen series of the data for each system.

Results:

The distillation data has been processed using the profile method. For alcohol systems the mass-transfer resistance is distributed between both phases while for the cyclohexane/n-heptane the gas phase mass-transfer resistance is clearly dominant. For larger loads the liquid-phase resistance in the C_6/C_7 system vanishes (see Figure 2).

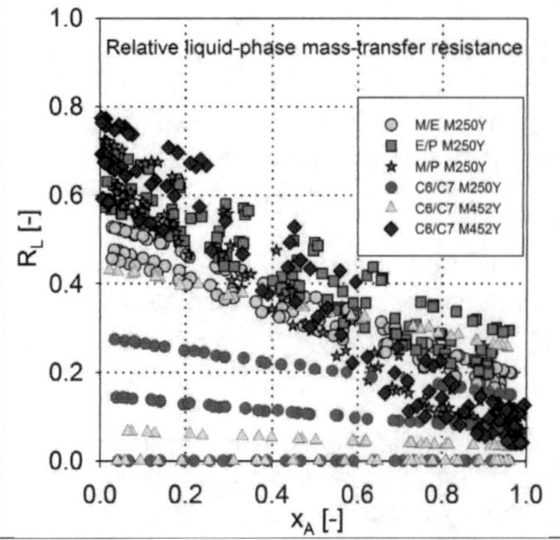

Figure 2: DISTILLATION relative mass-transfer resistance of the liquid phase

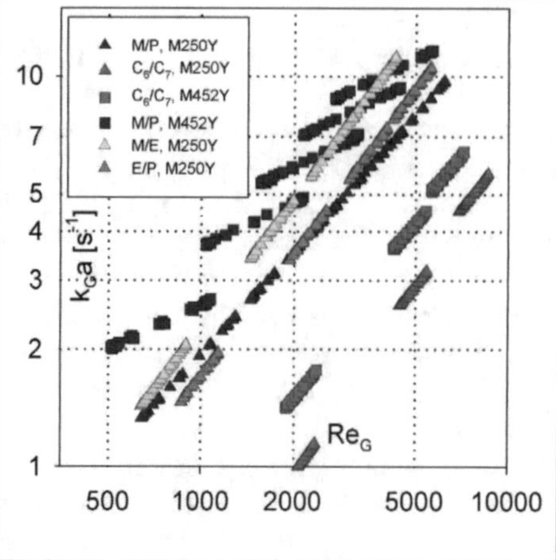

Figure 3: DISTILLATION k_Ga by the "profile method"

Evaluation of the volumetric mass-transfer coefficients is more secure and more precise for the phase with larger mass-transfer resistance, which is the gas-phase. These data (k_Ga) are therefore chosen for this presentation and together are presented in the Figure 3. As can be anticipated (and as is also predicted by the models) k_Ga values are higher for M452Y packing and for the alcohol systems. The data for individual packings and systems vary greatly and attempt to reduce them has been performed.

Dependence of the mass transfer coefficient k_G on the flows and physical properties of the system is assumed in the dimensionless form

$$Sh_G = C_1 \cdot Re^{\alpha} \cdot Sc^{1/3} \tag{6}$$

while the effective interfacial area is in the first approximation assumed to follow simple relationship

$$\frac{a}{a_G} = C_2 \cdot Re^{\beta} \tag{7}$$

Under such assumptions the k_Ga data obtained for various systems and packing dimensions can be expressed in the form

$$k_Ga = Sh_G \cdot \frac{D_G}{d_{eq}} \cdot \frac{a}{a_G} \cdot a_G = C_3 \cdot Sc^{1/3} \cdot \frac{D_G}{d_{eq}} \cdot a_G \cdot Re^{\gamma} \tag{8}$$

which is rearranged into

$$\frac{k_Ga}{Sc^{1/3} \cdot \frac{D_G}{d_{eq}} \cdot a_G} = C_3 \cdot Re^{\gamma} \tag{9}$$

The term in the denominator serves for recalculation of the k_Ga data from one system and packing size to another one, and the left hand side of the equation should be sole function of the Reynolds criteria – hydrodynamic function presented at the right-hand side of the equation.

The reduced data presented in the Figure 4 have been evaluated with the assumption of the plug flow in both phases. The data are clearly distributed into the three groups - those measured on the a) methanol/ethanol and ethanol/propanol system on the M250Y packing; b) methanol/propanol and C_6/C_7 system on the M250Y packing c) methanol/propanol and C_6/C_7 system on the M452Y packing. All the data measured on the methanol/ethanol, ethanol/propanol and C_6/C_7 system exhibit similar Reynolds criteria power γ (relationship 9) in the range 1.07-1.23, while the data measured on the methanol/propanol system exhibit significantly lower power (γ=0.93 for M250Y and γ=0.76 for M452Y). When methanol/propanol data are not considered, such behavior suggests that powers α and β could be constants and their sum equals to ~1.15. Due to the distribution of the data into groups according to the systems and packings the form of the recalculation factor

$$Sc^{1/3} \cdot \frac{D_G}{d_{eq}} \cdot a_G \tag{10}$$

doesn't seem to work well in this case, although the constant power γ of the hydrodynamic function at least suggests that the hydrodynamic of the phases under various conditions is analogical.

An attempt to remedy the situation has been performed through evaluation of the $k_G a$ data with the more realistic assumption of the plug flow with axial dispersion in the liquid phase. The fit of the experimental data is in average 42% better than in the case of the plug flow. Once again the reduced data doesn't form single function of the Reynolds criteria, but are dispersed in the broad (+/- 30%) region with methanol/ethanol data being high extreme and C_6/C_7 data the low one. With exception of the methanol/propanol data measured on the M452Y packing the data provide Reynolds criteria power γ (9) in the range 0.72-0.88, which is significantly lower than in the case of the plug flow. The absolute values of the (9) left-hand side remain almost the same, however. Important change in the behavior consists in the vanishing of the dependence on the packing type.

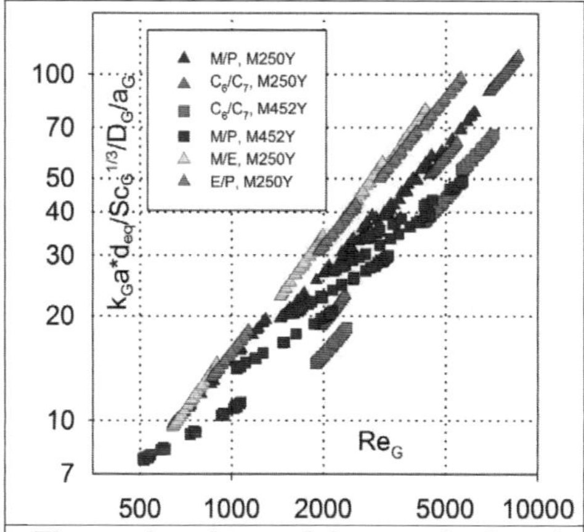

Figure 4: values of the recalculated $k_G a$ (9) evaluated under the PLUG FLOW assumption

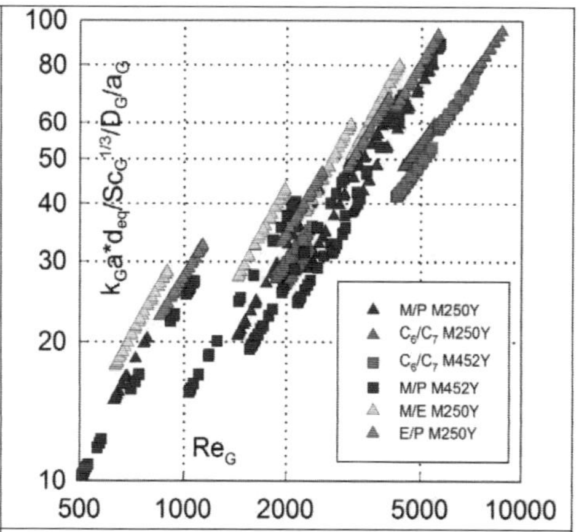

Figure 5: values of the recalculated $k_G a$ (9) evaluated under the PLUG FLOW +AXIAL DISPERSION assumption

The effective distillation interfacial area evaluated using absorption data recalculated to the distillation conditions according to the relations (4,5) are present in the Figures (6,7). In both cases the resulting effective area forms ~ 50-90% fraction of the geometrical area of the particular packing. Effective area evaluated for C_6/C_7 system is nearly independent of the mixture composition while that for M/E and E/P system changes dramatically. Dependence on the liquid load seems to be low when the plug flow with axial dispersion model is chosen for $k_G a$ evaluation. Surprisingly the results suggest that the effective interfacial of the C_6/C_7 system is lower than that of the alcohol systems.

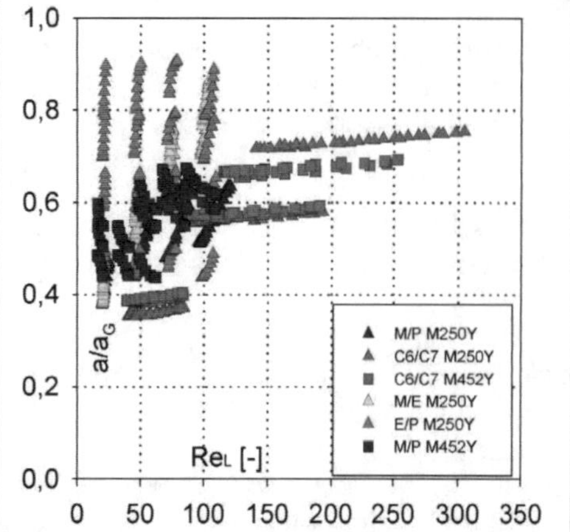

Figure 6: DISTILLATION interfacial area under the PLUG FLOW assumption

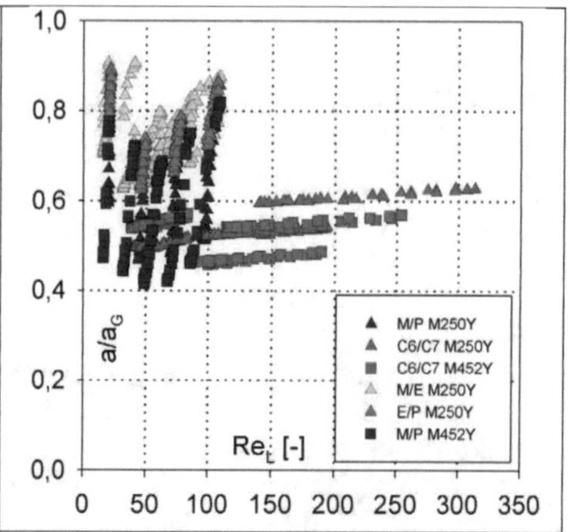

Figure 7: DISTILLATION interfacial area under the PLUG FLOW +AXIAL DISPERSION assumption

Conclusions:

The distillation is usually treated as a process with the dominant mass-transfer in the gas-phase. The data indicates that this is correct for the C_6/C_7 system, especially for larger loads, while for systems consisting of alcohols the mass-transfer resistance is distributed between the both phases.

The distillation k_Ga data are well described by the RBF and BS model for the C_6/C_7 system. For the methanol/ethanol system the k_Ga values are about 25% underestimated by these models. Common correlation of the distillation k_Ga data was not successful. While the hydrodynamics of the phases seems to be analogical, the recalculation of the data between the systems brings problems even when more sophisticated model for the liquid phase hydrodynamics is introduced for k_Ga evaluation.

The estimates of the effective interfacial area can't be conclusive due to weakly proven and actually rather challenged recalculation procedure from one system to another. As is, they support the idea of the effective interfacial area forming just about 50%-90% of the geometrical area of the packing although the packing is completely wet for all the conditions, as visually confirmed.

The system transfer procedure is only planned to be verified more thoroughly. Therefore the conclusions with respect to the distillation effective interfacial area should be viewed only as an attempt to shed some light on this perhaps most obscured mass-transfer parameter until reasonable direct measurement method is developed.

Does anybody have any idea?

Acknowledgements

Support of Grant Agency of the Czech Republic through the project No. 13-01251S is gratefully acknowledged.

Literature cited:

[1] Rocha, J.A., Bravo, J.L., Fair, J.R., 1996. Distillation columns containing structured packings: A comprehensive model for their performance. 2. Mass transfer model. Ind. Eng. Chem. Res. 35, 1660–1667.

[2] Billet, R., Schultes, M., 1999. Prediction of mass transfer columns with dumped and arranged packings. Updates summary of the calculation method of Billet and Schultes. Trans IChemE 77 (Part A), 498–504.

[3] Olujic', Ž., Behrens, M., Colli, L., Paglianti, A., 2004. Predicting the efficiency of corrugated sheet structured packings with large specific surface area. Chem. Biochem. Eng. Q. 18 (2), 89–96.

[4] Rejl, F.J., Valenz, L., Haidl, J., Kordač, M., Moucha, T. On the modeling of gas-phase mass-transfer in metal sheet structured packings. To be published in the Chemical Engineering Research and Design.

351006 Nagaoka Packing "Super H Pack"

Abstract:

NAGAOKA PACKING "Super H pack"

Tadayoshi Nagaoka

Laboratory of structured packing for distillation

Abstract

In the distillation and absorption processes of chemical plant, the structured packing was introduced instead of historical random packing and increased separation efficiency.

The structured packing, however, could not increase treating volume. When treating volume increases, speed of upward vapor also increases. In this circumstance, downward liquid, specially free liquid droplet at the bottom of distributor and structured packing can not go down due to high speed of upward vapor, and is accumulated in the column, resulting in start of flooding. These phenomena came from layout of all internals, because of separated layout, distributor, and structured packing.(Fig.2)

NAGAOKA PACKING “Super H pack” has connected a distributor, structured packing and collector by changing layout in the column. Super H pack realized increased speed operation of vapor and reached two-times larger treatment volume with high efficient separation.(Fig.1)

Structure and feature

NAGAOKA PACKING ”Super H pack” has five main features of unique structure as follows.(Fig.1)

1. Epoch-making connected structure with distributor, structured packing and collector pushes away the flooding point because of avoiding free liquid-droplet at the bottom of distributor and structured packing and expand operating range up to a higher F-Factor which enables high speed operation.
2. Super H pack is constructed with multiple vertical capillary wires which run with equal interval and in parallel in the full length.

Multiple linear liquid flow along the vertical capillary wires and upward organized vapor flow produce efficient vapor-liquid counter-contact.

Downward liquid flow which is led by the capillary wires can go down against high speed of upward vapor. Therefore high efficient separation and high speed operation which means larger treatment volume are achieved simultaneously.

1. Trough distributor is overflow type and has many equally spaced liquid ports and all capillary wires of structured packing hang from each of all liquid ports. Number of liquid drip points of distributor over 50,000/m2 can be available.

Combination of many equally spaced liquid ports and hanging capillary wires

produces uniform liquid distribution over the entire column.

1. Many equal liquid distribution starts at top of structured packing.

Elimination of the mal-distribution results from the constitution of vertical and parallel wires in the full length.

Therefore minimum specific surface area of Super H pack produces high efficient separation.

1. Constitution of vertical and parallel wires and minimum specific surface area achieve minimum pressure drop in operation.

Test result of distillation and valuation

Vacuum batch distillation.

(Photo 1)

Head pressure 100mmHg.

Fluid; Chlorobenzene and

Ethyl benzene.

Separation efficiency ;NTS/m

(Fig.3)

Pressure drop; mbar/NTS

(Fig.4)

Photo1

Indication	Remarks	Specific Surface Area m2/m3
HP-140	Super H pack	140m2/m3
HP-190	Super H pack	190m2/m3
HP-240	Super H pack	240m2/m3
MP-250	Conventional Sheet Metal	250m2/m3
BX-500	Conventional Wire Gauze	500m2/m3

u Comparison of vapor velocity, F-Factor(Load) m/s√kg/m3 (Fig.3 and 4)

Super H pack; --------------------------Max. F-Factor=5

Conventional structured packing

Sheet Metal and Wire Gauze; Max. F-Factor=3

After F-Factor=3,

Conventional structured packing----Pressure drop going up sharply

Flooding start. Impossible operation

Super H pack--------------------------------Pressure drop going up slowly

Stabilizing operation until F-Factor=5.

u Comparison of separation efficiency NTS/m (Fig.3)

After F-Factor=3

Conventional structured packing------Separation efficiency going down sharply

Impossible operation

Super H pack-------------------------------Separation efficiency going down slowly

Stabilizing operation until F-Factor=5

u Stabilizing operation of Super H pack in wide range is a main feature which separation efficiency(Fig.3) and pressure drop(Fig.4) graphs show .

u Separation efficiency of Super H pack in minimum specific surface area is superior to conventional structured packing with large specific surface area which show in Fig.3.

Conclusion

Super H pack is epoch-making structured packing which realized two-times larger treatment volume with high efficient separation simultaneously in condition of minimal specific surface area and minimal pressure drop.

Above features are proved in the distillation tests

In the revamping case, separation efficiency will be raised and treating volume will be increased two-times.

In the new plant case, column size will be reduced and cost-down will be realized in initial stage.

Column height------------Shortening by higher NTS/m

Column diameter---------Shortening by higher F-Factor/NTS

Running cost down will be added by higher separation efficiency, larger treating volume and lower pressure drop.

Large economic impact of initial stage and running cost down will be expected in the actual chemical plant.

Super H pack which is of a simple structure with multiple vertical and parallel wires will be applicable to processing of fluid which will condense in the column or processing of dirty materials.

Application field of Super H pack are not only distillation but also absorption, stripping, cooling and aeration etc.

Especially, application of Super H pack for absorption and stripping of CO_2 which will contribute to the earth will be highly expected.

.

Enhancing Fractionator Efficiency Using Properly Designed Vapor & Liquid Distribution Devices

Sounho Lee – GTC Technology US LLC
2014 AIChE Spring Meeting / Distillation Symposium
New Orleans LA April 1, 2014

GTC Technology – Engineered to Innovate

Introduction
How to achieve target distillation efficiency

GTC Technology – Engineered to Innovate

- Degree of separation of distillation column
  - Number of trays or packing type & height: primary parameter
  - Vapor/liquid distribution : also critical piece to achieve desired column performance but often overlooked in commercial distillation column design
  - More sensitive to vapor/liquid distribution in packed column
- Two (2) case studies
  - Actual revamp cases
  - Presenting "fixing small design flaws brings noticeable efficiency gain"
  - Packed column cases
  - Petroleum refining crude atmospheric column
  - Gas processing demethanizer

AIChE 2014 Spring Meeting / Distillation Symposium Page 2

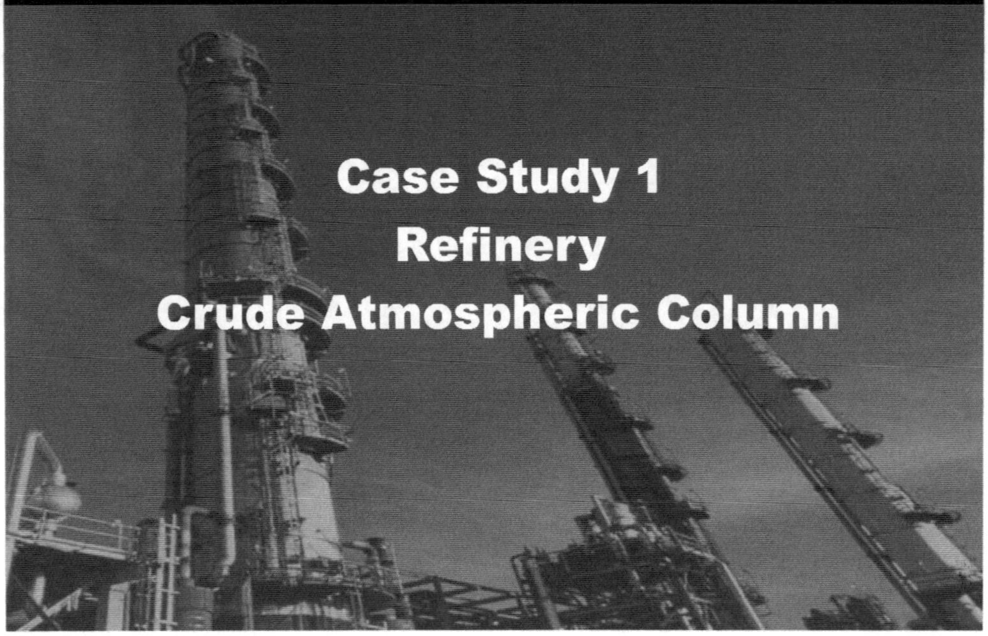

Case Study 1 - Crude Atm. Column Unit Configuration

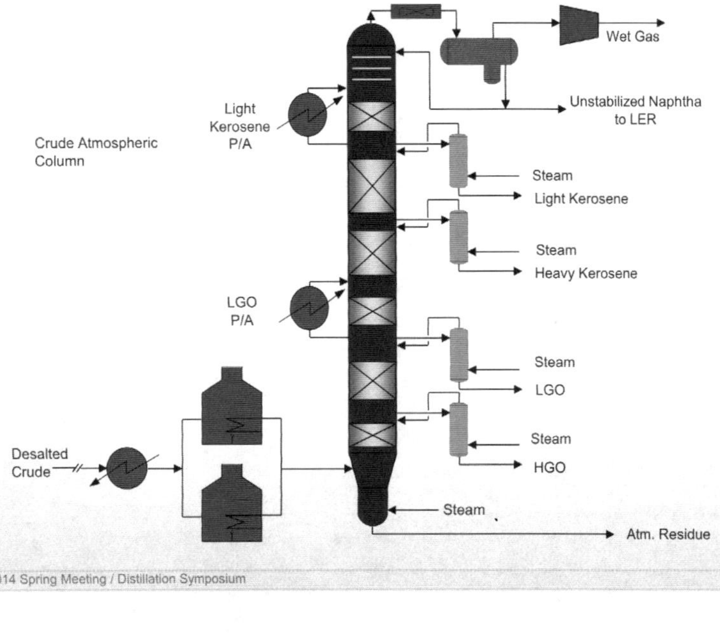

Case Study 1 - Crude Atm. Column Background

- Operating condition
  - Middle East Crude Slate
  - Heater Outlet Temperature: 635-653 °F (335-345 °C)
- Column Configuration
  - (6) structured packed beds
  - Naphtha-light kerosene fractionation & bottom stripping : trayed
- Pre-Revamp Limitation
  - Poor fractionation especially among light kerosene, heavy kerosene & LGO
  - Limited kerosene yield influenced refinery economics
  - Replacement with new packing did not improve fractionation performance

Case Study 1 – Crude Atm. Column Crude Assay vs. Product Yield Balance

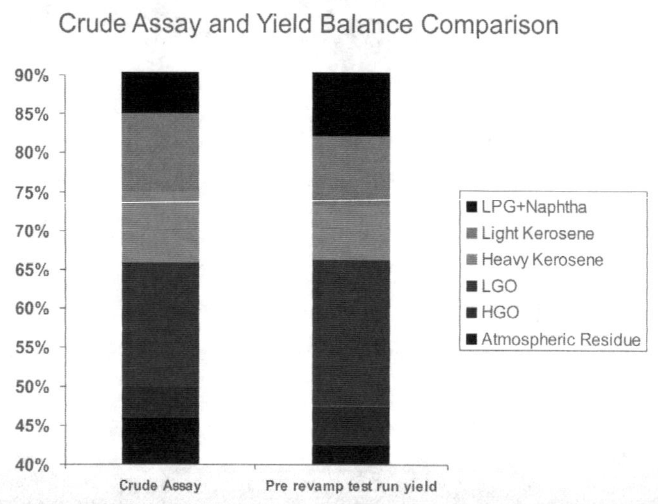

Case Study 1 - Crude Atm. Column
Root Cause Identification

- Packing type was not an issue
  - Selected packing types / bed heights were suitable to meet target efficiency
- Collector tray
  - Poor vapor distribution due to improper vapor riser/deck design
  - Poor liquid distribution between collector and liquid distributor
- Liquid distributor
  - Poor liquid quality
  - High flow deviation among troughs (crude atmospheric column)
  - Inadequate drip point arrangement (heavy kerosene side stripper)

Case Study 1 - Crude Atm. Column
Original Design Flaw

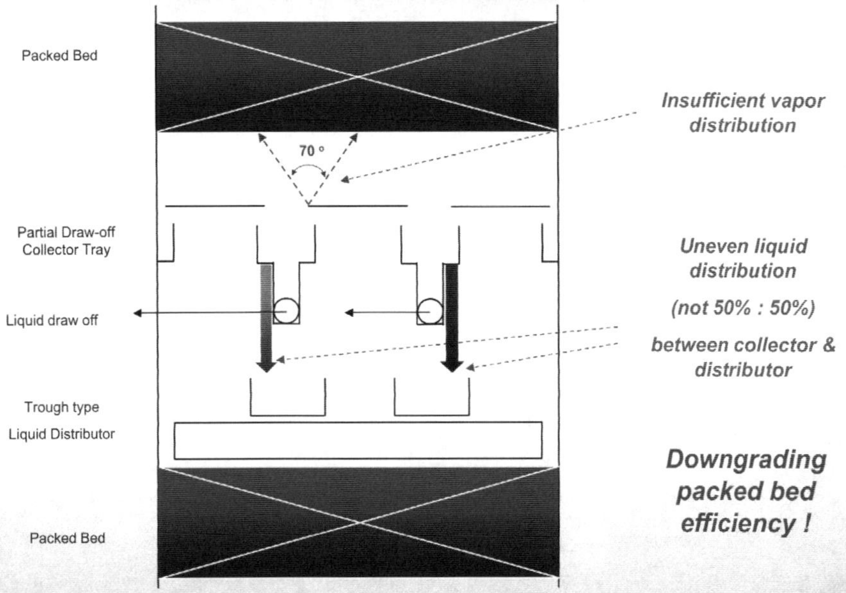

Case Study 1 - Crude Atm. Column
Modification Area

GTC Technology · Engineered to Innovate

- New collector trays and distributors are implemented for following sections
  - Light and heavy kerosene fractionation
  - Heavy kerosene and LGO fractionation
  - LGO pumparound
  - LGO/HGO fractionation
  - Wash section
  - Heavy kerosene side stripper (liquid distributor only)
- Existing structured packing for all pack beds was re-used

Case Study 1 - Crude Atm. Column
New Design Strategy

GTC Technology · Engineered to Innovate

- New collector tray (crude atmospheric column)
  - Chimney dimension enhancement to improve vapor distribution
  - Collector deck elevation optimized to improve vapor distribution
  - Gravity line hydraulic is evaluated and verified between the main column & side strippers
  - Improving downpipe design for uniform liquid split to two parting boxes of trough type liquid distributor
- Trough type liquid distributor (crude atmospheric column)
  - Flow deviation among troughs are minimized
  - Enhanced liquid distribution quality
- Pan type liquid distributor (heavy kerosene side stripper)
  - Enhanced liquid distribution quality

Case Study 1 - Crude Atm. Column
New Collector / Distributor - 3D View

GTC Technology – Engineered to Innovate

Case Study 1
Performance Summary

GTC Technology – Engineered to Innovate

Parameter	Pre-revamp Test Run	Post-revamp Test Run
Crude Charge (BPD)	BASE	Δ 0
Light Kerosene yield (LV%)	BASE	- Δ 0.3 LV%
Heavy Kerosene yield (LV%)	BASE	+ Δ 3.4 LV%
LGO yield (LV%)	BASE	+ Δ 1.0 LV%
HGO yield (LV%)	BASE	- Δ 0.9 LV%
Naphtha – Light Kerosene Gap, °F	BASE	+ Δ 2.0 °F
Light Kerosene flash point, °F	BASE	+ Δ 7.2 °F
Light Kerosene – Heavy Kerosene Gap, °F	BASE	+ Δ 11.0 °F
Heavy Kerosene flash point, °F	BASE	+ Δ 5.4 °F
Heavy Kerosene – LGO Gap, °F	BASE	+ Δ 95.4 °F
Unit Stripping Steam Light Kerosene Stripper, Lb/BBL	BASE	- Δ 89%
Unit Stripping Steam Heavy Kerosene Stripper, Lb/BBL	BASE	- Δ 26%
Overflash rate (% of total distillate)	BASE	- Δ 3%

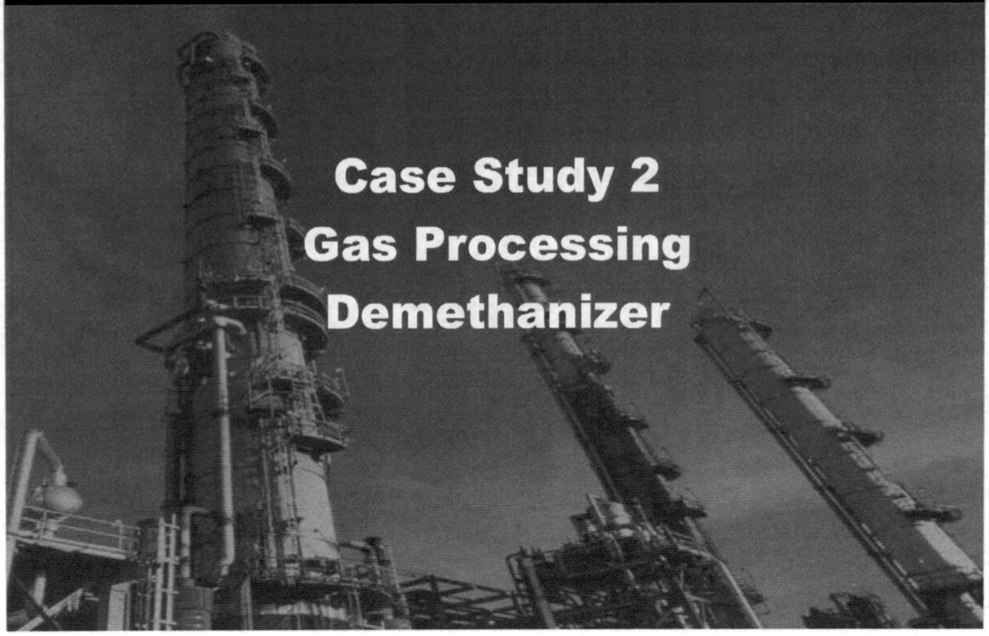

Case Study 2 – Demethanizer Background

- Demethanizer in gas processing plant
  - Single feed introduction @ bell section
  - Two reboilers for heat input: side and bottom reboilers
- Pre-retrofit column configuration
  - Third generation random packing installed for two beds
- Pre-retrofit limitation
  - C2+ component recoveries were downgraded at higher charge rate

Case Study 2 – Demethanizer Unit Configuration

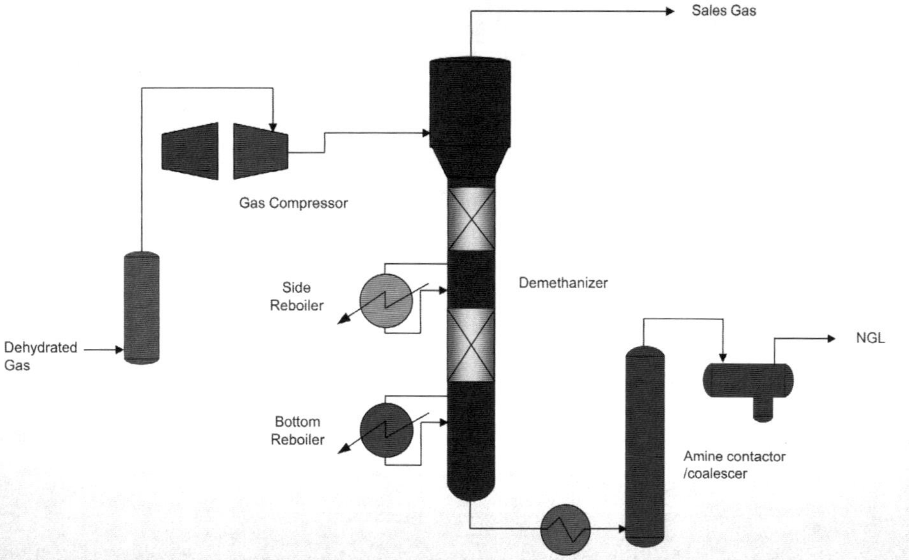

AIChE 2014 Spring Meeting / Distillation Symposium

Page 15

Case Study 2 – Demethanizer Pre-Retrofit Operating Condition

Operating Parameter	Pre-retrofit Operation 1	Pre-retrofit Operation 2
Plant Capacity (MMSCFD)	BASE	+ Δ 28%
Ethane Recovery (vol %) note 1	BASE	- Δ 16.8 %
Propane Recovery (vol %) note 1	BASE	- Δ 5.4 %
Column top temp, °F	BASE	+ Δ 23.5 °F
Side reboiler draw temp, °F	BASE	+ Δ 14.9 °F
Bottom reboiler draw temp, °F	BASE	**- Δ 63.3 °F**
Column top pressure, psig	BASE	+ Δ 3.1 psig
Simulated Side Reboiler Duty, MM BTU/hr	BASE	**- Δ 29 %**
Simulated Bottom Reboiler Duty, MM BTU/hr	BASE	+ Δ 10%

Note 1: recovery to bottom product

AIChE 2014 Spring Meeting / Distillation Symposium

Page 16

Case Study 2 – Demethanizer
Root Cause Analysis

GTC Technology – Engineered to Innovate

- Limited side reboiler duty and poor liquid/vapor distribution quality
  - Downgrade C2 recovery @ higher charge rate
  - Limited side reboiling
- Poor original collector tray design
  - Limited side reboiler circulation rate due to insufficient vapor riser height
  - Improper construction method: bolted construction
- Inadequate liquid distributor design
  - Poor liquid distribution quality
- Potential capacity limitation
  - Top bed capacity was reaching the limit with original packing size

Case Study 2 – Demethanizer
Modification Summary

GTC Technology – Engineered to Innovate

- Top bed (above side reboiling)
  - New collector tray design: taller riser height, seal welded construction
  - New distributor with improved distribution quality
  - Random packing size is increased by one step in order to increase packing capacity
  - New packing size is confirmed to achieve target degree of separation
- Bottom bed (below side reboiling)
  - New collector tray design: seal welded construction
  - New distributor with improved distribution quality
  - Maintained existing packing size

Case Study 2 – Demethanizer Performance Summary

GTC Technology – Engineered to Innovate

Operating Parameter	Pre-revamp Operation	Pre-revamp Operation	Post-revamp Operation	Post-revamp Operation
Plant Capacity (MMSCFD)	BASE	+ Δ 28%	BASE	+ Δ 36%
Ethane Recovery (vol %) note 1	**BASE**	**- Δ 16.8 %**	**+ Δ 25.3%**	**+ Δ 18.1%**
Propane Recovery (vol %) note 1	BASE	- Δ 5.4 %	**+ Δ 3.2%**	**+ Δ 1.2%**
Column top temp, °F	BASE	+ Δ 23.5 °F	- Δ 2.5 °F	- Δ 0.4 °F
Side reboiler draw temp, °F	BASE	+ Δ 14.9 °F	- Δ 4.8 °F	- Δ 1.6 °F
Bottom reboiler draw temp, °F	**BASE**	**- Δ 63.3 °F**	**- Δ 9.3 °F**	**- Δ 4.2 °F**
Column top pressure, psig	BASE	+ Δ 3.1 psig	- Δ 24.2 psig	- Δ 23.1 psig

Note 1: recovery to bottom product

AIChE 2014 Spring Meeting / Distillation Symposium Page 19

Case Study 2 – Demethanizer C2 Recovery Trend

GTC Technology – Engineered to Innovate

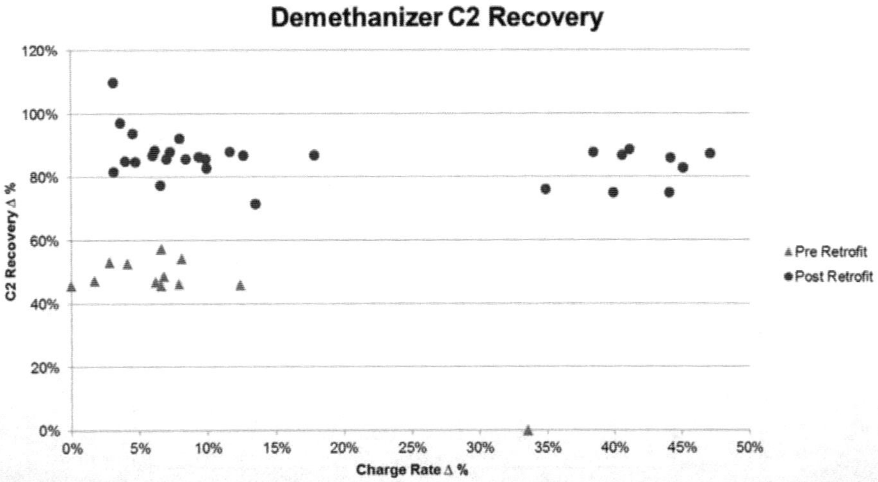

AIChE 2014 Spring Meeting / Distillation Symposium Page 20

References

1. Libermann N P, Process Design for Reliable Operations, *Gulf Publishing Company*, 2nd Edition

2. Kister H Z, Distillation Operation, *McGraw-Hill Comp*any, 1990

3. Kister H, et al, Troubleshoot chimney trays effectively, Hydrocarbon Processing, April 2001

4. Lee S H, et al, High-performance trays alone do not guarantee performance improvements, *Petroleum Technology Quarterly* Q2, 2012

5. Rukovena F, et al, Packed Tower Internals Important to Tower Efficiency, *AIChE Spring meeting*, April 2009

Thank you

Thank you

352098 A Submerged Particle Model for Liquid Holdup in Packed Beds

Abstract:

New packing elements emerge all the time in the quest for improved gas-liquid-solid contact in packed towers while minimizing pressure drop. Design of packed tower operations is still based on packing element specific empirical charts and it is still not possible to predict the transfer coefficients. Liquid holdup is a key parameter that influences wetting efficiency, pressure drop and transfer rates. A new model for liquid holdup in packed beds is developed considering gas to flow around particles enveloped by a film of trickling liquid. Equation for gas phase pressure drop is developed by adopting Ergun's equation incorporating the effect of presence of liquid phase on gas phase voidage and tortuosity for gas flow. The model equations are compared with the literature experimental data to evaluate the effect of parameters such as gas and liquid velocities, liquid properties, particle shape and size on liquid holdup and two phase pressure drop. The model equations compare reasonably well with experimental observations.

352135 A New Screening Tool for Designing Energy Efficient Distillation Systems

Abstract:

A New Screening Tool for Designing Energy Efficient Distillation Systems

Joshua Huff, Mohit Tawarmalani, Rakesh Agrawal

School of Chemical Engineering

Purdue University, West Lafayette, IN 47907

In the traditional industrial applications of multicomponent distillation, a relatively limited number of distillation column arrangements are repeatedly applied to a wide variety of separations. However, there are many separation process solutions differing significantly in cost and energy consumption that carry out the same overall process. Therefore a systematic method to identify and design optimal multicomponent separation sequences is needed that does not rely on the inventive activity of a few experienced engineers. Even for a commonly perceived "mature" technology such as distillation, a method to elucidate all possible multicomponent separation schemes has only been developed recently.

A matrix method has been developed to enumerate all distillation configurations which can perform a given separation. This generated "search space" includes both basic and thermally coupled configurations which complete an *n*-component separation in *n-1* columns. The size of this search space grows increasingly large as the number of components increases – for example, a 5 component separation has 6,128 distinct ways to perform it using 4 columns. Once we have this search space, we can ranklist all configurations in terms of their vapor duty, thermodynamic efficiency or total annualized cost. However, the results of this ranklist still take the form of a series of matrix outputs, which require detailed knowledge of the method to study and implement.

For this reason, a new tool has been developed to capture and visualize system information for distillation columns designed by the matrix method. The tool is capable of drawing flowsheets for all configurations in the search space and overlaying them with information on feed conditions, flowrates and compositions, section vapor flows, and reboiler and condenser vapor duties. The list of configurations is complete (for example, for a five component configuration it will provide drawings of all 6,128 configurations) and sortable by a number of criteria including vapor duty, thermodynamic efficiency, total annualized cost, and number of thermally coupled streams. Additionally, the ability to screen configurations based on a mix of structural criteria (such as presence of particular splits or columns) and operational criteria (such as vapor duty or number of thermally coupled streams) is built into the program.

This tool provides the ability to translate the complete search space screening of the matrix method to a form which is usable by practicing engineers to study, screen, and draw designs from a very large space of possible solutions. Due to the high global energy consumption of the distillation sector, this provides the opportunity for massive energy savings through both design of new energy-efficient distillation systems and strategic retrofit of existing systems for energy savings.

352226 More Operable Dividing Wall Columns

Abstract:

Industrial application of dividing wall columns (DWCs) for multicomponent separation has gained significance in recent years. Of predominant interest have been the dividing wall versions of the *n*-component Fully Thermally Coupled (FTC) configurations because of their low energy requirements. However, in such DWCs, to derive maximum energy/cost savings, the vapor split at the bottom of a vertical partition is often critical. But, in practice, the vapor split at the bottom of a partition is left unregulated. To overcome this operational challenge, we suggest new attractive multicomponent DWCs in this work that allow independent control of the vapor flowrate in each partitioned zone. Interestingly, we show that all the new ternary DWCs have the same minimum heat duty requirements as the FTC configuration.

351396 Dividing Wall Column as Energy Saving Retrofit Technology

Abstract:

Dividing Wall Column (DWC) is a fully heat coupled single shell column that can replace two or more conventional columns connected in series. As proven in many industrial applications, a three product DWC enables on average 30 % saving in energy and an equivalent saving in capital, accompanied by nearly halving the required plot area compared to conventional two or three column sequences. This however is related to new designs, and only few applications have been reported where an existing column has been revamped. These, however, have been originally single columns with a side product, which, by the virtue of nature of a DWC, are obvious candidates for a revamp.

Present paper uses an aromatics plant base case to show that existing two or three column sequences can be transformed in a cost-effective way into DWCs to allow energy saving equivalent to that achievable in new designs. This makes DWC to an interesting retrofit option for existing plants confronted with the need to comply in foreseeable future with CO_2 emissions reduction requirements imposed by legislation.

Consumer & Industrial Solutions

Process Intensification through Distillation Synthesis, Dividing Wall Columns, and Novel Recycle Methods

Mahesh Sawant, Daniel Trauth, John Pendergast, Ravi Dixit

2014 AIChE Spring Meeting

Dow.com

Overview

- Background on ANGUS business
- Vapor Phase Propane Nitration Technology and Limitations
- New Nitration Technology Process Intensification Opportunities
- Process Overview
- Process Intensification Improvements
- Conclusions

What is ANGUS Chemical?

- Wholly-owned subsidiary of the Dow Chemical Company
- Purchased in fall -1999
- Part of Dow Consumer & Industrial Solutions Group
- Plants in Louisiana and Germany
- Main technology is propane nitration
- Free-radical nitration
- Only large scale practitioner of the technology
- Specialty products based on nitroparaffin chemistry
- Applications include:
  - Paint additives
  - Pharmaceutical intermediates
  - Oil and gas
  - Process chemicals
- Process research that led to vapor phase propane nitration was started at Purdue University in the 1930's

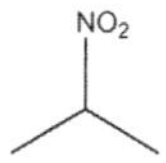

2-Nitropropane

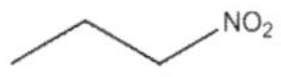

1-Nitropropane

Consumer & Industrial Solutions 3

Vapor Phase Propane Nitration

- Two or three stage reactor
- High temperature gas stream is fed from the top inlet
- Liquid acid at ambient temperature is introduced into the reactor via spray nozzles
- Droplets are heated, evaporate, and react with the gas feed
- Heat of reaction drives complete evaporation
- Delicate balance between water/nitric acid evaporation and exothermic reaction
- Fixed product selectivity
- Reactions take place in vapor phase only
- But, there are important limitations ....

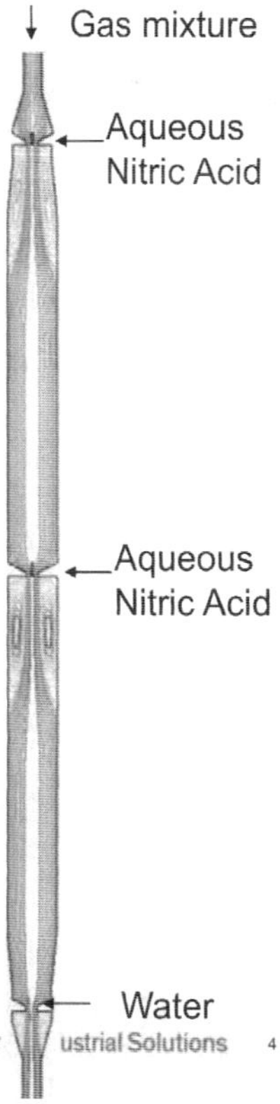

Consumer & Industrial Solutions 4

Mixed Phase Propane Nitration

- Downflow, co-current reactor setup
- Higher pressure/lower temperature process conditions
- Propane and nitric acid preheated and fed from the top inlet
- Boiling reactor; always maintain liquid phase
- Heat balance is insensitive to nitric acid concentration
- Simplified product slate
- Product selectivity control
- Biotreatable waste water
- Patented process
  - US 8324435, US 8519199, US 8558039

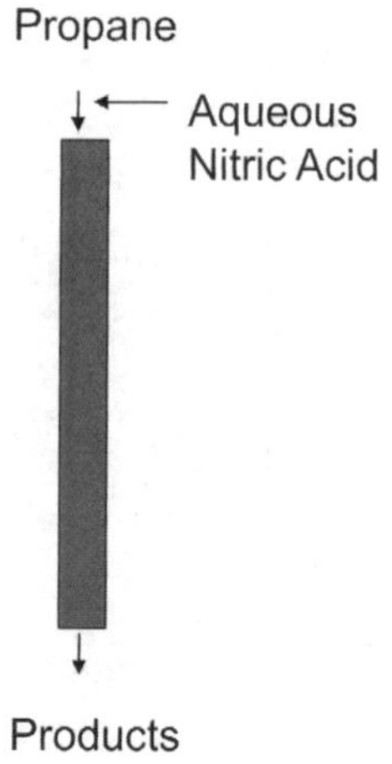

Drivers for Process Intensification

- Vapor phase process limitations
  - Fixed product selectivity
  - Low raw material conversions
  - Complex, energy intensive separation scheme

- Mixed Phase Propane Nitration
  - High selectivity allows for production of wide range of products
  - Improved raw material conversions and simplified product mix

A need exists for simplified, flexible and economical process for the manufacture of a variety of nitrated compounds

Vapor Phase Nitration: Simplified Downstream Process Sketch

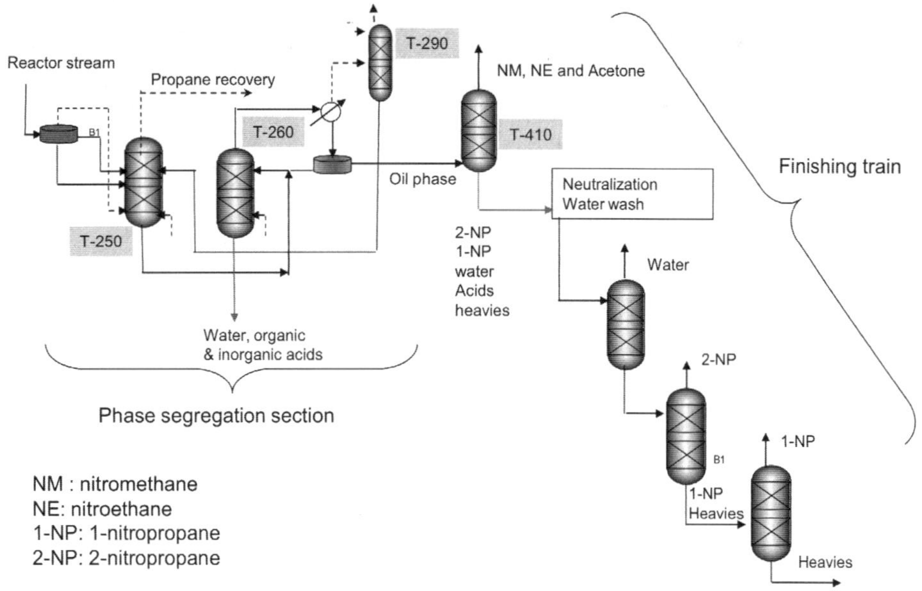

Intensification of Phase Segregation Section: Direct Recycle of Aqueous Stream

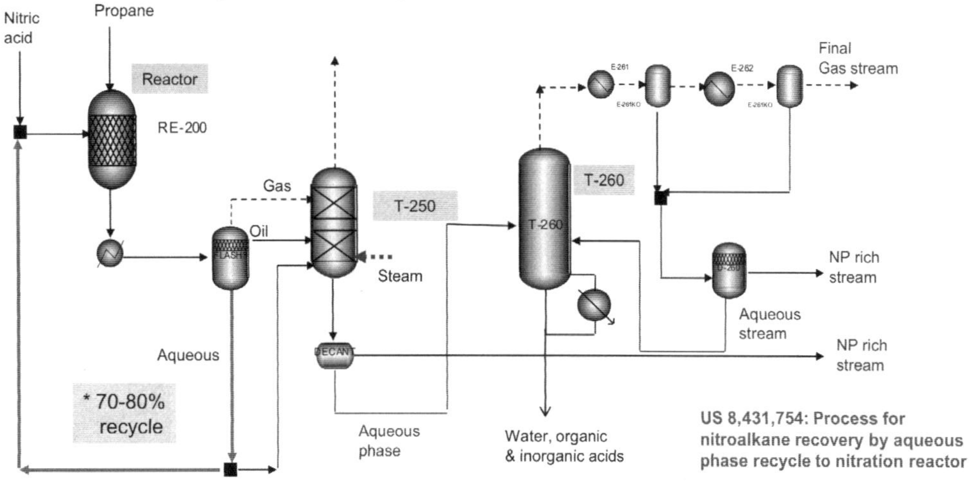

- ~70-80% aqueous stream recycled back to reactor depending on desired nitric acid feed strength
- No change in Nitroparaffin product selectivity or raw material yield
- Tail gas recovery column (T-290) eliminated
- Energy load on T-260 reduced by 30 – 40%

Intensification in Finishing Section – Dividing Wall Column

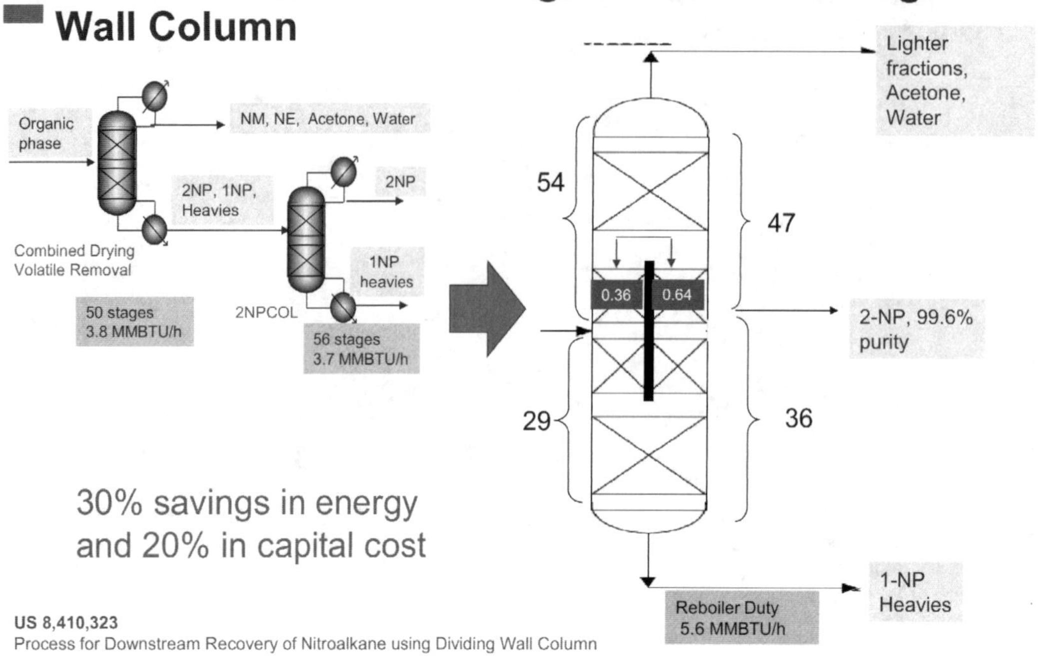

US 8,410,323
Process for Downstream Recovery of Nitroalkane using Dividing Wall Column

Consumer & Industrial Solutions

Intensification Reduces Columns from 7 to 4

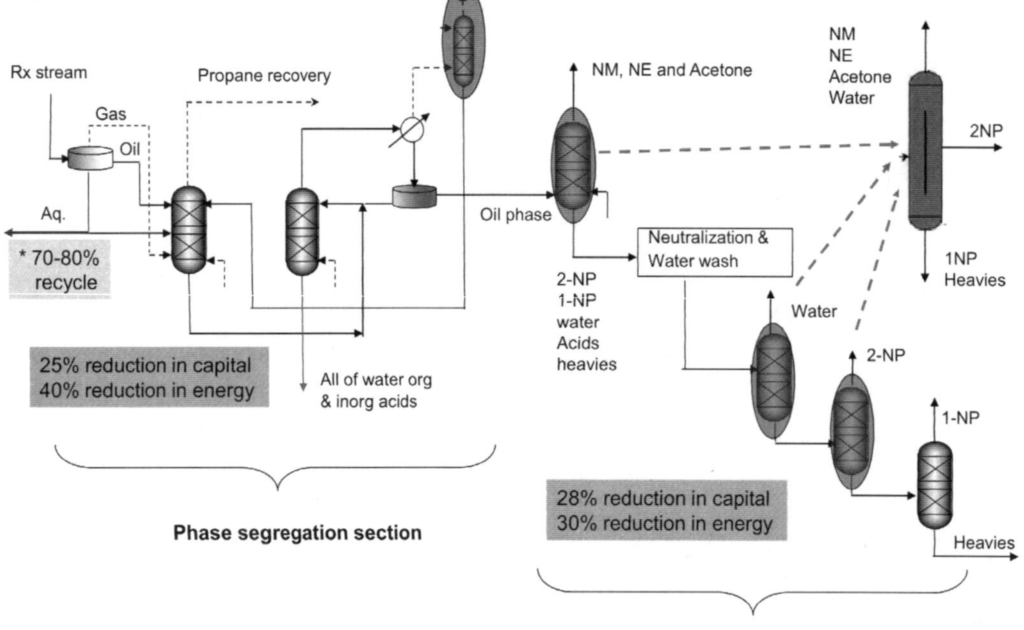

Consumer & Industrial Solutions
10

Conclusions

- Process intensification applied to propane nitration offers large opportunities:
  - 10 to 20% reduction in cost to manufacture
  - Reduced capital investment
  - Reduced environmental impact

- Pilot plant scale verification is complete

Parallel Distillation Calculations Using OPENMP

Spring 2014 AICHE Meeting New Orléans

Ron Bondy
Dr. David Bluck
Srividya Gummadi
David Van Peursem
Dr. Rajkumar Vedam
Schneider Electric/SimSci, Lake Forest, Ca

ron.bondy@invensys.com

Schneider Electric

Overview

- Introduction
- Multicore Computers
- Amdahl's Law
- Parallelization of Newton's Method
- OPENMP
- Results

Introduction

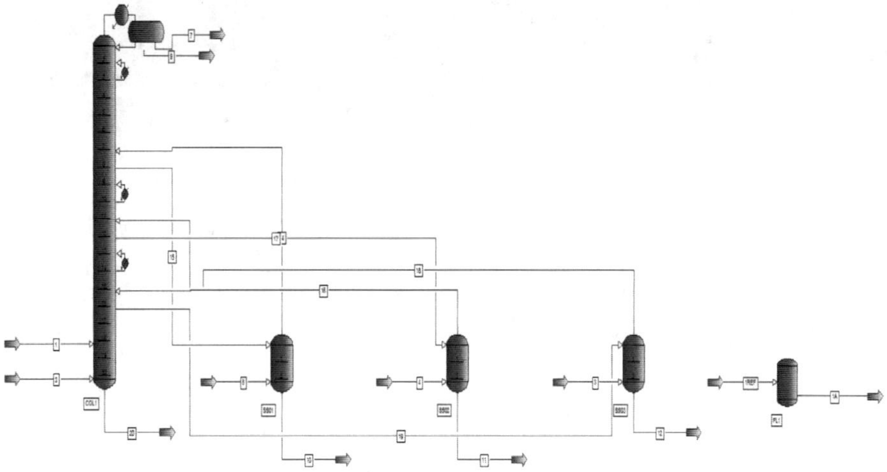

Introduction

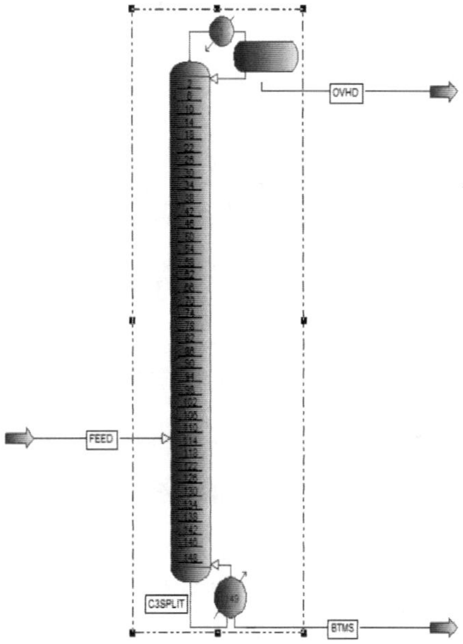

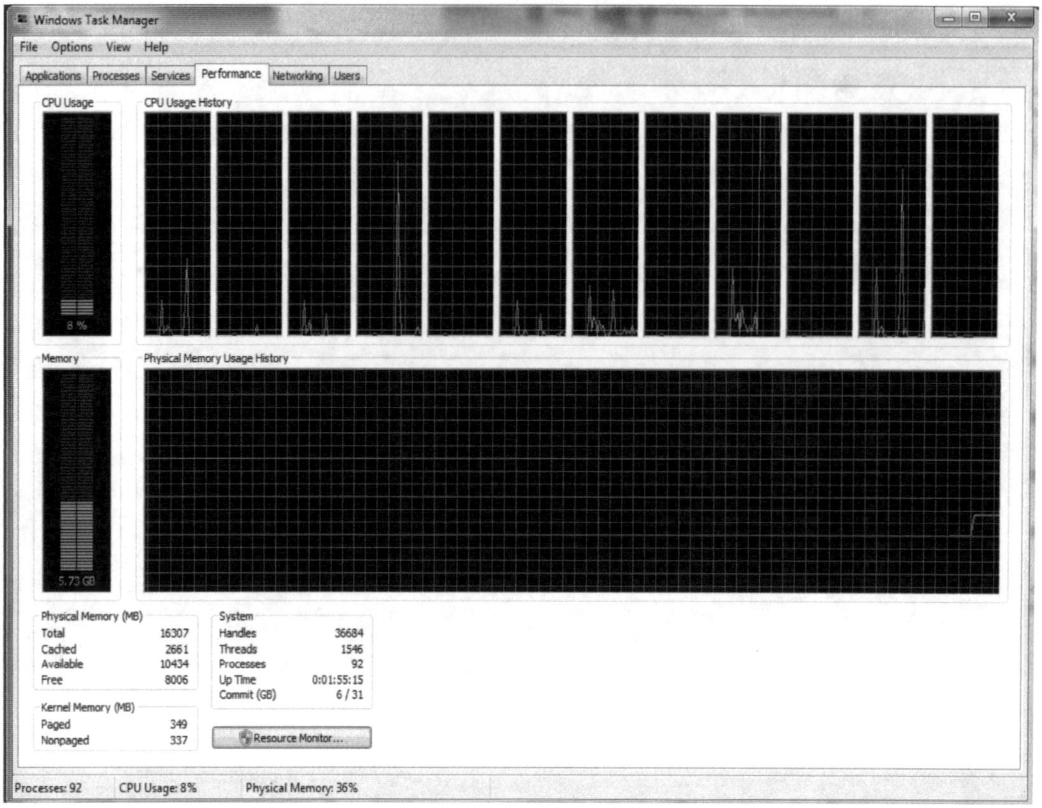
Windows Task Manager
File Options View Help
Applications Processes Services Performance Networking Users
CPU Usage
CPU Usage History
8 %
Memory
Physical Memory Usage History
5.73 GB
Physical Memory (MB)
Total 16307
Cached 2661
Available 10434
Free 8006
Kernel Memory (MB)
Paged 349
Nonpaged 337
System
Handles 36684
Threads 1546
Processes 92
Up Time 0:01:55:15
Commit (GB) 6 / 31
Resource Monitor...
Processes: 92
CPU Usage: 8%
Physical Memory: 36%

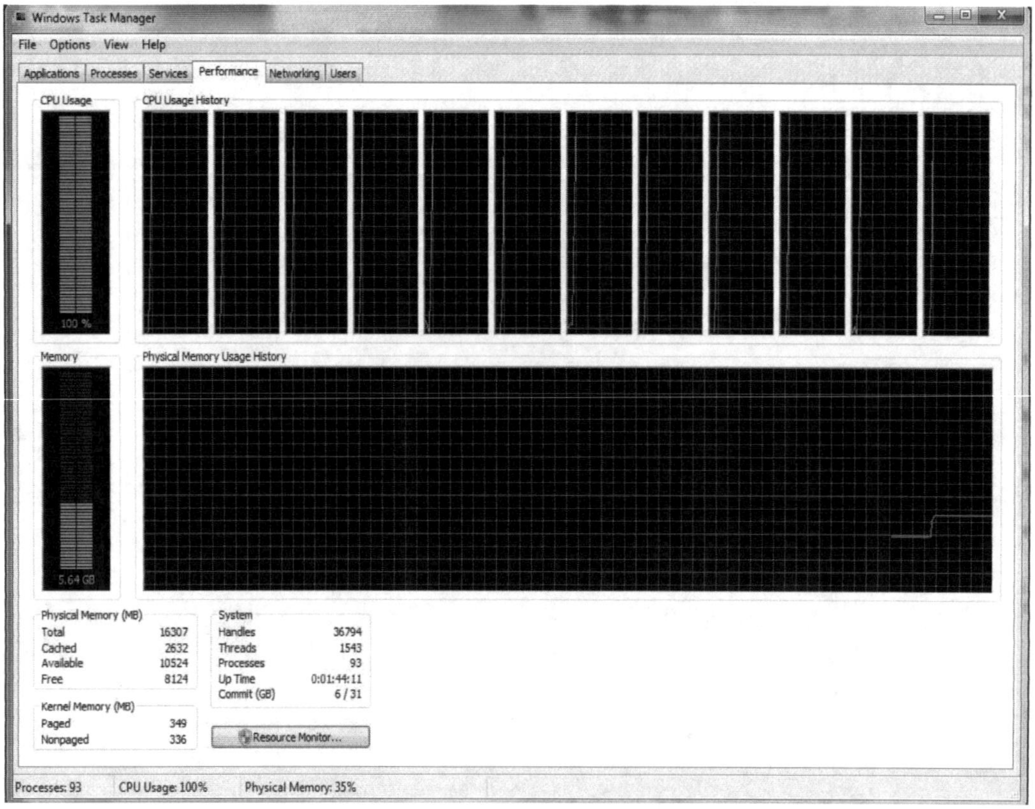
Windows Task Manager
File Options View Help
Applications Processes Services Performance Networking Users
CPU Usage
CPU Usage History
100 %
Memory
Physical Memory Usage History
5.64 GB
Physical Memory (MB)
Total 16307
Cached 2632
Available 10524
Free 8124
Kernel Memory (MB)
Paged 349
Nonpaged 336
System
Handles 36794
Threads 1543
Processes 93
Up Time 0:01:44:11
Commit (GB) 6 / 31
Resource Monitor...
Processes: 93
CPU Usage: 100%
Physical Memory: 35%

Introduction – CPU Speeds

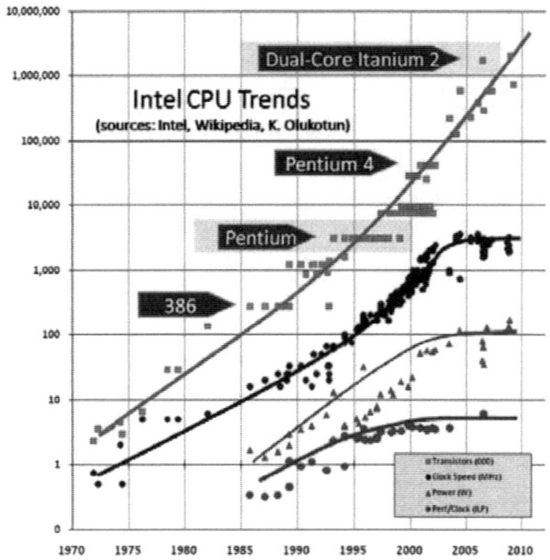

Herb Sutter,
Gotw.ca

Introduction – Multi Core CPU's

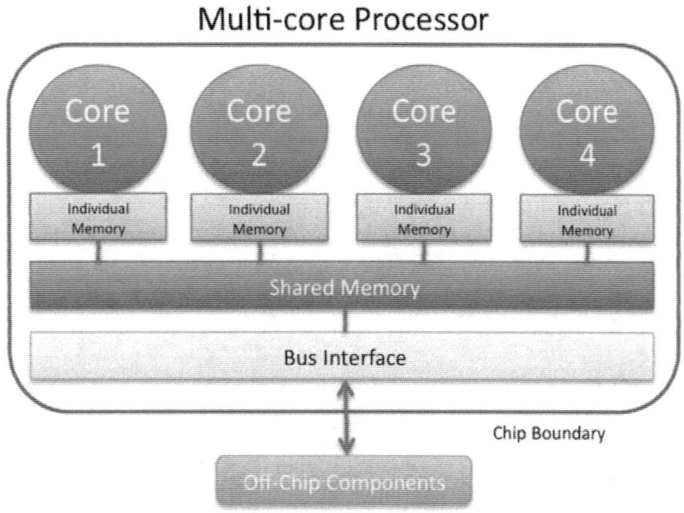

Dr. Raj Jain,
http://www.cse.wustl.edu/~jain/cse567-11/ftp/multcore/

Amdahl's Law (Theoretical Ideal)

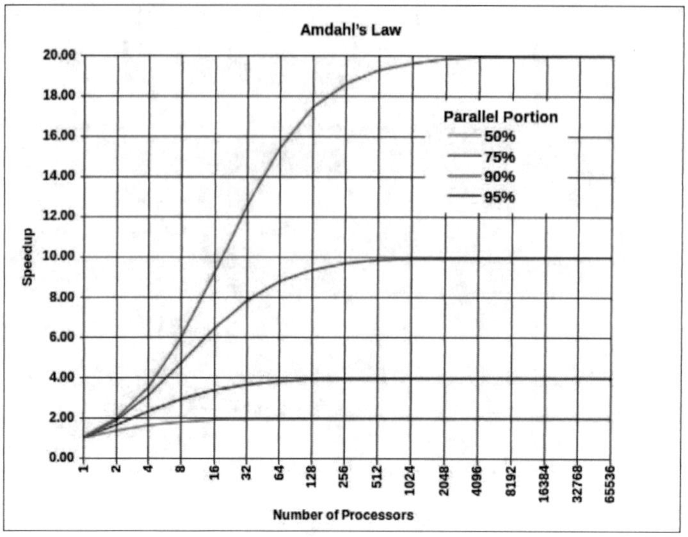

Newton's Method

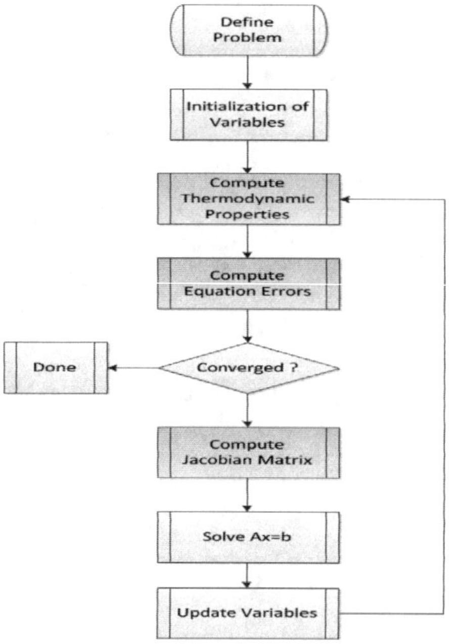

Code Development

- Code written in C++ using Visual Studio 2013 with OPENMP
- Object Model:

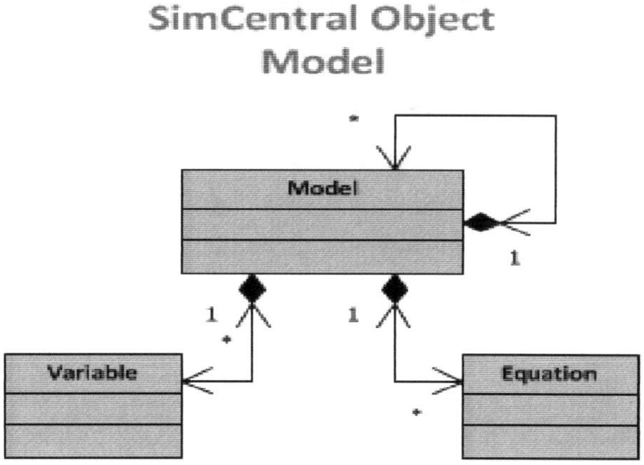

Schneider Electric - Division - Name – Date 11

Code Development

- Model objects are used for the column, distillation stage, fluid and fluid phase
- The model, equation and variable objects are the only 'domain' objects that the solver interacts with

Schneider Electric - Division - Name – Date 12

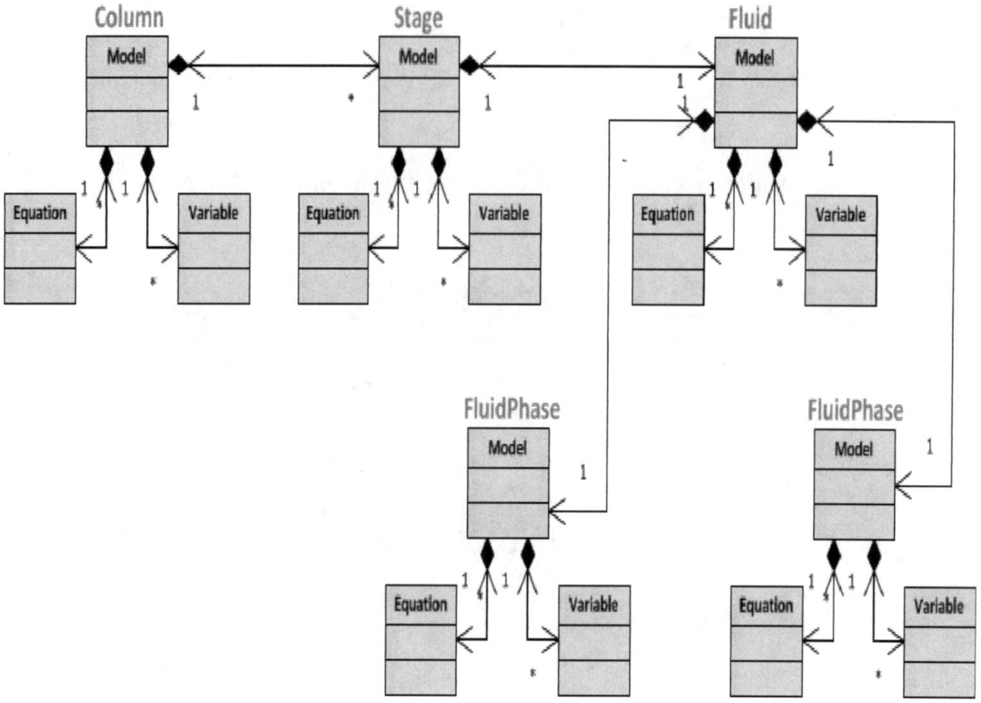

Schneider Electric - Division - Name – Date

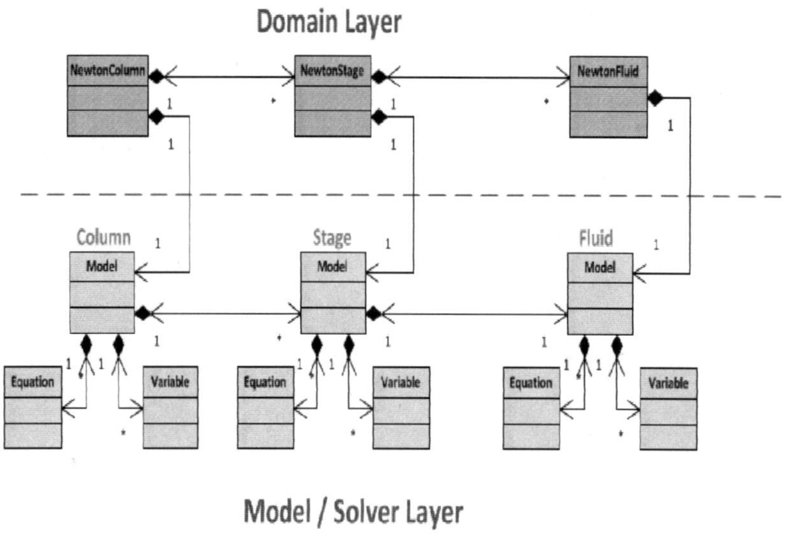

Schneider Electric - Division - Name – Date

Task Parallelism in Distillation

- Computing Thermodynamic Properties
  - Thermodynamic Calcs typically take 30% to 50% of a distillation column solution
- Computing Equation Errors
- Computing Jacobian Elements

- Each of the above is an excellent candidate to parallelize because these values on tray n are independent of other trays

Task Parallelism in Distillation

- Problem Definition
  - C3 Splitter
  - 150 Stages
  - 4 Components
    - Ethane
    - Propylene
    - Propane
    - Butane
  - SRK Thermodynamics

Parallel Thermodynamic Calculations

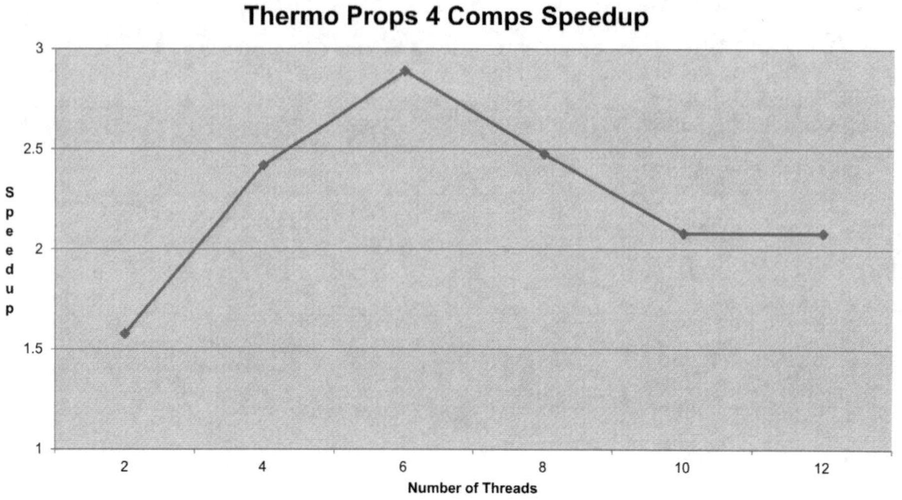

Schneider Electric - Division - Name – Date

17

Parallel Thermodynamic Calculations

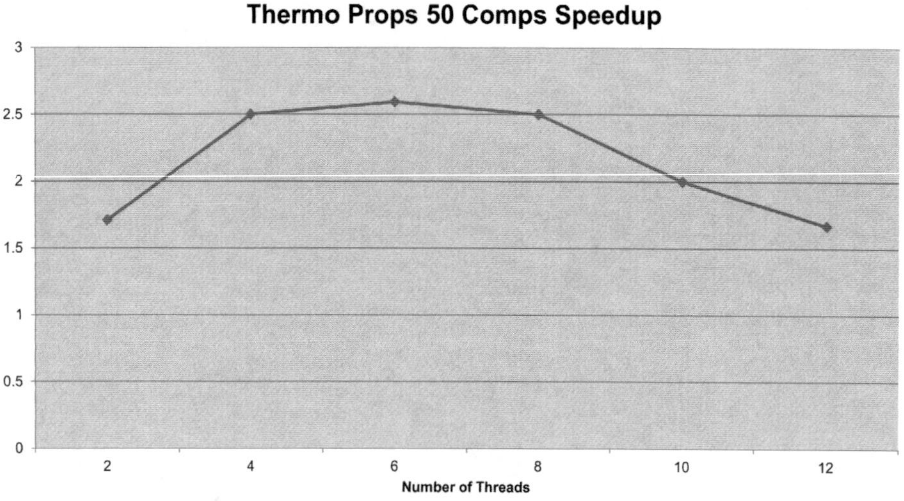

Schneider Electric - Division - Name – Date

18

Parallel Thermodynamic Calculations

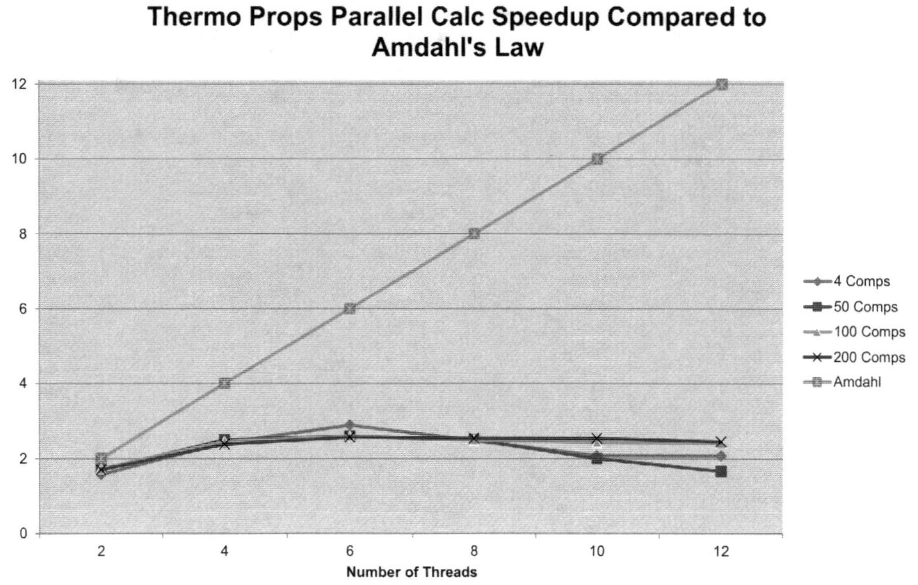

Schneider Electric - Division - Name – Date

19

Parallel Thermodynamic Calculations

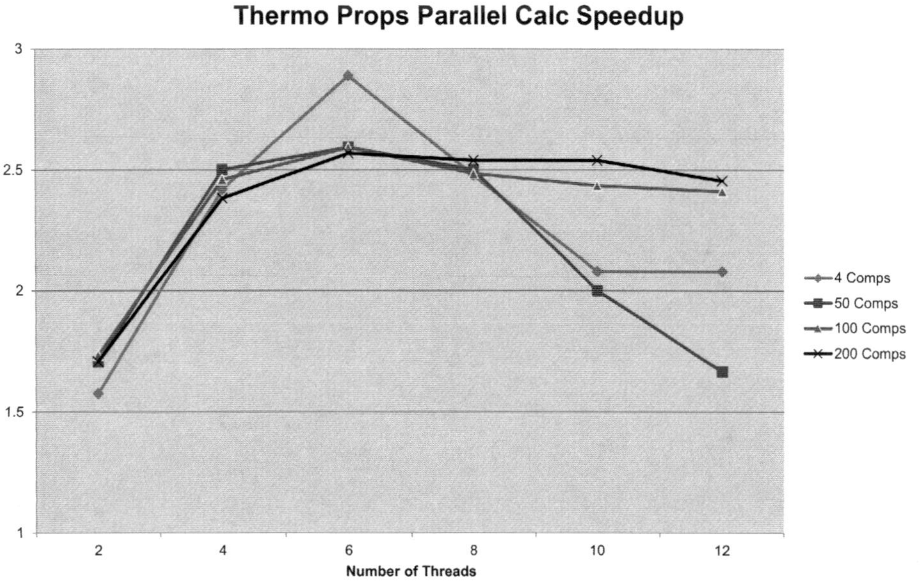

Schneider Electric - Division - Name – Date

20

Parallel Thermodynamic Calculations

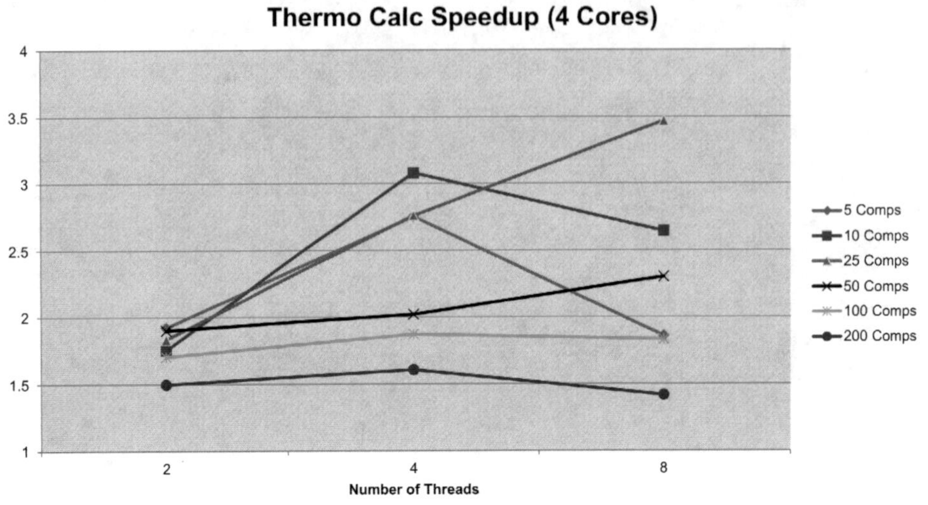

Parallel Thermodynamic Calculations

- Conclusion:
  - ***For numerically intensive calculations, use one thread per core***

Parallel Calculations

- Why does the performance of the parallel thermodynamic calculations not follow Amdahl's Law?

Multi Core CPU Architecture

- CPU cores contain a small amount of memory (32KB in this study)
  - This memory is the L1 cache
  - There is also an L2 cache that is larger than the L1 cache and slower
- CPU cores run at high speed
  - 3.2GHz in this study
- The CPU core communicates with the main RAM via a bus, that runs much slower than the CPU memory
  - 100 MHz in this study (32 times slower than the CPU)

CPU Bus

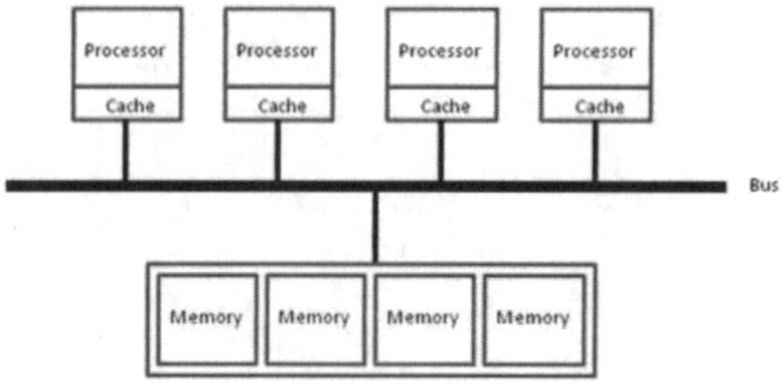

http://software.intel.com/en-us/articles/optimizing-applications-for-numa

Schneider Electric - Division - Name – Date

25

CPU Cache

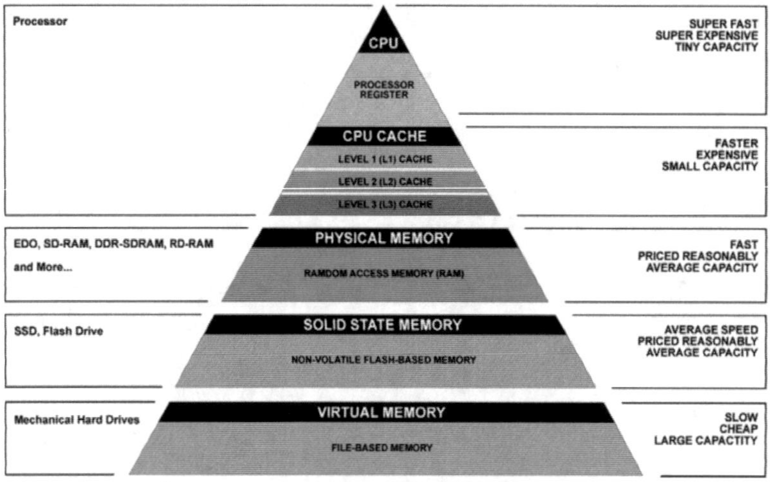

▲ Simplified Computer Memory Hierarchy
Illustration: Ryan J. Leng

Schneider Electric - Division - Name – Date

26

CPU-Z

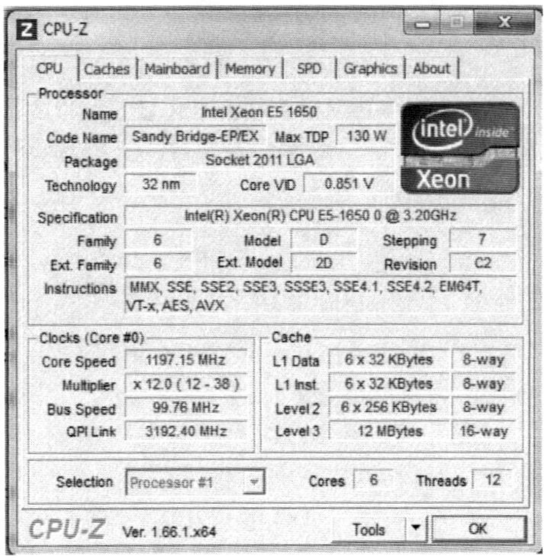

http://www.cpuid.com/softwares/cpu-z.html

SRK Data Requirements

- The approximate data storage requirements for the SRK equation of state is:
  - kija, kijb, kijT : 3*noc*noc
  - Pc, Tc, Pc, omega : 4*noc
  - ai,bi : 2*noc
  - Total : 3*noc*noc + 6*noc

SRK Data Requirements

3*noc*noc + 6*noc

NOC	Doubles	KB
5	105	1
50	7,800	61
100	30,600	239
200	121,200	947

Conclusion

- **Data Requirements when number of components is over 50 causes slow overall speeds due to the CPU bus speed being much slower than the CPU speed**

Questions?

- Thank you for coming
- Email ron.bondy@invensys.com with any followup questions or comments

Batch Distillation Strategy for Higher Purity Products

Z. Ted Fidkowski

AIR PRODUCTS

1

Air Products Performance Materials

Performance Materials Division in AP manufactures variety of specialty amines used in production of paints, coatings, sealants, personal care items (like soaps, shampoos, cosmetics) and also used as catalysts in production of rigid and flexible polyurethane foams.

2

AIR PRODUCTS

Problem Statement

- The following mixture of five components (some of them are pseudo-components or groups of pure compounds) is batch-distilled to give Product 1 and Product 2:
  - Light Impurities (**L**) – highest volatility
  - Product 1 (**P1**)
  - Intermediate Impurity (**I**)
  - Product 2 (**P2**)
  - Heavies (**H**) – lowest volatility
- Standard purities are 95% of **P1** and 97% of **P2**
- **Objective: Increase P1 purity to 99% for some fraction of P1 product**

3

Standard Distillation

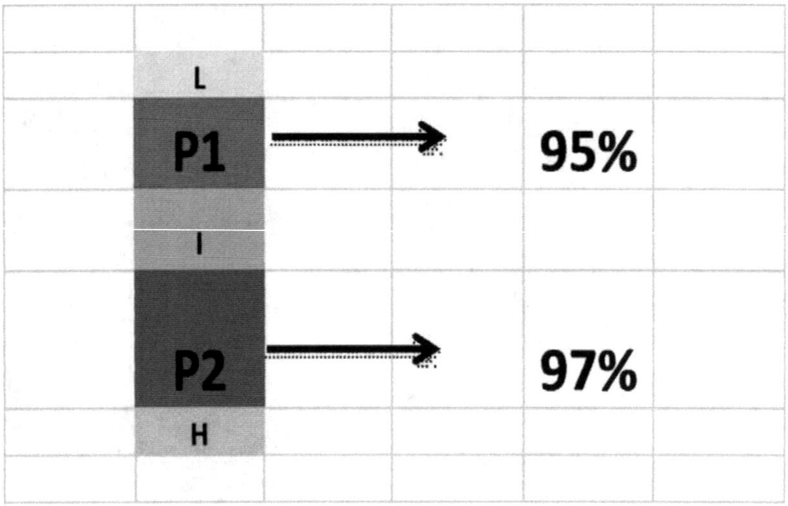

4

New Purities

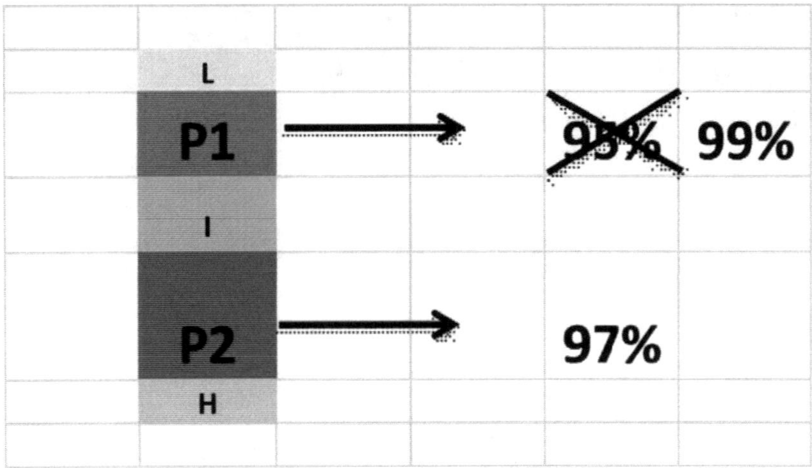

5

AIR PRODUCTS

Overview of Batch Distillation Strategies to Increase Product Purity

1. Standard Recipe
2. Increase Reflux in Pure P1 Cut
3. Rework Initial (Lower Purity) Product
4. Make High Purity Product from Front Cut

6

AIR PRODUCTS

Standard Recipe

1. Charge the pot, heat and degas
2. Low Boil Cut (remove L)
3. Front P1 Cut
4. Pure P1 Cut (make P1)
5. Intermediate Cut (remove I)
6. Front P2 Cut
7. Pure P2 Cut (make P2)

Each step has its own pressure, heat duty and reflux ratio.

7

AIR PRODUCTS

Standard Recipe

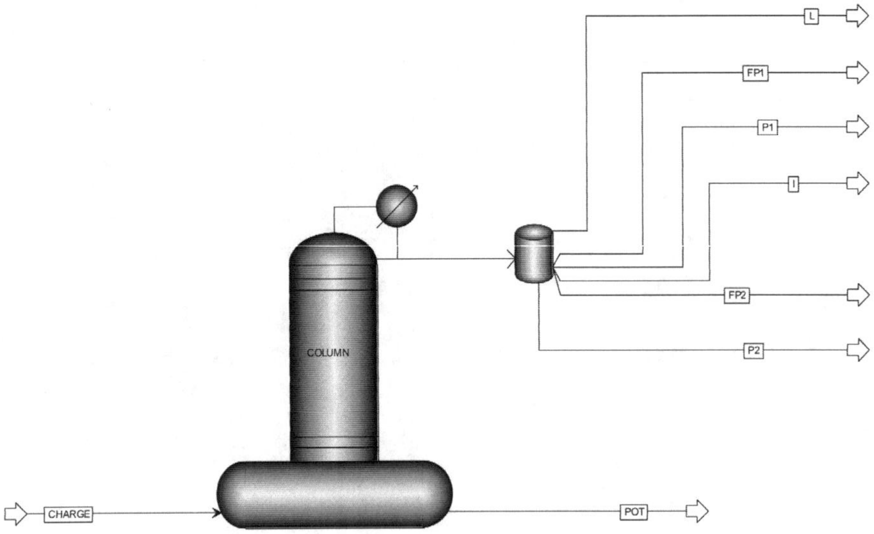

8

AIR PRODUCTS

Heat and Reflux

- Heat duty is maximized in each cut to speed up distillation; limiting factors are:
  - Column flooding (depends also on pressure)
  - Pot temperature (can be changed with pressure)
  - Heating utility temperature (steam pressure)
  - Cooling medium temperature
  - Reboiler size
  - Condenser size
- Reflux ratio is minimized, provided that desired purities are achieved.
- Establish points when to switch to and from a cut (depending on T, P, quantity, sampling)

9

AIR PRODUCTS

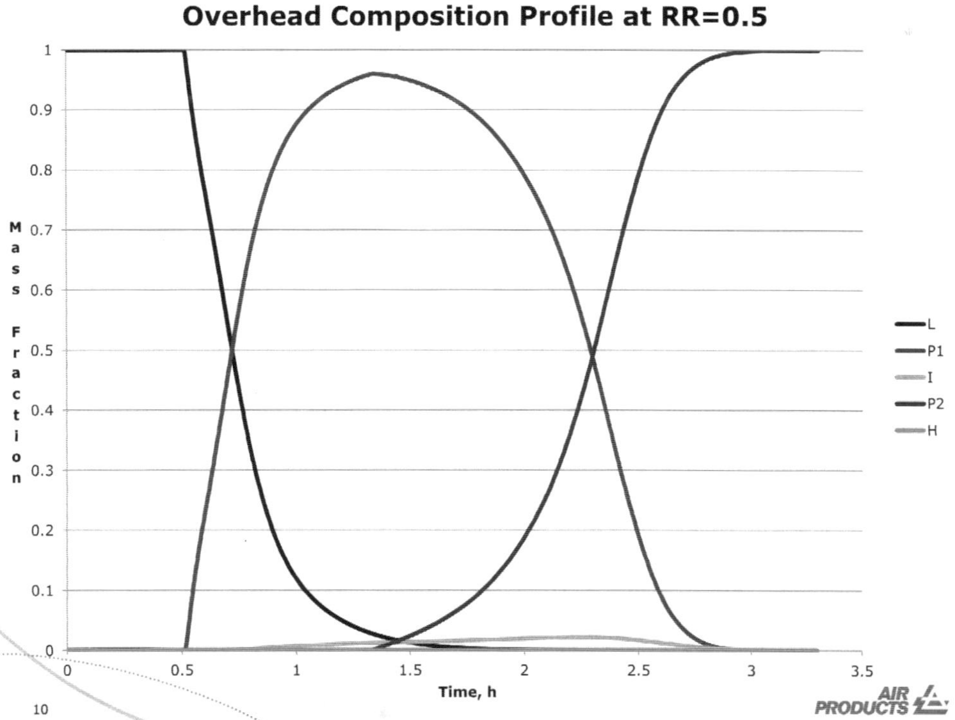

10

AIR PRODUCTS

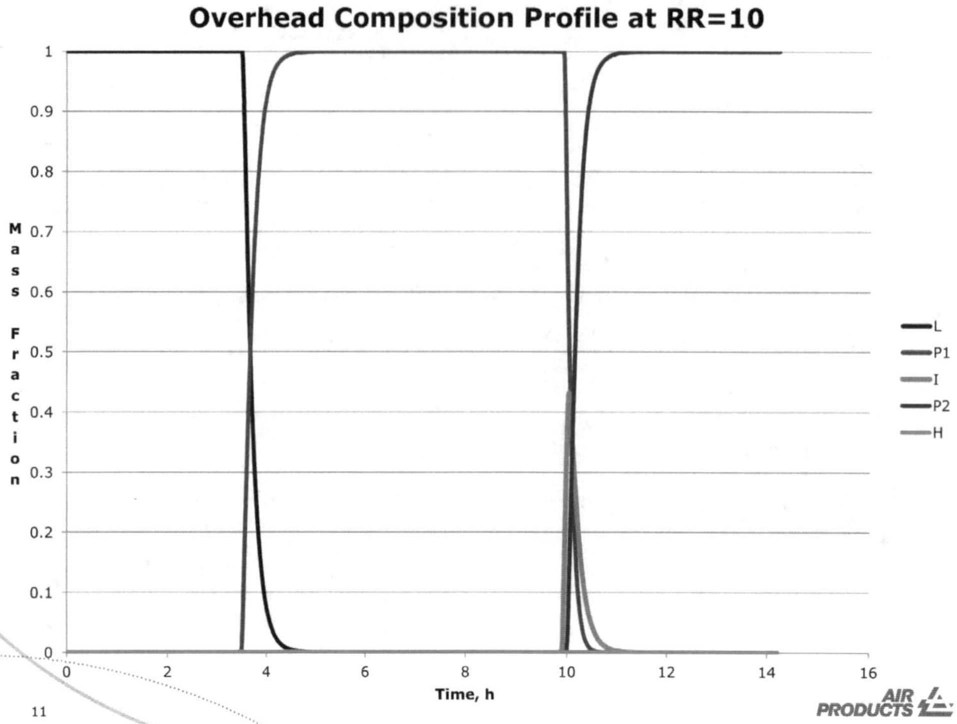

11

AIR PRODUCTS

Comparing High and Low Reflux Case

- Low reflux:
  - Distillation is quick
  - Purities are low
  - Recycled front cuts are big
- High Reflux:
  - Distillation is long
  - High purities
  - Less to recycle
  - Good start for the simulator – gives feasible products

12

AIR PRODUCTS

Strategy 1 – Increase RR in Pure Cut P1

- Run P1 Pure Cut at higher reflux and sample overhead over time
- Unsuccessful – couldn't reach high P1 purity for a long time
- Reason – look at the previous cuts (L, Front P1) that were run at lower reflux.
- Could go back and redo plant trials (cost)

13

AIR PRODUCTS

Strategy 1 – Increase RR in Pure Cut P1

- Problems:
  - Cut(s) before pure cut must run at high reflux, to remove Lights (additional cycle time)
  - We need to determine when to start P1 cut (easy to start too early or too late)
  - We need to determine when to finish this high purity P1 cut; cannot take too much, because next component (intermediate impurity I) will come out after P1 and spoil the P1 purity.
  - Major problem: AFTER distillation is done we will know if the product is in spec (or we need to take a lot of time for sampling)

14

AIR PRODUCTS

Strategy 2: Redistill Pure P1 Cut

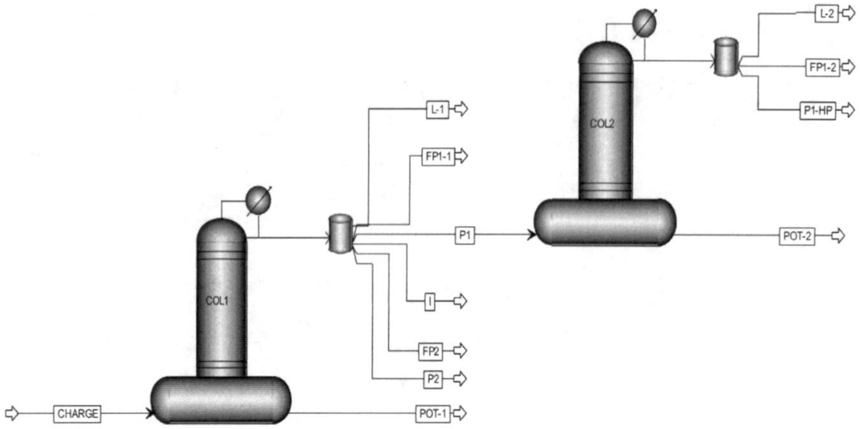

15

Strategy 2: Redistill Pure P1 Cut

- Seems straightforward
- Quite wasteful:
  - Additional energy needed
  - Additional time needed

16

Strategy 3: Redistill P1 Front Cuts

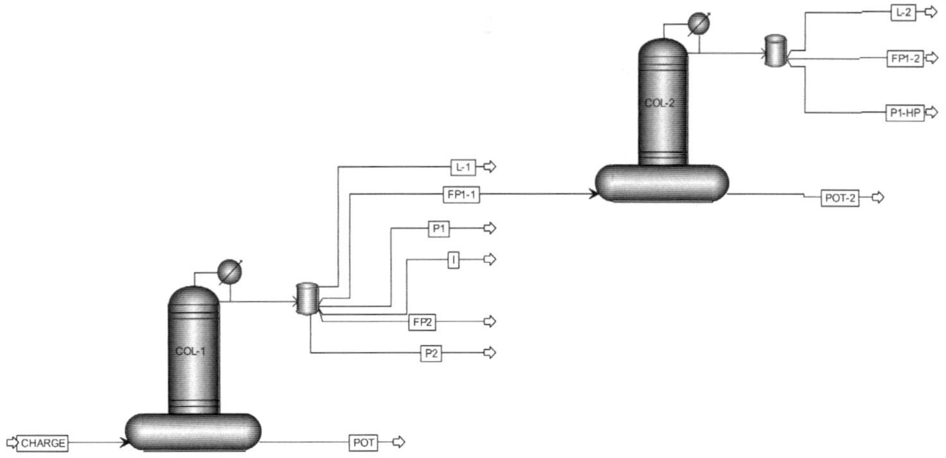

17

AIR PRODUCTS

Strategy 3: Redistill P1 Front Cuts

- P1 Front Cut contains only L and P1
- This Front Cut needs to be redistilled anyway
- Separation of Lights from P1 is quite easy
- There is no question when to finish: run distillation up to the end, since there is no heavier components coming after P1

18

AIR PRODUCTS

Conclusions:

- Three strategies to improve product purity in batch distillation were considered:
  1. Increase Pure Cut RR
  2. Rework Pure Cut
  3. Use Front Cut to make higher purity product
- The third method turned out to be the most convenient and economical.

19

AIR PRODUCTS

Thank you...
tell me more

www.airproducts.com

AIR PRODUCTS

20

351300 Troubleshooting and Rectification of a Giant C3 Splitter Tower Problem Part 1: Troubleshooting

Abstract:

The PetroLogistics Propylene dehydrogenation (PDH) 28 feet ID C3 Splitter started up in October 2010 and experienced low tray efficiencies and premature flood in its first year of operation. Due to the low tray efficiency it could not produce on-spec polymer grade propylene. PetroLogistics, Fluor (who was not involved in the tower design), and Sulzer formed a taskforce to conduct a troubleshooting investigation to determine the root cause of the poor performance and to propose a fix.

Our troubleshooting investigation combined hydraulic analysis and detailed multipass distribution calculations with the specialized technique of multichordal gamma scanning with quantitative analysis[1]. The hydraulic analysis and multipass calculations did not identify a reason for the low tray efficiencies, but confirmed that the trays are prone to channeling and maldistribution due to their large open areas. The gamma scans showed a maldistributed pattern on the trays, with high L/V ratios on the inside panels and low L/V ratios on the outside panels. The scans showed vapor cross flow channeling (VCFC) on the outside panels and reverse vapor cross flow channeling (RVCFC) on the inside panels. Flooding was observed on the inside panels well below the calculated flood point. The scans pointed at a combination of VCFC and multipass maldistribution as the root cause.

Our investigation identified the open slot area (15% of the active area) of the fixed valves to be the prime factor inducing the channeling and maldistribution. A likely initiator of the multipass maldistribution was liquid preferentially flowing to the inside panels from the false downcomers distributing the flashing reflux to the panels. The high ratios of flow path length to tray spacing (2.4 to 3.7), high weir loads, and integral trusses projecting a significant depth (4") into the vapor space were other conditions that promoted the channeling. Correlating liquid distribution with the effective weir lengths of the mod arc side downcomers (MOAD's) proved to be a challenge but our hydraulic calculations showed that the distribution ratios were quite robust to variation in the effective length of the MOADS.

To the best of our knowledge, this is the very first time that field measurements demonstrated interaction between VCFC with the inside-to-outside-pass maldistribution. A lesson learnt is that this interaction must be considered when designing and operating large-diameter towers. Finally, our investigation highlights that excessive open areas render trays prone to channeling and maldistribution, especially in large diameter towers containing multipass trays.

1. Kister, H. Z., "Use Quantitative Gamma Scans to Troubleshoot Maldistribution on Trays", Chem. Eng. Progr., February 2013.

351323 Troubleshooting and Rectification of a Giant C3 Splitter Tower Part 2: Rectification

Abstract:

The PetroLogistics Propylene dehydrogenation (PDH) 28 feet ID C3 Splitter started up in October 2010 and experienced low tray efficiencies and premature flood in its first year of operation. Due to the low tray efficiency it could not produce on-spec polymer grade propylene. PetroLogistics, Fluor (who was not involved in the tower design), and Sulzer formed a taskforce to conduct a troubleshooting investigation to determine the root cause of the poor performance and to propose a fix.

In Part 1, our troubleshooting investigation identified maldistribution pattern on the trays, with high L/V ratios on the inside panels and low L/V ratios on the outside panels, to be the root cause of the poor performance. Our investigation identified the open slot area (15% of the active area) of the fixed valves to be the prime factor inducing the channeling and maldistribution. A likely initiator of the multipass maldistribution was liquid preferentially flowing to the inside panels from the false downcomers distributing the flashing reflux to the panels of the top tray.

An unplanned shutdown of the PDH unit in June 2011 provided less than 30 days to engineer, fabricate, and install the fix. The most important and also most challenging modification was to blank the open slot area of the fixed valves from 15% to 11% of the active area. This was necessary to eliminate the channeling and the maldistribution between the passes.

Achieving an adequate blanking pattern turned out to be a unique challenge. The fixed valves are staggered, making it impossible to fully blank about half of the openings. Short flow path lengths and presence of support members created the risk of forming stagnant regions, where the froth may collapse. Downtime was to be minimized. It took innovative engineering to beat these challenges in the time frame given. The paper describes how this was achieved.

The task force also modified the piping entrance into the reflux false downcomers and increased the false downcomer height to prevent liquid pouring out from the false downcomers and preferentially flowing into the inside panels. Other less important modifications included adding anti-jump baffles to prevent cross movement of liquid across downcomers and adding downcomer blocks on the inboard side of the off-center downcomers to improve liquid distribution to the passes. Even though these were considered less important, they were easy to implement and therefore incorporated.

Upon restart, the modified tower achieved tray efficiencies comparable to those obtained in a well-operated, smaller-diameter, low pressure C3 Splitter. Flooding and instability were fully eliminated. The tower successfully produces on-spec polymer grade propylene.

351413 A High Capacity Re-Tray That Wasn't - a Practical Lesson in Counter-Intuitive Thinking

Abstract:

A plant upgrade project involving a partial column re-tray produced an unexpected result, with actually a performance deficit afterwards. This study details the practical thinking and application of fundamental ideas of tray flooding employed to understand the root cause of the problem and the simple fix used to return the column to acceptable operation. The lesson from this experience also led to a more fundamental and generally applicable moral for distillation practioners everywhere.

AIChE 119d

Evaluation and Testing of a Two Phase Feed Distributor

Veronica R. Saunders
Research Engineer
NOVA Chemicals Research and Technology Centre, Calgary, AB, Canada

Eric Clavelle
Principal Scientist
NOVA Chemicals Research and Technology Centre, Calgary, AB, Canada

Andre Bernard
Process Design Engineer
NOVA Chemicals (Canada) Ltd, Corunna, ON, Canada

Prepared for Presentation at the 2014 Spring National Meeting
New Orleans, LA, March 30th – April 3rd, 2014

Evaluation and Testing of a Two Phase Feed Distributor

Veronica R. Saunders
Research Engineer
NOVA Chemicals Research and Technology Centre, Calgary, AB, Canada

Eric Clavelle
Principal Scientist
NOVA Chemicals Research and Technology Centre, Calgary, AB, Canada

Andre Bernard
Process Design Engineer
NOVA Chemicals (Canada) Ltd, Corunna, ON, Canada

Abstract

Feed locations in a distillation tower represent transition points where liquid and vapor volumetric rate change significantly on each side of the distributor. Poor distribution may cause pre-mature flooding of the trays directly above and/ or below the distributor, which can result in a capacity shortfall. Flashing feeds add to the complexity of the design and have been known to significantly impact the ability of distributors to achieve the expected capacity. This paper discusses the evaluation, testing and re-design of feed distributors intended for the re-tray of a depropanizer in an olefins facility. Physical space constraint and a significant vapor fraction in the feed streams were challenges imposed on the design.

A full scale test of the originally proposed distributor, tested at the volume fraction expected for the design fluids, proved unsuccessful. Using flow regime mapping techniques and momentum equations for stratified flow, the system was scaled down to lab scale with air-water test fluids. Testing of various distributor designs at lab scale with air-water was completed and the performance of the successful design was simulated with CFD at lab scale with air-water for validation and at full scale with the process fluids.

Introduction

NOVA Chemicals Corunna site is an olefins plant with a front-end crude unit and back end aromatic unit. The plant was originally designed as a naphtha cracker but has since developed feed flexibility through expansions and revamps. The direction and ability in processing higher volume of naphtha and gas oil feedstock in the 2005 – 2008 timeframe became constrained by the depropanizer. A debottleneck project involving a 1 for 1 re-tray of the tower was then initiated. Re-tray of the depropanizer was to be executed during a plant turnaround scheduled in 2011. A drastic change in market conditions for olefins plant feedstock in 2010 drove a desire to crack more ethane and significantly less liquid feeds. The re-tray project was therefore canceled just before the execution stage. Nonetheless, a fair amount of knowledge was gained throughout the design and implementation stages, especially around the feed distributors. A significant vapor fraction in the tower feeds along with physical limitations for accommodating the feed distributors' hardware made it challenging to effectively distribute the liquid on the feed trays. An air/ water test on the initial distributors design proved unsuccessful and required the design team to step back and re-design the distributors. Even though the project didn't reach the execution stage, the learning gained from the re-design effort will serve future initiatives related to trouble shooting or designing similar feed distributors.

Depropanizer Debottleneck

Description

The depropanizer is a 147 ft tower with a diameter of 7 ½ ft and a total of 62 trays. In its current configuration, the first 33 trays are conventional 2 pass sieve trays. Trays 34 to 62 are high capacity trays. Heat to the bottom reboiler is supplied by low pressure steam. Propylene refrigeration (3°C) provides cooling to the overhead condenser. The tower has 2 main feed points at tray 34 and tray 39. A third minor feed point, which accommodates a small recycle stream from the C3 rerun tower, is located on tray 22.

As shown on Figure 1, the tower receives feed from the deethanizer bottom. Because of feed distributor limitations, the feed is split between two feed nozzles, one above tray 34 and one above tray 39. Cross exchange with the depropanizer overhead product provides cooling to the feed stream entering the tower at tray 34. A cooling water exchanger provides cooling to the stream that feeds the tower at tray 39.

The plant is equipped with a condensate stripper, which could strip light ends from the charge gas compressor (CGC) 4th and 5th stage liquid knock out pots. The bottom of the condensate stripper then feeds tray 39 of the depropanizer. The condensate stripper provides pre-fractionation to the depropanizer and reduces its hydraulic load. The condensate stripper, which had been taken off stream to accommodate a different process scheme referred as liquid re-injection[1] was re-commissioned as the depropanizer became a constraint to the heavier plant feed slate.

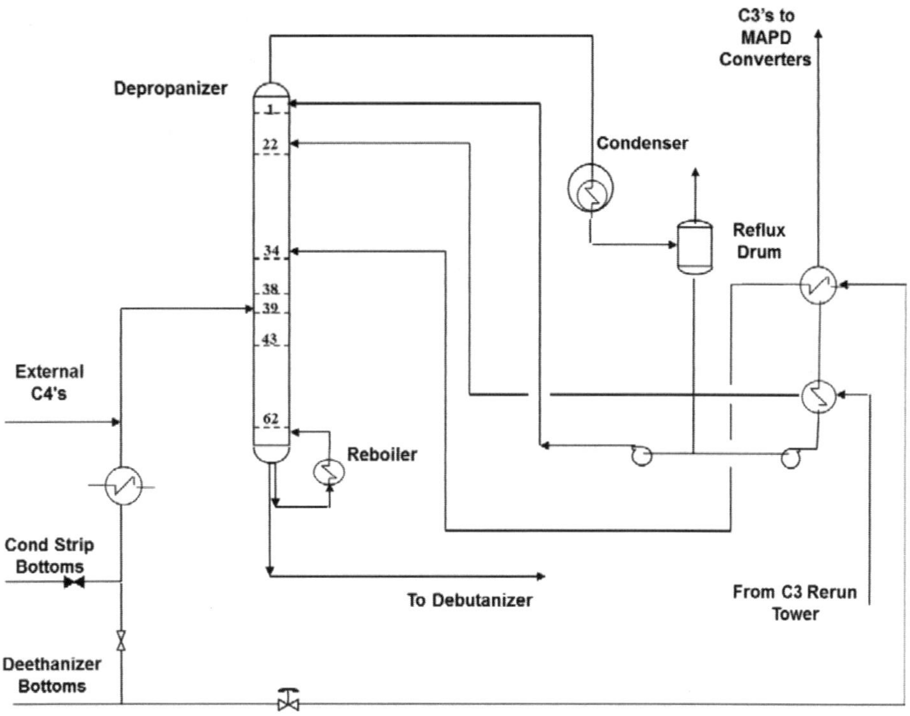

Figure 1: Depropanizer

Depropanizer Re-Tray

Because of the energy benefits, the depropanizer re-tray was based on the liquid re-injection process scheme while maintaining the flexibility of operating the condensate stripper if needed. Based on results from process simulations, a single feed point at tray 34 would provide more capacity than splitting the deethanizer bottom stream between the two feed points. However, because of the desire to maintain a condensate stripper operating mode, a feed distributor was still required at tray 39.

A 1 for 1 full re-tray of the tower with [1]HiFi™ high capacity trays was selected for the re-tray option. The feed conditions and tray design would dictate the feed distributor characteristics.

Feed Distributor Requirements

The feed distributors were to accommodate the two operating modes described above at the expected design rate. A turndown case for each of the two operating modes was also established and needed evaluation. As provided in Table 1, this led to four cases for the feed

[1] **HiFi is a trade mark of Shell Global Solution**

distributor located above tray 34 (nozzle 12) and two cases for the feed distributor located above tray 39 (nozzle 13). The liquid and vapor volumetric flows are expressed as a percentage of the design case.

Table 1: Case Descriptions

Distributor above tray 34

Case	Description	Feed		
		Liquid % of design	Vapour % of design	Vapour Fraction (volume base)
1	Liquid Re-Injection – Design	100	100	0.872
2	Condensate Stripper	58	0.11	0.013
3	Turndown – Liquid Re-Injection	87	17	0.56
4	Turndown – Condensate Stripper	38	76	0.932

Distributor above tray 39

Case	Description	Feed		
		Liquid % of design	Vapour % of design	Vapour Fraction (volume base)
2	Condensate Stripper - design	100	100	0.817
4	Turndown – Condensate Stripper	57	30	0.703

As indicated in Table 1, the feed distributor above tray 34 design case is based on Case 1. The feed distributor above tray 39 design is based on Case 2. Cases 1 and 3 do not apply to the feed distributor above tray 39 as they represent a no flow condition.

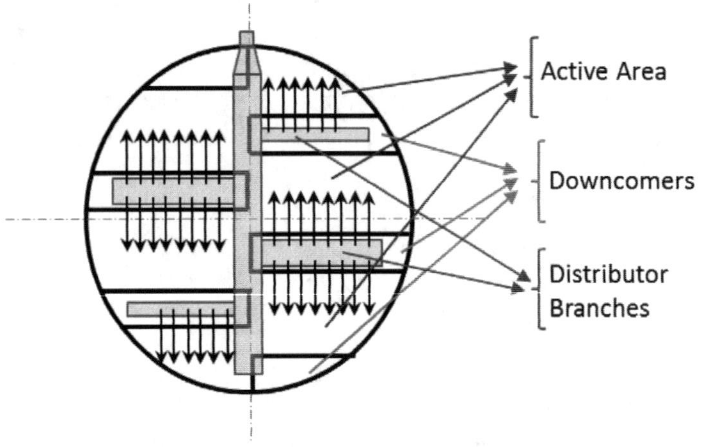

Figure 2: Distributor Schematic

The common practice with HiFi™ trays is to distribute the liquid on the active area of the tray located beneath the distributor. As illustrated in Figure 2, a distributor header, oriented perpendicular to the downcomers, travels across the center axes of the tower. Branches from the main header are located over the downcomers and project the liquid sideways on the active area of the tray.

HiFi™ trays differ from other multi-pass trays as there are no physical barriers isolating the active areas from each other. This provides more robustness and makes the HiFi™ trays less sensitive to mal-distribution. Regardless of the characteristics, good design practice would aim at evenly loading the downcomers of the tray located beneath the feed distributor.The liquids from the distributor and the tray above are considered in the design process.

For the design case, the liquid flow from each of the distributor branches was calculated such that the downcomer loadings (gpm/ft^2) and downcomer velocity (ft/sec) are identical for each downcomer on the tray below the feed distributor. As shown on Figure 3, F1, F2 and F3 refer to liquid coming from downcomers located above the feed distributor. F1, F2 and F3 flowrates have been calculated from the design simulation of the tower, where the total stage flow is distributed amongst the downcomers proportionally to the weir length of each downcomer. Given the symmetry of the HiFi™ tray, only 1 half of the tray needs to be calculated. The total liquid rate of the feed stage is obtained from the tower simulation profile.

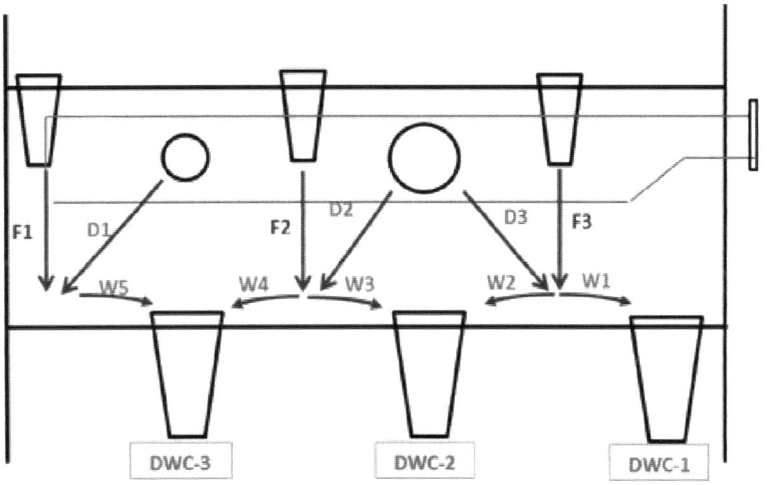

Figure 3: Distributor Flow Calculation

Assuming liquid distribution proportional to the weir length, the individual weir flow can be calculated from:

W_i = (total stage liquid rate) * $weir_i$ length/(tray total weir length)

With i being the weir number as illustrated in Figure 3. The flow from each of the distributor branches can then be calculated by material balance:

$D1 = W5 - F1$

$D2 = (W3 + W4) - F2$

$D3 = (W1 + W2) - F3$

D1, D2 and D3 being the distributor liquid flow contributions as shown on Figure 3. Table 2 provides the required liquid distribution for the feed distributors.

Table 2: Feed Distributor Liquid Distribution

Flow Path	Above Tray 34 (% Volume)	Above Tray 39 (% Volume)
D1	11.15	11.81
D2	21.07	21.07
D3	17.78	17.12
D1'	11.15	11.81
D2'	21.07	21.07
D3'	17.78	17.12

Design Objectives, Criteria, and Physical Constraints

Objective

The objective of the header design is to:

1. compliment the tray geometry already specified, and its lack of symmetry,
2. adequately distribute the flow according to the downcomer loading specifications for the given design flow rates, and
3. fit within the physical boundaries of the tower, manhole, and the trays above and below.

Original Design and Testing Method

A full scale test of the originally proposed distributor, tested at the volume fraction expected for the design fluids, proved unsuccessful. One of the four branches delivered only 2% of the flow, thus starving 1/6th of the tray and providing a surplus to the tray's central section. Strong oscillations were apparent in the branch flow through the pulsing of the liquid phase at the outlets. The three furthest downstream branches experienced alternating flow of 100% liquid and 100% vapor exiting the holes on the branches closest to the header.

Design Downcomer Loadings

As described above, the design downcomer loadings were determined by summing up the perimeter of each downcomer and weighting it against the total tray area. See Table 2. The percentages listed in Table 2 are used to judge the flow performance of the designed headers.

Design Criteria

The design criterion is the flow distribution performance to the downcomers, i.e., overall downcomer balancing. The design achieving the closest to ideal distribution in terms of (1) percentage of flow to each downcomer compared with its ideal loading, and (2) tray left-right balancing (50%-50% being the ideal) is the most successful design. In terms of priority, criteria (2), the left-right flow balancing, ranks highest; re-establishment of flow distribution over one side of the tray is considered easier than re-establishing balance between the left and right sides of the tray[2].

Physical Constraints

The maximum space available for Nozzle 12 (above tray 34) and Nozzle 13 (above tray 39) was taken to be no greater than the space required by the originally designed headers. As a result, Nozzle 12 and 13's maximum pipe diameter for the main header was set at a 12" Sch10, although a larger diameter would have been beneficial in slowing the liquid and vapor speeds. The largest branch diameter was increased from an 8" SCH10 to a 10" SCH10 as room was available for the slight adjustment in branch diameter. The 6"-Sch10 dogleg inlet of Nozzle 13 was incorporated into the design of Nozzle 13, and positioned as previously

[2] A central beam between the left and right side of the tray and the direction of flow on the trays makes re-establishment of balancing over the tray more difficult than between downcomers on one side of a tray

designed. The heights of the main header and dog leg above the tray were also maintained as a design requirement.

Overview of the Design Methodology

The Stages of the Design Process

The design process is split into four stages. The first involves scaling the flow rates to a 3:1 scale and with alternative fluids (water and air) to facilitate in-house experimental testing of the header designs due to the risk associated with the depropanizer process fluids and the size of the full scale header. In-house experimentation was deemed necessary due to the computational expense in terms of time and processor needs to complete two-phase multifluid models at full-scale. The experiments allow the computational expense to be focused on the designs and modifications with the most potential. Stage 2 involves validating the computational fluid dynamic (CFD) simulation method by replicating one of the experimental trials with CFD using the experimentally scaled apparatus, air and water, and the scaled flow rates. The third stage involves testing the flow distribution of the proposed designs with the scaled air-water system and scaled apparatus. When potential is noted, the designs are tested with CFD at full scale and with the depropanizer process fluids and rates, which is stage four.

Stage 1: Scaling the Process Fluids

The steps involved in Stage 1, scaling the process, are outlined below:

1. Predict the flow regime of each design flow rate for the full scale depropanizer tower riser by creating a flow regime map of the flow system (dependent on density, viscosity, surface tension, pipe orientation, diameter, and pipe roughness),
2. Calculate the holdup of the system,
3. Create a flow regime map for the experimentally scaled riser (3:1 scale) with air and water properties (dependent on air-water density, viscosity, surface tension, pipe orientation, diameter, and pipe roughness),
4. Iterate the mass flow rates of air and water that provides a similar holdup value to the depropanizer flow system and a similar location on the flow regime map.

The location of the mass flow rates of the liquid and vapor phases on a flow regime map specify the regime and the proximity of the flow case to transition lines (to other regimes). The proximities of the flow conditions for both full scale and experimental scale to transition lines were maintained.

Stage 2: Testing the Computational Process

The process used to test the computational fluid dynamics method is as follows:

1. Create a scaled model with acrylic tubing complete with:
    a. a riser of 2.5" inner diameter and 60 diameters in height,
    b. a 90° long radius bend, and
    c. a reducer to connect to the 3.75" inner diameter header.
2. Complete three experimental trials at one of the scaled flow rates with the scaled manifold, and record the distribution of liquid to each tray section (Left D1, Left D2, Left D3, Right D1, Right D2 and Right D3).

3. Complete CFD analysis of the scaled header and riser with the flow properties and flow rates of the experiment while sampling the liquid mass flux out each outlet hole and summing appropriately to the Left D1, Left D2, Left D3, Right D1, Right D2 and Right D3 tray areas.
4. Assess CFD predicted performance against the experimental performance in terms of liquid distribution and visual observation.

Stage 3: Testing the Designs Experimentally

Stage 3 uses the flow rates scaled in Stage 1 to test the headers, hole sizes, hole layouts and internals experimentally and, therefore, more quickly than CFD allows, thus reducing the total number of CFD multiphase simulations required.

The following procedure outlines the steps involved in Stage 3:

1. Decide feature/modification to test,
2. Construct scaled manifold with design feature incorporated,
3. Complete three experimental trials for each scaled flow rate with the scaled manifold and record the distribution of liquid to Left D1, Left D2, Left D3, Right D1, Right D2 and Right D3,
4. Assess performance of design feature with liquid distribution analysis (left-right performance and distribution performance),
5. Repeat Steps 2-4 until satisfied with the result,
6. Complete CFD analysis with full scale header and process fluids,
7. Return to Step 1 if required.

Only the designs with the most potential, as determined by the experimental results, were modelled with CFD. As the depropanizer process fluids are significantly different then the experimental fluid in terms of density ratio and viscosity, CFD was used to assess the top two design headers and any fine tuning modifications that were deemed necessary.

Details of Design Process

This section describes in detail the steps of Stage 1: Testing the Scaling Process, Stage 2: Testing the Computational Process, and Stage 3: Testing the Designs.

Stage 1: Scaling the Process Fluids

In multiphase flow scaling of process fluids is more complicated due to the physical phenomena referred to as holdup and the presence of flow regimes. Considerable differences between the physical properties of the liquid and vapor phases (density specifically in our case) cause the vapor to flow at higher speed than liquid. The result being that the "volume fraction (holdup) of liquid at any point in [the] pipe [is] greater than that in the mixture entering or leaving the pipe" [2, p169]. The liquid holdup was deemed the primary scaling factor, as opposed to matching the volume fraction. Flow regime prediction was carried out using various mechanistic models, such as Barnea's *Unified Model for Predicting Flow-*

Pattern Transitions [3] and Taitel and Dukler[4]. Mechanistic models predict very different transitions between flow regimes for fluid systems[3] with different density ratios, viscosities, pipe inclination and pipe diameter. Therefore, to ensure the flow regime was consistent between the scaled system and the full-scale system, the position of the full scale and scaled system on their relative flow regime maps was set as the second scaling factor. More specifically, the relative location of the liquid and vapor superficial velocity pairs to the transition line were matched as best as possible.

Stage 2: Testing the Computational Process

Computational simulations of the favored designs were completed using the transient, Euler-Euler multi-fluid model (ANSYS 12.0). The 3D model that was input into the computational domain included all minor offsets, inconsistencies in spacing, and ridges at transitions resulting from the construction process and the materials used to build the scaled header. The k-e realizable turbulence and symmetric drag models were used for the validation test, and were used on all subsequent CFD models. A user-defined function summed up the liquid mass flux and vapor mass flux exiting each outlet and stored the information at every time step in order to provide time averaged data, as provided by the experimental testing.

Stage 3: Testing the Designs Experimentally, and Scaling the Design Features to Full Scale

This section outlines the method and criteria applied to size the holes for the header and to determine the number of holes required for each tray area (and therefore branch) for both the scaled and full scale designs, thus ensuring the design process was scaled appropriately, as opposed to simply the size of the holes being scaled. For example, a scaled, air-water header with n number of holes and diameter would not have the same theoretical dimensionless, stratified liquid height as a full-scale depropanizer process fluid header with *n* holes and diameter *d*. The headers at this stage in the design process had both bottomside and topside holes in each branch to provide separate paths for the liquid and vapor exiting the branch.

The following method is used to size the liquid holes and set their number per tray area:

1. Set branch diameter, number and size of liquid and vapor holes for each branch.
2. Designate percentage of flow to each branch.
3. Enter calculation with the assumption of stratified liquid at a set height.
4. Calculate the velocity out each bottomside hole via Bernolli's principle (dependent on the liquid height within the branch assuming stratified flow).
5. Sum up flow rates out each bottomside hole using velocity from step 3 and bottomside hole outlet area,
6. Assuming stratified flow, iterate the liquid dimensionless height in the branch using the stratified flow equations [3] until the liquid mass flow rate entering branch equals the liquid mass flow rate exiting branch.

---

[3] The term system refers to the particular fluids, temperature, pressure, pipe orientation, pipe inner diameter and surface roughness, as all are fixed in order to generate a flow regime map, i.e., mechanistically based flow regime maps are dependent on density, viscosity, surface tension, pipe orientation, pipe inner diameter and surface roughness.

7. Calculate the cross sectional area remaining for the vapor phase.
8. Assess header hole details against criteria provided below.

The iteration is repeated by changing the size and number of holes and the distribution of holes per tray area or branch until the design criteria are met.

Design Criteria for Hole Sizing

The design criteria are listed below:

1. Stratified liquid level dimensionless height between 0.2 and 0.5. Height range selected to attempt to keep liquid away from vapor holes, yet retain a liquid level for flow cases with reduced liquid flow rates. The stratified liquid height was determined for the flow case with the most and least liquid.
2. Mass flow rate of liquid deviation from the ideal < 1% of ideal mass flow rate. The flow out the branch is determined by the liquid height within the branch via Bernoulli's principal.
3. Vapor velocity < 2m/s. Due to the high vapor flow rate of the design conditions and restrictions on pipe diameter, this value was selected to prevent over perforating the top surface of the header, which would allow liquid to splash without constraint.
4. Predicted maldistribution along branch < 5%, as stated for exit manifolds by Tilton [5].

Distributor Fluid Velocities

Design criteria for strictly single phase distribution headers are well established. As suggested by Tilton [5] the average pressure drop across the holes was kept large compared to that over the length of the branch. The multiphase header designs adhere to Tilton's <5% maldistribution criteria by ensuring that the pressure drop over the hole is 5-10 times greater than the pressure drop down the branch. For this multiphase application, the pressure drop in the branch was calculated for the liquid phase using true velocity with a slip surface at the liquid and vapor interface.

Kister [6] provides the following rules of thumb for the liquid velocity, v_L = 5 ft/s or 1.5m/s.

Analysis, Experimental Results and Discussion of Stages 1, 2, and 3

Introduction

The following section includes a quick header velocity calculation for each flow case, droplet size characterization of liquid droplets carried by the vapor phase within the main header, and the results of Stages 1, 2 and 3. A discussion summarizes the findings.

Simple Vapor and Liquid Velocity Calculations

Calculations of both the liquid and vapor speeds in the main headers of Nozzle 12 and Nozzle 13 were calculated using liquid holdup. All but one case exceed the recommended liquid speed within the distributor. In addition, the vapor speeds, which range from 0.1 to 9.7m/s in Nozzle 12 and from 1.2 to 2.2m/s in Nozzle 13, highlight concerns of liquid entrainment and high momentum. Consequently, if opportunity existed for a larger manifold (i.e., if there existed acceptable space), a larger pipe diameter would have been incorporated into the design.

Maximum Stable Droplet Size and Stokes Flow Analysis

A quick calculation of the maximum stable droplet size and stokes flow[4] for each flow case was completed. Ideally the maximum stable droplet size in the vapor phase would be large and Stokes Number >> 1.0, thus resulting in liquid droplets being less likely to be carried indefinitely with the vapor phase. The characteristics length for the Stokes number calculation was the topside branch holes, thus representing the turning radius required for a droplet to exit the header through a topside hole. The average vapor speed was taken from the full scale CFD simulations. For all cases, the Stokes number is greater than 1.0 within the header and in the vapor vents, thus entrained droplets are expected to be knocked out at elbows, tees, and other changes of direction required by the flow. The designs proposed made use of multiple changes in flow direction.

Stage 1 Result: Process Scaling and Experimental Application

When comparing the flow regime maps of the experimental scale, air-water system to the full scale, depropanizer system, a disparity in superficial velocities is evident. The depropanizer system density ratio, viscosities, surface tension and riser diameters cause transitions between regimes to occur at lower superficial velocities than for the air-water system. The result of the scaling method is higher velocities for the scaled air-water system than for the full scale system, which is a requirement to match flow regimes. Thus, the scaled system is a conservatively scaled system, as the full-scale system is likely an easier system to distribute due to its lower momentum.

The cases tested included a bubbly flow, an annular flow near the intermittent/slug flow transition, intermittent/slug flow near the annular transition, and an intermittent/slug.

Stage 2 Result: Testing the Computational Process

The average distribution from three experiments and the time averaged results of the computational trial to each downcomer tray area was compared for an early proposed design. Comparing the experimental results and CFD results shows good agreement between downcomer distributions with deviations ranging between 0.5% and 2.3% of the total mass flow. The average deviation[5] was 1.3% with a standard deviation of 1%. This validates our computational method as an effective predictive tool of flow distribution, as the deviation falls well within the reasonable bounds required for adequate performance. See Figure 4.

---

[4] Stokes number is defined as the relation between the particle response time and system response time.

[5] The average was not flow weighted; each downcomer was an equal weight in the average calculation.

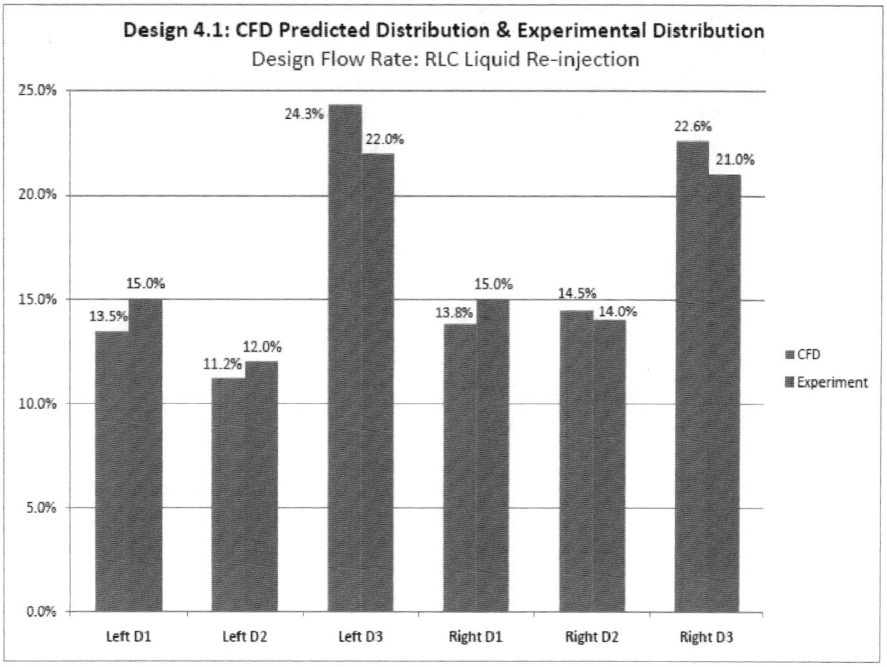

Figure 4 CFD & Experimental Results of Design 4.1 & RLC Liquid Re-Injection Flow Rate: Distribution of Flow

Stage 3: Experimental Findings of Interest

The velocity calculations and droplet size analysis helped shape the header designs in two important ways. The former highlighted the RLC Condensate Stripper as a potentially challenging case to distribute, and therefore, resources were not wasted attempting to better the already drastic improvement sought and found with Design 5 for this flow case.

The concern regarding the RLC Condensate Stripper flow was confirmed by the flow testing of Design 4 and 5; the RLC Condensate Stripper was the poorest performer despite drastic improvements made for the remaining flow cases. For Design 5.0, 5.1 and 5.2, overall maldistribution is ≤ 5% with the exception of the RLC Condensate Stripper flow case. After both experimental and computational testing, it is believed that unless the header can be made larger to slow down the vapor speed, the less than ideal distribution of this flow case will persist.

The droplet size calculations help support the usefulness of two design features: exterior liquid shields[6] and vapor vents[7], installed in Design 5. The exterior liquid shields were incorporated into the design and the vapor vents left for further investigation. These features both work to knock out droplets from the vapor phase and to direct any liquid leaving through the branch topside holes downwards.

---

[6] A single horizontal plate placed above each branch.
[7] Small pipes inserted into each vapor hole (topside hole on the branch).

During the design phase an opportunity to better match the header layout (branch take offs) to the tray division (D1, D2 and D3) was identified, that would extend the left-right header symmetry to the entire header. The result is Design 5, the off centered 3-way tee to 3-way tee. Design 5 includes flow aligners installed between the 3-way tees for both the right and left side of the tray, and a "wave breaker" to knock the riser annular flow down to stratified upstream of the first tee. See Figure 5 for the performance witnessed experimentally. Further modifications were made to Design 5.0 during the computational simulation phase.

Design	Flow Case	Overall Maldistribution		
		$\bar{x}$	s	Left %
Design 5.0	Heavy 1 Liquid Re-injection	3%	2%	2%
Design 5.0	Heavy 1 Condensate Stripper	3%	2%	1%
Design 5.0	RLC Liquid Re-injection	5%	3%	9%
Design 5.0	RLC Condensate Stripper	12%	7%	25%

Figure 5 Experimental Performance of Design 5.0

Stage 4: Description of Computational Fluid Dynamics Simulation Models

Two designs (Designs 4 and 5) were found successful during experimentation and further testing was completed with CFD with the depropanizer process fluids and at full scale. For each design, subsequent smaller modifications were tested to better performance.

Stage 5: Computational Fluid Dynamics Results and Discussion

For Design 5, two modifications performed best, Design 5.4.1 and Design 5.1. Average maldistributions were less than 4% with Design 5.1 and 1% with Design 5.4.1. All Design 5 modifications were successful in terms of left-right distribution performance with a left-side maldistribution less than 4% for all simulations. Design 5.1, 5.4.1 and 5.2 performed best with a left maldistribution of < +0.3%, -0.5% and +1.0% respectively.

Of the two best performing designs, Design 5.4.1 was favored due to its lower outlet speeds than its leading competitor. Pressure oscillations and average outlet velocities were recorded by CFD and found acceptable.

Figures of Design 5.4.1 are provided below.

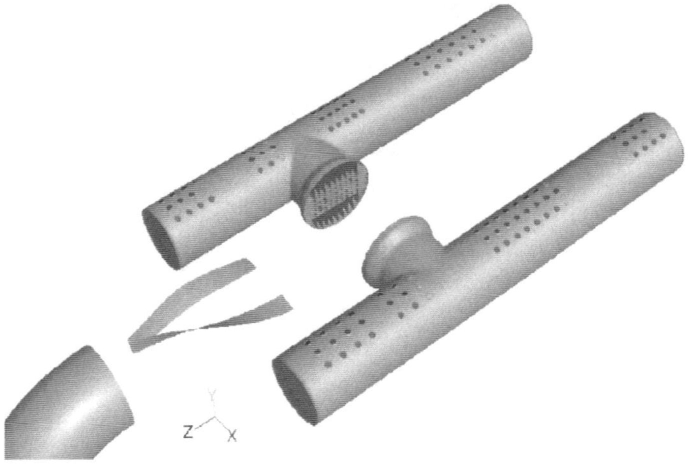

Figure 6 Design 5.4.1 Top View Isometric with Flow Straightener and Wave Breaker

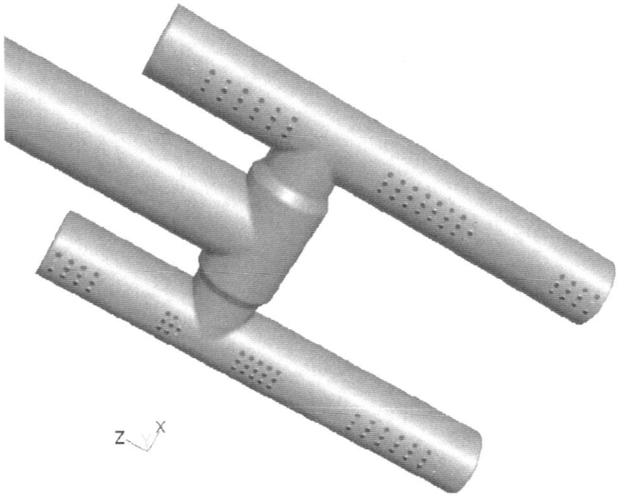

Figure 7 Design 5.4.1 Bottom View Isometric

Summary

A multiphase (liquid and vapor) header was designed that compliments the existing tray geometry and distributes the liquid flow according to the downcomer loadings.

The design process, including a novel scaling technique, in-house experimental testing and computational fluid dynamic modeling, is described in detail. For both Nozzle 12 and Nozzle 13, an "Off-center H" header design is recommended. Nozzle 13's dog leg entry makes attaining left-right balance more challenging and requires an additional flow straightener. Overall maldistribution to the downcomers is approximately 1% for Nozzle 12 with a left-right maldistribution of 0.5%. The final Nozzle 12 design has a predicted pressure drop of 31.9kPa; the predicted pressure drop for Nozzle 13 is 32.2kPa from the outlet of the dog leg pipe to the nozzle outlets.

Included in the design process is the validation of the CFD model against a scaled air-water experiment conducted in-house, which showed impressive ability of CFD to predict the distribution of a challenging, high Reynold's number, multiphase flow. The average deviation to each downcomer between the experimental header test and the computational fluid dynamic model was 1.3% of the total mass flow with a standard deviation of 1%.

References

1. Bernard, A. and R. Hayden "Planning and Designing the Modernization of the Recovery Area of a Flexible Cracker", 16th Ethylene Producer's Conference, AIChE 2004 Spring Meeting.
2. Chhabra, R. P. and Richardson, J. F. Non-Newtonian Flow in the Process Industries: Fundamentals and Engineering Applications. Oxford: Butterworth Heinemann, 1999.
3. Barnea, D. 1987. A unified model for predicting flow-pattern transitions for the whole range of pipe inclinations. International Journal of Multiphase Flow 13.1, 1-12.4
4. Taitel, Y. and Dukler, A. E. 1976. A model for predicting flow regime transitions in horizontal and near horizontal gas- liquid flow. AIChE J. 22, 47-55.
5. Tilton, J. N. Fluid and Particle Dynamics. In R. H. Perry, D. W. Green, & J. O. Maloney (Eds.), Perry's Chemical Engineers' Handbook 7th Ed. (pp.6.1-6.54). New York: McGraw-Hill. 1997.
6. Kister, H. Z. Distillation Operation. Boston: McGraw Hill. 1990.

Steam Stripping of Aniline Process Wastewater – A Tutorial, Review, and Critique

ROBERT G. KUNZ, Ph.D., P.E.

RGK Environmental Consulting, L.L.C.
Hillsborough, North Carolina, USA

Paper 119e Advances in Distillation Tray Technology

Presented at the 2014 AIChE Spring National Meeting

March 30 – April 3, 2014 New Orleans, LA

ABSTRACT

The aniline-water system is a classic illustration of a so-called heterogeneous azeotrope (heteroazeotrope). The liquid phase separates into two immiscible layers, an oily aniline-rich organic layer and a water-rich aqueous layer, in equilibrium with each other and both in equilibrium with a single vapor phase. Aniline-water vapor-liquid equilibrium (VLE) data, including little known results, are reviewed in this presentation.

When aniline is manufactured from nitrobenzene, an understanding of phase equilibria is important in the separation of the aniline product from the water of reaction and their subsequent purification. Soluble water must be removed from the aniline oil, and the aniline soluble in the water phase must be treated prior to discharge to minimize environmental contamination. The present discussion concentrates on treatment of the water phase.

A proposed design and cost estimate for a steam stripper of aniline process wastewater to reduce the load on downstream wastewater treatment facilities is contained in a 1973 US EPA development document. In such a system, the overhead composition must approach the lower-boiling azeotrope, while the higher-boiling bottoms composition approaches pure water. However, the overhead composition in the stripper material balance appears to be incorrectly stated and impossible to achieve regardless of operating conditions. A pressure of 20 psig and the temperatures listed appear to be inconsistent as well. Your author is not aware of any corrections having been issued, and the design as published is called into question.

To attain the same bottoms composition with an overhead composition approaching the azeotrope, one must increase the overhead flow and decrease the bottoms flow. The number of theoretical trays required to achieve the desired separation is then calculated from the Kremser-Souders-Brown Equation. The number of actual trays is estimated using a range of tray efficiencies derived from published correlations plus hard-to-find aniline steam-stripper laboratory data. This analysis confirms the number of trays shown in the EPA diagram to be adequate at atmospheric pressure and 20 psig, with a higher tray efficiency needed at 20 psig.

INTRODUCTION

The subject of this paper started out to be a simple textbook example of a heterogeneous azeotrope – one type of vapor-liquid equilibrium (VLE) diagram that may be encountered in environmental control calculations. This was to join previous examples in the author's book[1] and subsequent paper[2] involving Raoult's Law, positive and negative deviations therefrom, and maximum- and minimum-boiling homogeneous azeotropes.

For a heterogeneous azeotrope to occur, the liquid phase breaks up into two immiscible solutions of differing composition, each in equilibrium with the other and both in equilibrium with a single vapor phase. The x-y VLE curve is flat in the area of mutual insolubility, and with few exceptions, the y=x 45° line intersects this constant y-value within the insoluble range.[3] Solutions of this type cannot be separated by distillation to obtain two products on opposite sides of the intersection. Distillation of a minimum-boiling azeotrope results in an overhead as close as desired to the azeotropic composition and a bottoms composition as close as desired to either of the pure components, depending on which one is richer in the feed.

The aniline-water system exhibits such behavior. It seemed a likely candidate to illustrate this phenomenon, with a steam-stripper design contained in a 1973 US EPA development document[4] providing a practical application to environmental control. Validation of the design was to be simple and straightforward.

Instead, this exercise turned into an extended project since the operating temperature and pressure of the steam stripper as presented appear to be inconsistent. Furthermore, contrary to expectations, its overhead composition was eventually discovered not to approach an attainable azeotropic composition at any reasonable set of operating conditions.

Nevertheless, once the perceived discrepancies are corrected, the design turns out to be capable of producing the stated aniline bottoms concentration at both atmospheric pressure and 20 psig, the elevated pressure assumed. However, a higher overall tray efficiency is necessary at 20 psig to maintain the same number of actual plates calculated for the atmospheric case. Well known published data, newly developed information, and recently uncovered little known experimental results from the past were used in the evaluation.

After consideration of some background information immediately below, this paper makes use of the aforementioned data, as necessary, along with widely accepted thermodynamic principles to deduce a technical basis for the wastewater stripper design.[5]

BACKGROUND INFORMATION

Nomenclature, Appearance, and Physical Properties

Aniline ($C_6H_5NH_2$) – a.k.a. *aminobenzene, aniline, aniline oil, benzidam, benzenamine, benzoamine, blue oil, cyanol,* and *phenyl amine*, among others[6,7] – is an oily organic liquid with a molecular weight of 93.1265 [6] and an atmospheric boiling point of 184.13 °C.[8] It is colorless when pure but gradually darkens to a reddish brown upon exposure to light and air.[7,9-10] The aniline molecule contains both a benzene ring, making it an aromatic chemical, and an amine group (–NH_2), making it a typical amine having the characteristic pungent odor of dead fish.[7] Safety and toxicology will be discussed later.

Solubility. Aniline is soluble in alcohol, ether, acetone, and benzene in all proportions [8-9,11] but exhibits limited miscibility with water.[12=14] Sufficiently concentrated solutions of aniline and water break up into two immiscible liquid layers at temperatures below an upper critical solution (consolute) temperature of 167.5 °C (Figure 1).[12-15] The water-rich aqueous phase containing some dissolved aniline is termed *aniline water* or *the water layer*; the organic phase consisting of aniline with water dissolved in it is called *aniline oil* or *the aniline-rich layer.*[14]

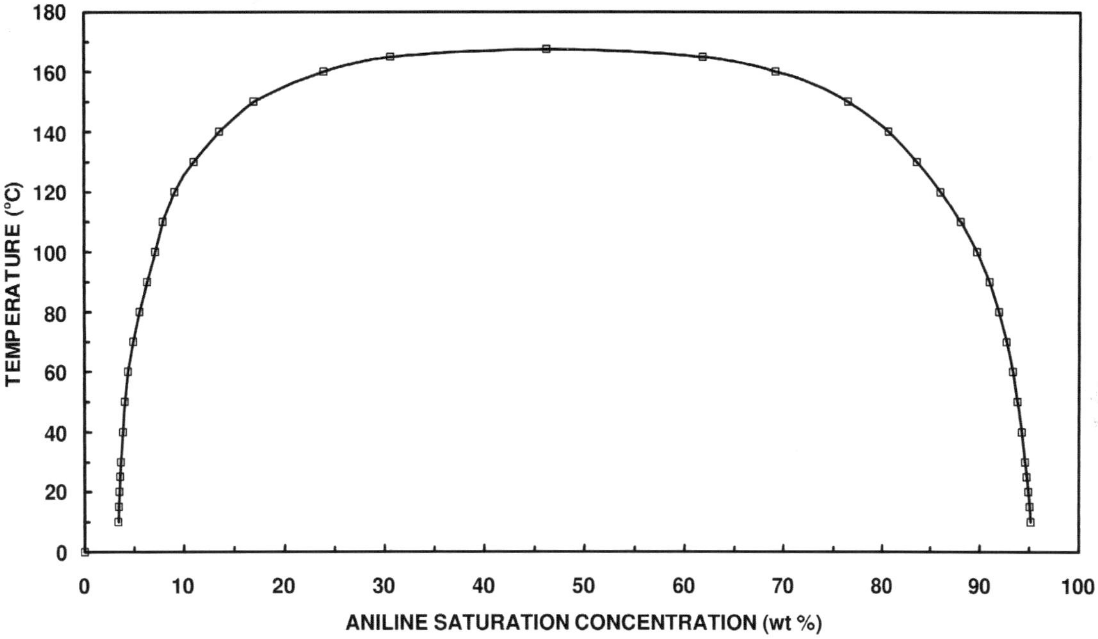

FIG. 1 Mutual Solubility of Aniline and Water in the Liquid Phase

The data of Figure 1 are plotted here in weight percent (wt %) as originally reported and are converted, as needed, to mole % for the VLE calculations discussed in a subsequent section. The left-hand branch of the curve in Figure 1 defines the locus of compositions for the saturated water phase, that is, the maximum solubility of aniline in water. Aniline solubility in the water layer is decreased by the presence of inorganic salts,[16] a reduction assumed for engineering purposes in one study to be 20%.[14]

The right-hand branch displays the maximum solubility of water in the saturated aniline phase. The area under the dome-shaped curve denotes effective concentrations for composite mixtures within the two-phase region. A single liquid phase exists outside the dome. At any given temperature, the two saturated liquid-phase compositions are in equilibrium with each other and in equilibrium with a single vapor phase. A diagram showing the liquid-phase relationships of Figure 1 in conjunction with VLE behavior is depicted in Ref. **[15]**.

Specific gravities of the two liquid layers[14,17] are plotted against temperature in Figure 2 between 32 to 160 °F (0 and 71.1 °C). Values are close to the respective densities of the pure components.[14,18-19] At these temperatures, the less dense aqueous phase rides on top; the aniline layer lies on the bottom.

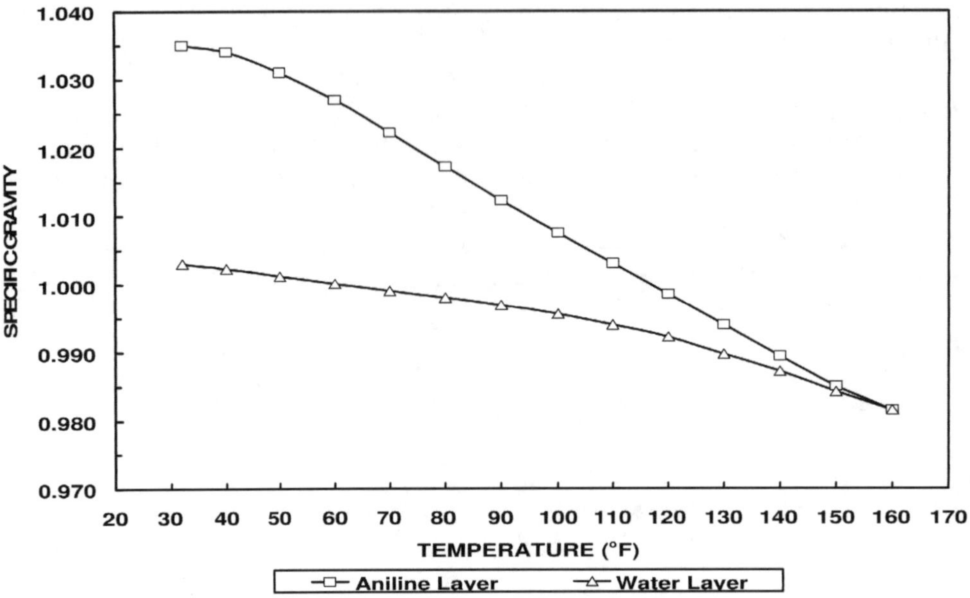

FIG. 2 Specific Gravities of Saturated Aniline-Water Solutions

Commercial Production

Commercial production of aniline today is primarily by catalytic hydrogenation of nitrobenzene in the gas or liquid phase in a fixed or fluidized bed, [4,9,11,17,20-21] with lesser production from reaction of iron filings and HCl.[11] The nitrobenzene is prepared by nitration of benzene.[17,20,22]

Aniline is an important industrial chemical used primarily as an intermediate in the manufacture of methyl diphenyl diisocyanate (MDI), itself an intermediate in the production of polyurethane foams.[23] Other applications include chemicals used to vulcanize rubber, agricultural chemicals, dyes and pigments, resins, pharmaceuticals, and miscellaneous uses.[23]

About 97% of the nitrobenzene produced in the U.S went into the production of aniline,[11,24,27] and approximately 2/3 of the aniline production went to produce MDI.[23]

Safety and Toxicology

Benzene, nitrobenzene, and aniline are dangerous chemicals and must be treated with great caution.[9-11,14,22-28] All are flammable and therefore potentially explosive. They are readily ingested by the body through the lungs, skin, and mucous membranes. Acute exposure to any of them can be seriously incapacitating or fatal. Chronic exposure to benzene is known to cause leukemia. Chronic exposure to nitrobenzene may cause anemia and damage to the spleen and liver.[27] Aniline is a suspected carcinogen[11] because of significant evidence of increased bladder cancer in humans.[10] Health, safety, and toxicology issues are discussed further in Appendix A.

Chemical Reactions to Produce Aniline

Reduction of nitrobenzene:[4,9,11,27]

$$C_6H_5NO_2 + 3\ H_2 \xrightarrow[\text{fluidized bed at 270 °C and 2.3 atm (~ psig)}]{\text{copper-silica catalyst}} C_6H_5NH_2 + 2\ H_2O \quad (1)$$

Liquid-phase reduction with iron filings (the historic Béchamp Process, overall reaction as follows):[9,20-21]

$$4\ C_6H_5NO_2 + 9\ Fe + 4\ H_2O \xrightarrow{FeCl_2 + HCl} 4\ C_6H_5NH_2 + 3\ Fe_3O_4 \quad (2)$$

and by ammonolysis of phenol[11,17] or chlorobenzene:[17,20-21]

$$C_6H_5OH + NH_3 \xrightarrow{SiO_2\text{-}Al_2O_3 \text{ catalyst}} C_6H_5NH_2 + H_2O \quad (3)$$

$$C_6H_5Cl + NH_3 \xrightarrow{Cu^+ \text{ reducing agent}} C_6H_5NH_2 + HCl \quad (4)$$

Of these, the chief means of aniline production is by hydrogenation of nitrobenzene,[11] followed by iron-filing reduction. The nitrobenzene in turn is formed by nitration of benzene at approximately 50 °C,[7,17,27,29] using a mixture of nitric and sulfuric acids.[4,11,17,20,27-28] Sulfuric acid is used as a dehydrating agent, which drives the nitration reaction to completion.[20]

$$C_6H_6 + HNO_3 \xrightarrow{H_2SO_4} C_6H_5NO_2 + H_2O \quad (5)$$

The 1973 US EPA development document contains a flow sheet for hydrogenation of nitrobenzene starting with benzene plus mixed nitric and sulfuric acids and ending with aniline, redrawn here as Figure 3. Similar overall flow sheets are presented in Refs. **[20-21]**. Flow sheets for some of the other processes can be found in the cited references.[4,9,11,17,20-21,28]

Steam Stripping of Aniline Wastewater

The aniline water from the final liquid-liquid separator in Figure 3 contains 3.1 wt %[4] to 3.5 wt % aniline.[9] This stream can be extracted with benzene or nitrobenzene to recover aniline or can be steam stripped. Steam stripping will minimize aniline contamination in the effluent if discharged directly, or will reduce the load on downstream wastewater treatment facilities if further treated. Aniline concentrations as low as 440[30] and 25[31] ppmw have been measured in different multi-plate laboratory steam-stripping columns. Long-term mean treatment performance concentrations for aniline in the pharmaceutical industry are 1240 mg/L (≈ 1240 ppmw) from steam stripping and 0.030 mg/L from biological treatment.[32]

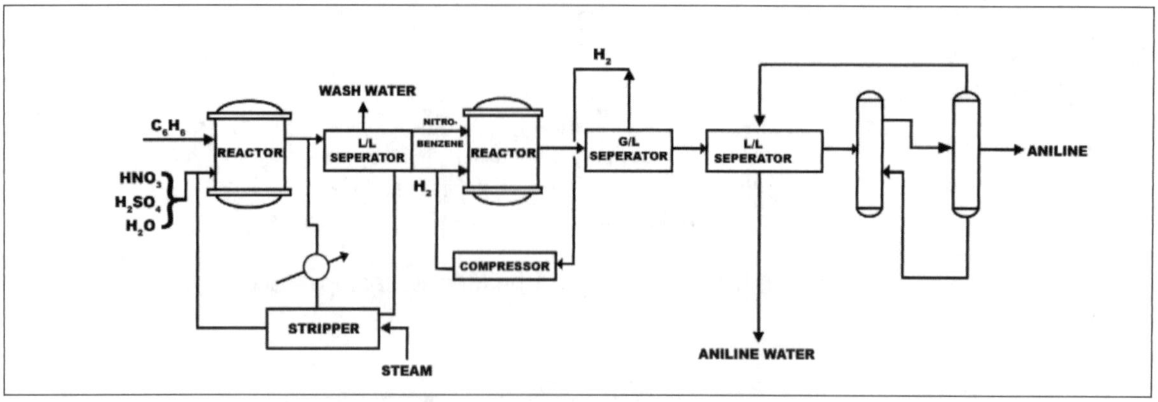

FIG. 3 Benzene to Aniline Flow Sheet

1973 EPA DEVELOPMENT DOCUMENT STEAM STRIPPER

The flow diagram from a proposed design and cost estimate for a steam stripper of aniline process wastewater contained in the 1973 EPA development document [4] and reproduced in a later book[33] is also redrawn here (Figure 4). The feed, with no recycle, is stated to be 17 gallons per minute (gpm) of 3.1 wt % aniline, that is, the limit of solubility in salt-free water at 100 °F (37.8 °C). The steam stripper in the diagram produces an aqueous liquid effluent containing about 2000 ppmw (≈ 2000 mg/L) aniline plus an overhead vapor mixture of approximately 52 wt % aniline and 48 wt % water, which is sent to incineration. The stripper bottoms is used to preheat the feed and is further heat-exchanged with cooling water to 95 °F (35 °C). The design uses a reboiler to provide non-contact steam, presumably to avoid adding contaminated steam condensate to any wastewater to be treated or discharged.[34] Our mission in the remainder of this paper will be to verify that the conditions of Figure 4 are indeed reasonable to achieve such an effluent concentration.

Problems with the Design

It is not entirely clear at first whether this design is based on theory or reflects an actual operating unit. A later EPA document[35] alludes to steam-stripper installations at two named US aniline plants, but, unfortunately, no specifics are given. However, the overhead composition should be at or near that of the aniline-water heteroazeotrope.

The composition of an azeotrope varies somewhat with pressure (or temperature).[36] From below atmospheric pressure up to several atmospheres, the azeotropic composition for this system is said to be on the order of **17 wt %±** in some publications[14, 37-40] and an equivalent **0.03 to 0.05+ mole %** in others.[41-43] Because the overhead composition of 52 wt % aniline derived from Figure 4 works out to be about **17 <u>mole %</u>** aniline, and this numerical value is suspiciously close to the accepted azeotropic composition expressed in **wt %**, it seems as if someone confused the units when creating the diagram. The basis of the diagram therefore appears to be incorrect, and your author is not aware of any corrections that may have been issued. Furthermore, since the stated vapor-phase composition cannot be achieved in practice below the consolute temperature, the design appears to be based on a simulation rather than on an operating unit.

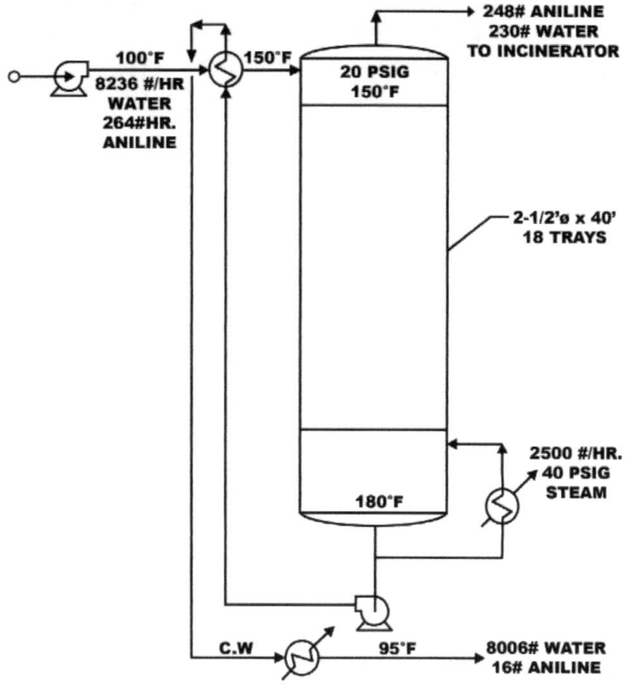

FIG. 4 Steam Stripper

In addition, the pressure and temperatures listed in Figure 4 appear to be inconsistent – 20 psig (34.696 psia, 2.36 atm, 1794.295 mmHg, 2.33 bar, 239.2 kPa), with temperatures top to bottom of 150 °F (65.6 °C) and 180 °F (82.2 °C). Since almost pure water is being distilled, the indicated temperatures would correspond to operation under vacuum, but with no vacuum-producing equipment shown. The present evaluation has chosen to believe the stated pressure of 20 psig as the more reasonable, rather than a vacuum installation; the design will be re-evaluated at 20 psig and at atmospheric pressure as well.

Temperature at the top, corresponding to the azeotrope at 20 psig, is estimated to be 125 °C (257 °F), closer to 250 °F than the 150 °F shown in Figure 4. The liquid leaving the bottom of the column is essentially pure water, and its temperature at 20 psig from Steam Tables[19] should be 258.8 °F (126 °C). Temperature of the separate 40-psig stripping steam is 286.7 °F, closer to 280 °F than the 180 °F shown in the diagram. At atmospheric pressure, temperature of the azeotrope is 98.5 °C,[38,43] and the temperature of 100 % water is 100 °C.

Corrected Material Balance

In the distillation/stripping of the aqueous phase, it makes sense that the overhead in Figure 4 should approach the (minimum-boiling) azeotropic composition, while the bottoms composition approaches that of pure water, boiling somewhat higher. As indicated above, the difference amounts to about 1.5 °C at atmospheric pressure and 1 °C at 20 psig.

Since the original overhead composition is impossible to attain, the material balance around the stripper must be revised. To continue to achieve the same bottoms concentration of 2000 ppmw of aniline at the same feed rate (F) and composition, the flow of overhead vapor (D) must be increased and the bottoms flow (B) must be decreased. This would require incineration of more distillate in the flow scheme of Figure 4. By material balance:

$$F = D + B \tag{6}$$

and

$$X_F\,F = Y_D\,D + X_B\,B \tag{7}$$

with Original wt Fraction (X_F) Aniline in Feed (F) = 264 / 8500 = 0.03106

Original wt Fraction (X_B) Aniline in Bottoms (B) = 16 / 8022 = 0.0020

The 2500 lb/hr of 40-psig steam added to the reboiler (~30% of the feed)[34] is assumed to raise an equal amount of steam in the bottom of the stripper by indirect heat exchange. The resulting vapor flow (V) is assumed constant from plate to plate all the way to the overhead distillate if one neglects the minor heat- and mass transfer effects in dilute solutions.[34]

Therefore, B = 8500 – 2500 = 6000 from Equation (6). This is L and is also assumed constant.

Then Wt Fraction (Y_D) Aniline in Distillate (D) =

$$Y_D = (X_F\,F - X_B\,B) / D \tag{8}$$

[(0.03106 × 8500 = 264) – (0.0020 × 6000 = 12)] / 2500 = 252 / 2500 = 0.1008 by rearrangement of Equation (7). This corresponds to a mole fraction of 0.021. less than azeotropic compositions of 0.036 (16.18 wt%) at 98.5 °C/1 atm and 0.042 (18.47 wt%) at 125 °C/20 psig. Calculations at 20 psig are exactly the same. A summary is shown in Table 1.

TABLE 1. Summary of Aniline Steam-Stripper Material Balance Calculations

Condition	Y_D Mole Fraction Aniline	Y_D Weight Fraction Aniline	F Feed Rate (lb / hr)	D Distillate Rate (lb / hr)	B Bottoms Rate (lb / hr)
1 atm and 20 psig	**0.021**	**0.1008**	**8500**	**2500**	**6000**
Original Diagram	**0.1726**	**0.5188**	**8500**	**478**	**8022**
X_F = 0.03106 wt fraction aniline and 0.0062 mole fraction aniline for all cases					
X_B = 0.002 wt fraction aniline and 0.0004 mole fraction aniline for all cases					

INPUTS TO STEAM STRIPPER DESIGN

Vapor-Liquid Equilibrium (VLE) Data. Extensive VLE data covering both phases of the aniline-water system were published in 1940 in a single comprehensive study by Griswold and coworkers.[43] Data were obtained at 100 °C and at 745 mmHg. near standard atmospheric pressure. Other, less extensive constant-temperature data at 20,[44] 40,[45] 50,[16] 56.3,[46] and 75 °C [46] are available in the literature; additional data at 742 mmHg,[47] 760 mmHg, [48-49] an unspecified atmospheric pressure,[30] and higher pressures (6, 11, and 16.4 atm) [47] have also been reported.

Aniline-Phase VLE Data. The Griswold aniline-phase x-y data [43,50] are plotted in Figures 5 and 6, along with points at ~ 99 °C and 760 mmHg from two of the other investigations. [51-52] The major difference between the data in the two figures occurs at low water concentrations.

The Margules Equations, [1,5,15,36] DECHEMA's best-fit activity coefficient (γ) formulation for these cases,[50]

for water (component 1) $$\ln \gamma_1 = x_2^2 \, [A_{1\text{-}2} + 2\, x_1 \, (A_{2\text{-}1} - A_{1\text{-}2})] \tag{9}$$

for aniline (component 2) $$\ln \gamma_2 = x_1^2 \, [A_{2\text{-}1} + 2\, x_2 \, (A_{1\text{-}2} - A_{2\text{-}1})] \tag{10}$$

were used to draw the curve correlating the data in each figure, with DECHEMA's parameters $A_{1\text{-}2} = 1.7569$ and $A_{2\text{-}1} = 3.1726$ for the 100 °C data and $A_{1\text{-}2} = 1.4748$ and $A_{2\text{-}1} = 3.1920$ for the 745-mmHg data. Coefficients for the van Laar, Wilson, NTRL, and UNIQUAC models are also given by DECHEMA. A fit of activity coefficients to the van Laar Equations [1,5,15,36]

for water (component 1) $$\ln \gamma_1 = A_{1\text{-}2} / [1 + (A_{1\text{-}2}\, x_1) / (A_{2\text{-}1}\, x_2)]^2 \tag{11}$$

for aniline (component 2) $$\ln \gamma_2 = A_{2\text{-}1} / [1 + (A_{2\text{-}1}\, x_2) / (A_{1\text{-}2}\, x_1)]^2 \tag{12}$$

with different values of $A_{1\text{-}2}$ and $A_{2\text{-}1}$ is shown in Ref. **[15]** for the 100 °C data. The coefficients $A_{1\text{-}2}$ and $A_{2\text{-}1}$ for both the Margules and the van Laar Equations correspond to the logarithms of γ^{∞}, the activity coefficient at infinite dilution for water and aniline, respectively.

Partial pressure (p) was computed from equations applicable at low total pressure (P_T). These relationships can also be employed here at 20 psig without serious error (Appendix B).

$$p_1 = \gamma_1 \, x_1 \, p_1^{vap} \tag{13}$$

and $$p_2 = \gamma_2 \, x_2 \, p_2^{vap} \tag{14}$$

with $$p_1 + p_2 = P_T \tag{15}$$

For the constant-temperature data at 100 °C, calculations are straightforward. For the constant-pressure data, a trial-and-error procedure was employed to choose the proper temperature (t) at each value of water mole fraction (x_1) such that the sum of the individual partial pressures from Equation (15) equals the observed total pressure of 745 mmHg.

Rather than using the Antoine Vapor Pressure Equation

$$\log_{10} p_2^{vap}\,(\text{mmHg}) = A - B / [t\,(°\text{C}) + C] \tag{16}$$

with handbook values of constants A = 7.96681, B = 1668.21, and C = 228.0 valid for water between 60 and 150 °C,[53] vapor-pressure (p_1^{vap}) of water was calculated from the relationship given in the Keenan and Keyes Steam Tables as valid from 50 to 374.11 °C.[54]

$$\log_{10} \frac{p_c}{p} = \frac{x}{T} \left[\frac{a + b\,x + c\,x^3 + e\,x^4}{1 + d\,x} \right] \quad (17)$$

Where								
p	=	p^{vap} (int atm)	x	=	647.27°K – T	c	=	7.515484×10^{-9}
p_c	=	218.167 int atm	a	=	3.3463130	d	=	1.3794481×10^{-2}
T	=	t(°C)+273.16	b	=	4.14113×10^{-2}	e	=	6.56444×10^{-11}

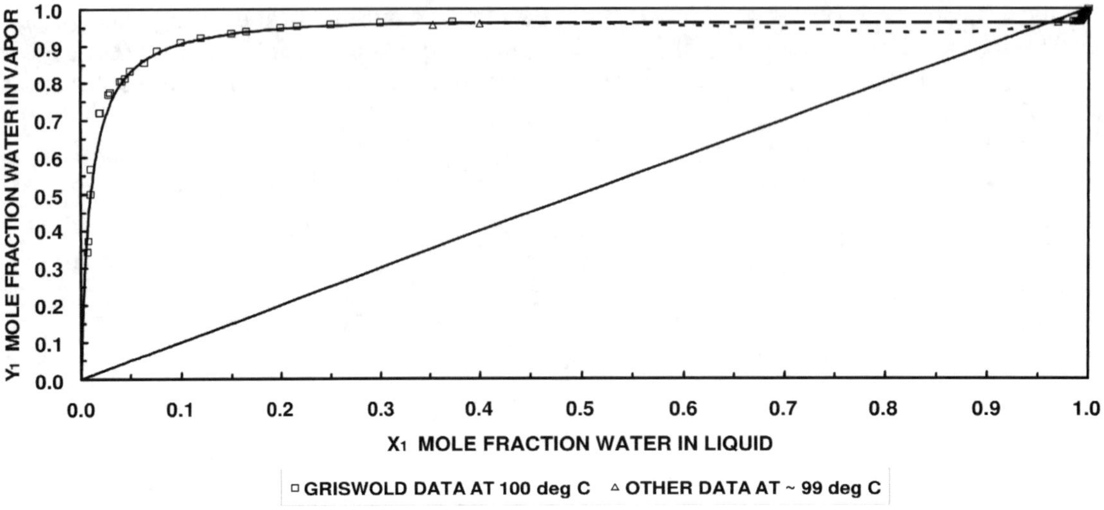

FIG. 5 VLE x-y Diagram for Aniline and Water at 100 °C

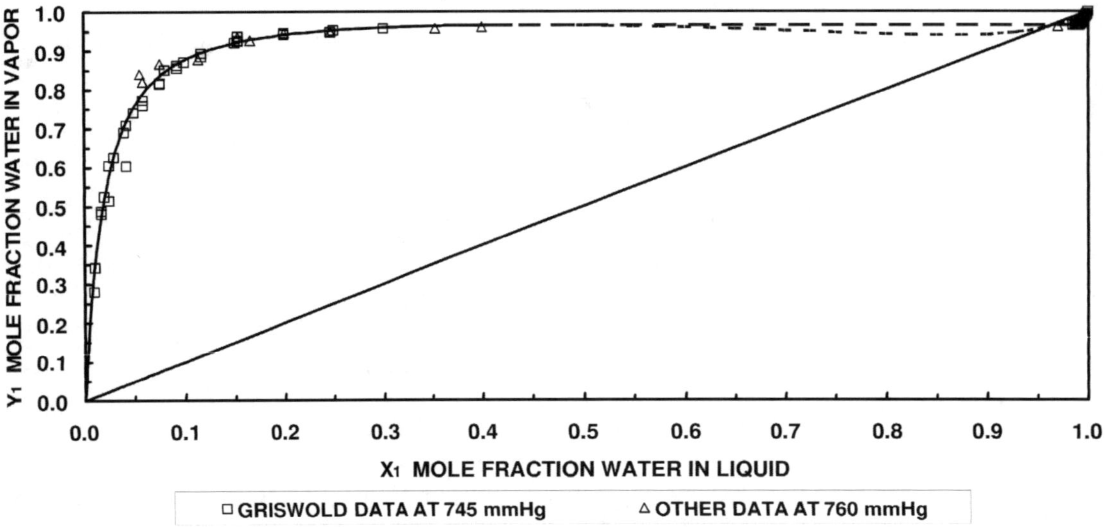

FIG. 6 VLE x-y Diagram for Aniline and Water near Standard Atmospheric Pressure

This relationship was employed in the event that the vapor pressure of water would be needed for a case at an elevated total pressure and temperatures beyond the range of validity of other water vapor-pressure equations commonly employed.

Temperature–dependent vapor pressure for aniline (p_2^{vap}) comes from the Antoine Equation with constants A = 7.24179, B = 1675.3, and C = 200.[53] (The DECHEMA fits[50] of the Margules Equations to the Griswold aniline-water VLE data[43] had used Antoine constants A, B, and C of 8.07131, 1730.630, and 233.426 for water (valid 1 °C–100 °C) and 7.46441, 1840.790, and 216.923 for aniline (valid 35 °C–184 °C).)

The Antoine Equation with constants from Lange's Handbook[53] agrees well with vapor pressures for aniline listed in Ref. **[8]** taken from the extensive Stull Tables [55] and to within 0.14 – 2.1 mmHg with a newer, more complicated vapor pressure equation noted in Ref. **[56]**.

At low pressure, vapor-phase mole fractions (y) are given by

$$y_1 = p_1 / P_T \quad \text{and} \quad y_2 = p_2 / P_T \tag{18}$$

The broken horizontal line smoothly joining the two sections of the heavy solid curve in Figure 5 and in Figure 6 denotes the two-phase region of limited miscibility, where vapor-phase composition and temperature are constant. Liquid mole fractions at the ends of the broken-line segments coincide closely with the liquid-liquid equilibrium data of Figure 1 at the appropriate temperature. Calculated vapor composition agrees well with experimental VLE data. The dotted section of the curve in the two-phase region of Figures 5 and 6 is not real but an artifact of the Margules activity-coefficient calculations. The solid diagonal line is y = x.

Water-Phase VLE Data. An exploded view of the x-y diagram of the aqueous phase at 745 mmHg is depicted in Figure 7. It shows the data of Griswold et al.[43,51] along with data points obtained later by another researcher[30] in order to verify the Griswold data. Once again, the dashed horizontal line extends over the region of mutual insolubility, and the dotted line reflects the vapor-phase values of Figure 6 calculated from the Margules Equations. The Margules model, which works so well for the aniline phase in Figures 5 and 6, is not even close to the water-phase experimental data in Figure 7.

Instead, the Griswold data are newly fitted here by means of the van Laar Equations with $A_{1\text{-}2}$ = 1.40 and $A_{2\text{-}1}$ = 4.51. As shown by the solid curved line in Figure 7, this fit starts at the composition of pure water (x_1 = 1.0, y_1 = 1.0) in the upper right-hand corner and proceeds downward toward the extension of the dashed horizontal line from the two-phase region, with a short transition section in between. The van Laar fit smoothly fills the gap at the upper end of the data points, approaching the composition of pure water with a reasonable curvature. Average deviation of the absolute value between calculated and experimental results is 0.00048 mole fraction, of the same order of magnitude as the accuracy to 0.0002 mole fraction aniline in the water phase claimed by the original investigators.[43] Use of absolute values prevents positive and negative deviations from canceling out one another.

Calculated values of temperature along the van Laar curve-fit at 745 mmHg range from 98.3 °C to 99.4 °C; measurements at 745 mm Hg for the two-phase region are reported at 98.5 °C and 99.2 °C. This helps to explain the "99 °C and 745 mm" caption on the water-phase data in the top part of the original investigators' Figure 1. Temperature of the azeotrope itself is estimated from the published data to be 98.5 °C at 745 mmHg and 100 °C at 750 mmHg. At a liquid-phase solubility limit of about 0.985 mole fraction water (0.015 mole fraction aniline), calculated vapor-phase composition of the azeotrope at 745 mmHg is 0.964 mole fraction water (0.036 mole fraction aniline). This compares to an aniline mole fraction of 0.030 to 0.036 estimated directly from the original data. These azeotropic properties lie within the ranges listed in various compilations cited above.

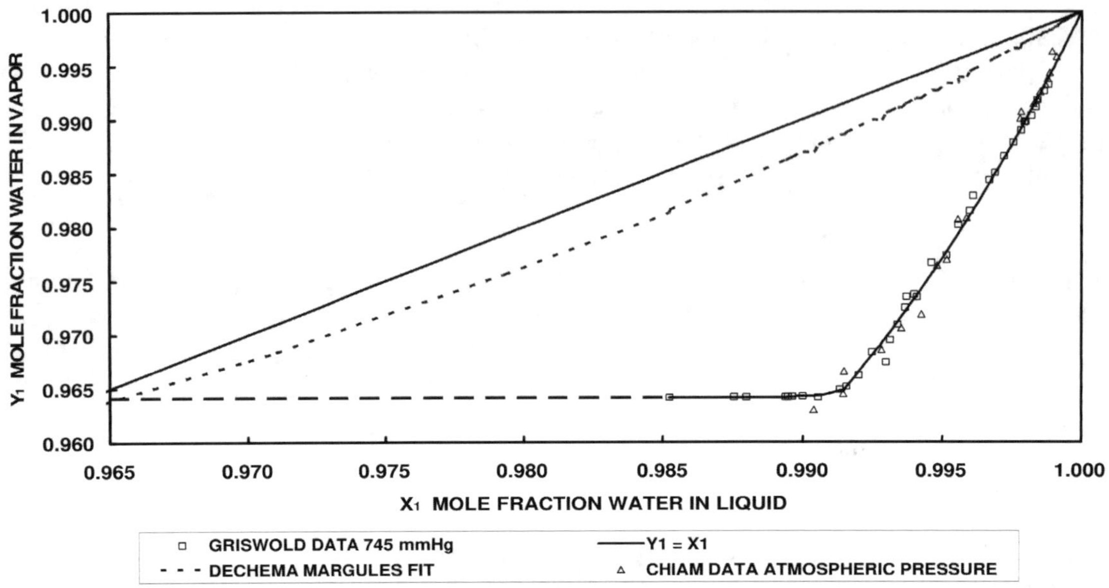

FIG. 7 VLE x-y Diagram for Aniline and Water at 745 mmHg Pressure (Exploded View of Aqueous Phase)

Estimation of Activity Coefficients from Liquid-Liquid Solubility Data. In addition to a fit to VLE data, activity coefficients can be derived from liquid-liquid mutual solubility data. This is an accepted method to extend one's information,[5,14-15,36] as further explained in Appendix C. The procedure makes use of the fact that the single vapor phase and both liquid phases are all in equilibrium with one another. For each component

Aniline Phase Water Phase

$$y = \gamma_A x_A p^{vap} / P_T = \gamma_W x_W p^{vap} / P_T \tag{19}$$

The ratio p^{vap} / P_T, the same for both liquid phases, cancels out, and one is left with

$$y = \gamma_A x_A = \gamma_W x_W \tag{20}$$

Equation (20) allows the establishment of an activity-coefficient formulation valid at least in the vicinity of the liquid miscibility gap.

To evaluate this possibility of extending VLE "data" to other conditions such as 20 psig, the technique was first tried out at atmospheric pressure, where the Griswold data are available to provide a check. Accordingly, constants for the Margules Equations were recalculated based on the liquid-liquid solubility data at 100 °C. As expected, the resulting fit of aniline-phase data is good in the immediate vicinity of the equilibrium conditions; it deteriorates, however, as one moves farther and farther away. Similarly, the van Laar Equations provide a good fit for the aniline phase over a somewhat wider range, but they also fail to reproduce all the data.

Fit of the water-phase data is markedly improved by the use of the van Laar Equations, but calculated and experimental values still lie a significant distance apart. The final van Laar fit of the water-phase data shown in Figure 7 comes about after an empirical adjustment of $A_{2\text{-}1}$ to a value of 4.51. This value of $A_{2\text{-}1}$, ln (γ^{∞}) for aniline, turns out to be the same as that found for ln (γ^{∞}) at 100 °C in an extensive investigation by others.[56] The best-fit value for $A_{1\text{-}2}$ with $A_{2\text{-}1}$ = 4.51 is 1.40 (rounded), as originally found directly from the solubility data. The conclusion in this case is that constants for the Margules and van Laar Equations obtained from liquid-liquid solubility data may not be reliable except in the immediate vicinity of the equilibrium points.*

The only thing the van Laar Equations were used to estimate for the 20-psig case is the azeotropic point. As in the atmospheric-pressure case, the minimum-boiling heteroazeotrope at 20 psig (34.696 psia) is expected to be slightly below the boiling point of water at this pressure, which is roughly 126 °C.[19] Starting with solubility data at 120 °C of x_1 = 0.98345294 and x_2 = 0.01654706 in the water phase, van Laar constants were determined, allowing estimation of the azeotrope temperature and vapor-phase composition. Linear interpolation between 120 °C and 130 °C of the solubility data at this new temperature allows a second determination of constants, temperature, and vapor-phase composition – and so on. The method converges within five iterations to almost exactly 125 °C, x_1 = 0.9823318 (x_2 = 0.0176682), and y_1 = 0.95776 (y_2 = 0.04224) for $A_{1\text{-}2}$ = 1.150847 and $A_{2\text{-}1}$ = 4.018355. Experimental results from a single study at pressures of 0.98, 6, 11, and 16.4 atm show an aniline vapor-phase mole fraction (y_2) for the azeotrope between 0.044 and 0.057.[47]

Henry's Law Constants for Aniline in Water at Various Temperatures. An estimate of comprehensive VLE data at 20 psig is somewhat uncertain, but fortunately it is not necessary to generate a complete x-y diagram to design a steam stripper for the water phase. Following the method outlined in Ref. **[14]**, the procedure to be employed will estimate the Henry's Law Constant (H) from mutual solubility data. Expressing the VLE relationship in the form of Henry's Law will then allow the shortcut Kremser Equation [15,36,57-59] to be used for stripping-column design,[60] as explained later.

---

* Treybal[15] indicates that the Margules Equations are best at following the observed data when $A_{1\text{-}2}$ and $A_{2\text{-}1}$ are nearly equal (possibly less than 2:1). The van Laar Equations can handle high values of $A_{1\text{-}2}$ and $A_{2\text{-}1}$ and greater dissymmetry in constants and are better in following the data than the Margules Equations at a ratio of constants in the vicinity of 2:1. The ratio of constants for the DECHEMA fits to the Margules Equations is approximately 2:1, but the fits from solubility data show a ratio of 3:1 or even greater.

Henry's Law relates liquid and vapor mole fractions

$$p_2 = H\, x_2 \tag{21}$$

$$\text{and} \quad y_2 = H\, x_2 \,/\, P_T \tag{22}$$

where H in units of pressure per mole fraction (m.f.) is a constant at constant temperature.

We note that the corresponding atmospheric data can be approximated by Henry's Law (linear in x_2 with a slope of H / P_T and a zero intercept) for aniline at dilute concentrations. We further note that Raoult's Law [Equation (13), with $\gamma_1 = 1.0$] applies for the water component as the solution composition becomes closer to that of pure water and the temperature approaches the boiling point of pure water at the prevailing pressure. There is a general rule that when one component of a binary obeys Henry's Law, the other in almost pure solution follows Raoult's Law.[61] Therefore, for the water phase,

$$P_T = p_1 + p_2 = p_1^{vap}\, x_1 + H\,(1 - x_1) \tag{23}$$

Ref. **[14]** states that this system follows Henry's Law for aniline and Raoult's Law for water for concentrations up to the solubility limit of aniline in the water phase, and that it follows Raoult's Law [Equation (14), with $\gamma_1 = 1.0$] for aniline and Henry's Law for water for concentrations up to the solubility limit of water in the aniline phase. This approximation works well enough for the water phase but becomes noticeably less accurate as water concentration increases in the aniline phase, as shown by the data of Griswold et al.[43] Maximum activity coefficient for water (γ_1) in the water phase calculated from the several activity-coefficient models tested is 1.0045; activity coefficient for aniline (γ_2) in the aniline phase at the two-phase point is on the order of 1.2.

Estimation of Henry's Law Constants from Solubility Data

Ref. **[14]** outlines a method to estimate the Henry's Law Constant (H) for the aniline-water system based on mutual solubility data. Equation (19) above written for aniline, component 2, is modified to include H:

Aniline Phase — Water Phase

$$y_2 = \gamma_{2A}\, x_{2A}\, p_2^{vap} / P_T = \gamma_{2W}\, x_{2W}\, p_2^{vap} / P_T = H\, x_{2W} / P_T \tag{24}$$

From which

$$H = \gamma_{2A}\, p_2^{vap}\,(x_{2A} / x_{2W}) \tag{25}$$

or if $\gamma_{2A} = 1.0$,

$$H = p_2^{vap}\,(x_{2A} / x_{2W}) \tag{26}$$

At 100 °C, $H = (1)\,(45.44)\,(0.62757) / (0.01475) = 1933$ mmHg / m.f.

and $H / P_T = (1933/760) = 2.54$ at 760 mmHg

This number is in agreement with the data-regression line for the *calculated* entries of equilibrium aniline vapor concentrations over the water phase at 100 °C, as listed in Griswold's Table I, and must have been how they were calculated. Those values and the least squares regression line are plotted in Figure 8. The liquid composition at the highest point plotted, listed as "2 phases" in the Griswold table, was estimated here from aniline water-phase mutual solubility at 100 °C (Figure 1).

In a similar manner, values of H from Equation (26) at 98.5 °C, 125 °C, and 126 °C are calculated to be 1861, 3594, and 3650 mmHg / m.f., respectively, using the corresponding aniline vapor pressures of 42.60, 122.2, and 126.7 mmHg. These Henry's Law Constants could be up to 20 % higher if a value of $\gamma_2 = 1.2$ in the aniline phase were taken into account [Equation (25)]. In addition, Ref. **[14]** advises for design that the solubility of aniline in the water phase should be decreased by 20 % because of the salting-out effect of inorganic salts from the manufacturing process.† (One does not take this effect into account when evaluating experimental laboratory data obtained with purified solutions.) A 20% decrease in x_{2W} appearing in the denominator of either Equation (25) or Equation (26) would then increase the calculated Henry's Law Constant by another 25%, and both of these factors together would result in a combined 50 % increase. Final values of H from this estimating technique are listed in Table 2, Column 4.

Another technique to estimate H for aniline in water is through its activity coefficient at infinite dilution (γ^{∞}) since

$$H = \gamma^{\infty} p^{vap} \qquad (27)$$

Values of γ^{∞} for aniline in water between 273 °K and 373 °K have been determined by others from original VLE measurements and related thermal data for highly dilute solutions, combined with values from the literature.[56] Henry's Law Constants were then calculated by those authors from the relationship of Equation (27). Results are presented in tables, charts, and correlating equations

$$\ln(\gamma^{\infty}) = 19.6708 - 14.6501 / \tau - 15.3761 \ln(\tau) \qquad (28)$$

$$H = 48.9325 - 46.4403 / \tau - 24.6450 \ln(\tau) \qquad (29)$$

where $\tau = (T\ °K) / 298.15$ and H is in units of kPa / m.f.

Values of γ^{∞} and H (mmHg/m.f.) calculated at 100 °C and 126 °C from the above equations are listed in Table 2, Columns 2 and 3. Natural logarithm of γ^{∞} matches almost perfectly the $A_{2\text{-}1}$ obtained from a curve-fit of the water phase data of Figure 7 to the van Laar Equations.

† This has to be based on practical plant experience. A decrease in solubility is consistent with laboratory tests when various salts were added to determine the "salting-out" effect on aniline solubility in the water phase at 50 °C.[16]

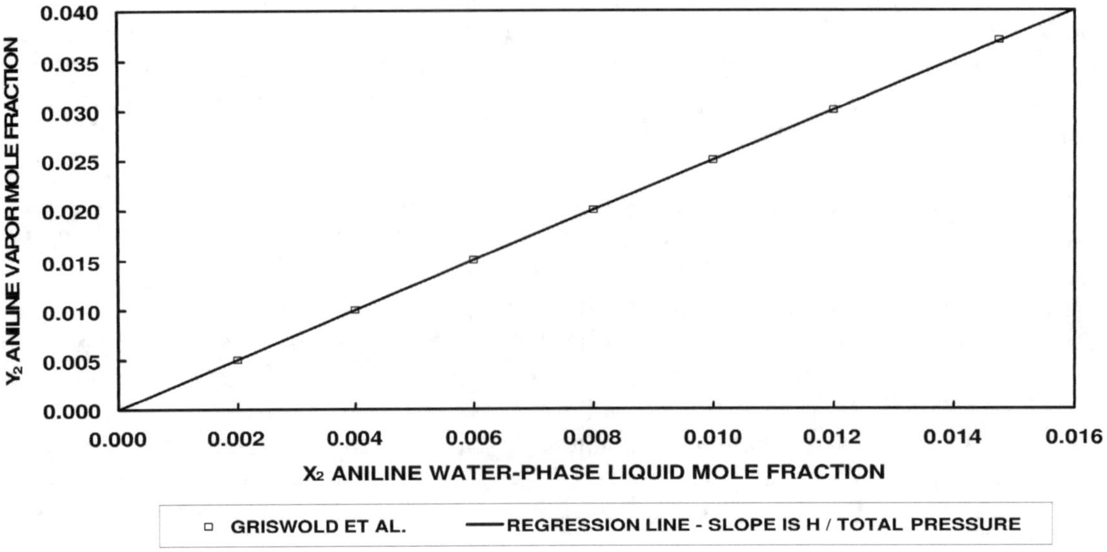

FIG. 8 Griswold Vapor-Phase Calculations for the Aniline-Water System at 100 °C

One can expect activity coefficients from a system having an upper critical solution temperature to decrease with increasing temperature, as shown by γ^{∞} for aniline.[56] This decrease is overwhelmed by the marked increase in vapor pressure with increasing temperature to cause an increase in H,[56] but at a slower rate.

TABLE 2. Gamma Infinity (γ^{∞}) and Henry's Law Constant (H) for Aniline in Water

Temperature (°C)	γ^{∞} Dimensionless	H (mmHg / mole fraction) From $\gamma^{\infty} \times p^{vap}$	From Solubility
98.5	—	—	*2792*
100	91.7	4070	2900
125	—	—	*5392*
126	69.5 (extrapolated)	6877 (extrapolated)	5475

Comparison of Henry's Law Estimates

The calculated aniline-water x-y diagram for the water phase is plotted in Figure 9, using the van Laar coefficients from the constant pressure function of Figure 7. The constant pressure data points from Figure 7 are reproduced as well. At the scale of the figures, the curves of Figures 7 and 9 are virtually identical. The solid y=x 45° diagonal extending from lower left to upper right intersects the dashed horizontal line denoting the two-phase region.

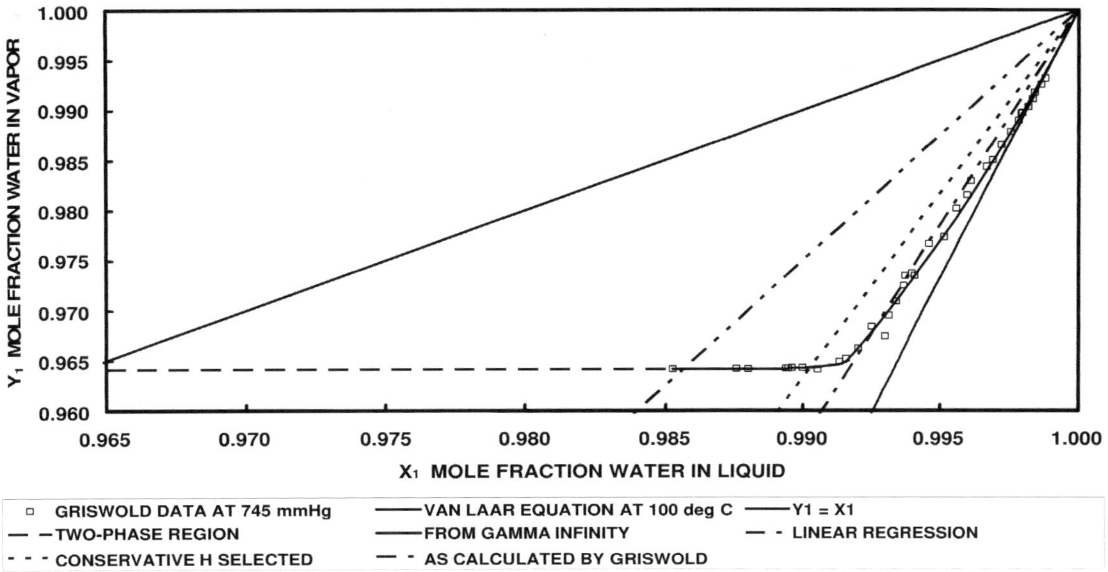

FIG. 9 Calculated VLE x-y Diagram for Aniline and Water at 100 °C
Exploded View of the Water Phase with Comparison of Henry's Law Values

The other straight lines emanating from the point x, y = 1.0 (100 % water, 0 % aniline) denote by their slopes several values of the Henry's Law Constant (H) for aniline in water estimated by various techniques. A Henry's Law Constant is needed to allow the use of a shortcut method to compute the required number of trays for the distillation column of Figure 4 to achieve the desired separation.

The solid slanted straight line farthest to the right (H = 4070 mmHg/m.f.) is calculated from γ^{∞} and Equation (27). It corresponds to the tangent to the VLE x-y relationship at $x_2=0$ (100% water). This value of H is thus the highest to be obtained at this temperature when that relationship is curved rather than a straight line throughout. Such values are appropriate when stripping parts per million (ppm) concentrations of contaminants from wastewater.[62]

The second straight line to the left (dash-dot) is the linear least-squares fit with a zero constant of the first 25 data points in Griswold's Table II[43] before the data begin to taper off significantly. (Refer again to Figure 9.) Numerical value of H is ~ 3300. It constitutes a straight-line average approximating the curved x-y equilibrium relationship between the azeotropic composition and pure water. One could use this value of H along with the Kremser Equation discussed below when higher contaminant levels in this range are to be stripped.[60]

Similar comments apply to the next line to the left. This one is obtained from solubility data via Equation (25), corrected for the salting-out effect on aniline solubility in the water phase, and calculated at the temperature of the azeotrope. It produces a more conservative version of H (2792, the lesser value shown in italics in Table 2) and for this reason will be used here to evaluate the steam stripper design.

In contrast, the last line (dash-dot-dot) farthest to the left is nowhere near the equilibrium relationship at 100 °C. This line is the regression of calculated entries in Griswold's

Table I.[43] The regression line and calculated points are plotted against aniline mole fraction (x_2) in Figure 8 above. Those tabular entries were apparently calculated from Equation (26) assuming $\gamma_{2A} = 1.0$. However, they are in fact a "straw man," and Griswold and co-workers point out the inadequacy of indiscriminately assuming that Raoult's and Henry's Laws apply to solution equilibria for the single-phase regions of the aniline-water system. That inadequacy plus the lack of data are indicated to have prompted their experimental investigation. The resulting Henry's Law Constant (1900 at atm pressure and 100 °C is not representative of the VLE relationship at this condition and is too low to be of use in the Kremser Equation.

EVALUATION OF STEAM STRIPPER DESIGN

Number of Theoretical Equilibrium Stages via the Kremser Equation

Stripper design was assessed using the Kremser Equation, also known as the *Kremser-Souders-Brown*, *Kremser-Brown-Souders*, or the *Souders-Brown* Equation.[15,36,57]

$$\frac{x_B}{x_F} = \frac{S-1}{S^{N+1}-1} \quad \text{(for } S > 1.0\text{)} \qquad (30)$$

where x_B, x_F = Liquid phase mole fraction of the solute in the bottoms and feed, respectively, for the stripper configuration pictured in Figure 4

N = Number of theoretical stripping stages

S = stripping factor = $(H/P_T)(V/L)$

and V & L = molar flow rates of vapor and liquid in the column

If one neglects the moles of solute (aniline) in the water and steam, V/L is given approximately by the ratio of the weight of steam injected into the stripper to that of the treated water in the underflow leaving the stripper.[57]

When rearranged to solve for N, the equation becomes

$$N = \frac{\log\left[(x_F / x_B)(S-1) + 1\right]}{\log S} - 1 \qquad (31)$$

Either natural logarithms or common logs can be used but must be the same throughout.

These equations assume that Henry's Law is valid to provide a linear VLE relationship, both V and L are constant throughout the stripper, dilute solutions, adiabatic conditions (no heat loss), a saturated feed preheated to its boiling point, overhead and bottoms streams leaving the stripper at equilibrium conditions[57] – plus no heat- and mass transfer effects.[34]

The next step is to calculate N from Equation (31) for the atmospheric pressure (760 mmHg) case. The factors x_F and x_B are from Table 1. Henry's Law Constant (H) is from Table 2. Vapor rate (V) of 2500 lb_m/hr comes from the stripping-steam rate of Figure 4 or the distillate rate of Table 1. Liquid rate (L) equals total bottoms flow of 6000 lb_m/hr from Table 1. Inserting the numbers into Equation (31),

$$N = \frac{\ln \{(0.0062/0.0004)\ [(2792/760)(2500/6000) - 1] + 1\}}{\ln [(2792/760)(2500/6000\ )]} - 1$$

$$N = \frac{\ln [(15.5)\ (0.5307) + 1]}{\ln [1.5307]} - 1$$

$$N = \frac{\ln(9.2259)}{\ln(1.5307)} - 1 = 4.2194 \quad \text{Call 5}$$

The reboiler is considered to be a theoretical stage, resulting in 5 – 1 = 4 theoretical stages needed in the 18-tray column shown in Figure 4.

Similar calculations for the 20-psig case produce an H/P_T of 3.005, the same V/L of 0.4167, an S of 1.2522, and 8 – 1 = 7 – 1 = 6 theoretical stages necessary for the separation.

Stripping Column Efficiency

Because of poor vapor-liquid contact, insufficient contact time, inadequate mixing, inhibited mass transfer, fluid flow effects, etc., a VLE system does not achieve equilibrium on each plate. The influence of system properties, fluid flow, and plate geometry is further discussed in Ref. **[63]**. It is therefore is necessary to define the synonymous *stage*, *plate*, *tray*, or *overall tower efficiency* [E_o] to compensate for our lack of knowledge of the approach to equilibrium

$$\text{Efficiency } (E_o) = 100 \times N / N_{\text{actual}} \text{ (number of actual trays, plates, or stages)} \qquad (32)$$

The Murphree plate efficiency [E_M] [64-65] applies to a single stage; it is defined as the ratio of the actual change in the average vapor composition accomplished on a tray to the theoretical change in average vapor composition if the vapor leaving the tray were in equilibrium with the liquid leaving the tray.[34] This measure has a more theoretical basis than E_o but is less convenient to use because it must be applied to each individual plate.[34]

By Correlation. Without specific experimental data at hand, one can make use of a correlation to estimate the efficiency (E_o) and from it the number of actual physical stages required to bring about the desired separation. The two correlations mentioned below by way of example are by no means exhaustive.

The O'Connell correlation for absorbers and strippers is found in graphical form in Ref. **[36]** and in several editions of Perry's Handbook.[66-68] This relates plate efficiency [in percent on the linear vertical y-axis to the product of several physical properties – liquid density (ρ_L) in lb_m/ft^3, the dimensionless equilibrium ratio ($K = H/P_T$), molecular weight (M), and liquid viscosity (μ_L) in centipoises (cP) – on the logarithmic horizontal axis. Since the liquid on the trays is almost pure water, properties of water have been used here as inputs.

Liquid density of 59.81 lbm/ft^3 has been obtained from Steam Tables[19] at 100 °C for the atmospheric case and 58.55 lbm/ft^3 at 126 °C for a design at 20 psig. Viscosities of liquid water (0.42 cP at 100 °C and 0.34 cP at 126 °C) were read from a nomograph in Perry's 6th edition.[66] The value of the grouping $\rho_L/[(K)(M)(\mu_L)]$ on the x-axis calculates to (59.81)/[(3.674)(18.02)(0.42)] = 2.1509511 for the atmospheric case and to (58.55)/[(3.005)(18.02)(0.34)=3.1801579 at 20 psig. These figures translate into efficiencies of 46 % and 50 % as read from the O'Connell curve.

The Drickamer-Bradford correlation,[36] which may or may not be valid for steam stripping, was derived from petroleum refinery hydrocarbon distillation columns and relies solely on the average viscosity of the liquid phase. That correlation yields plate efficiencies in the range of 40 % to 50 %. The EPA Design Manual for Stripping of Organics[34] cites a number of correlations where efficiency varies inversely with liquid viscosity.

Using Data. As noted in general in Perry's,[68] the best procedure to correct for tray efficiency would be to employ experimental data from a steam stripping column on a large enough scale to be representative of an actual commercial unit. Steam-stripping data for aniline in water from 3-in i.d.[30] and 10-in i.d.[31] laboratory columns are available in the open literature. Even better would be to have data from a pilot plant or a full-size aniline-water steam stripper; unfortunately, no such data could be found.

In one laboratory study using a 3-in i.d. column, overall plate efficiencies [E_O] varied from 61.8 % to 77.6 % for 14 runs over a range of operating conditions;[30] the simultaneously derived Murphree efficiencies [E_M] averaged over the 3 plates of the column were much the same (57.8 % to 83.4 %). The lowest efficiency [E_O] obtained was about 60 %. The investigator deduced that the mass transfer is controlled by gas-phase resistance. Perry's[66-68] indicates that gas-film-controlled systems may exhibit plate efficiencies of as much as 50 % to 100 %. Another study of stripping aniline from water solution, briefly reported, showed results using a Murphree efficiency [E_M] of 58 % for plates 5 through 10 of an 8-in i.d. 10-plate laboratory column.[31]

Number of Actual Plates Required

Based on the correlations and laboratory data considered, a conservative range of 40 % to 60 % overall efficiency has been selected to estimate the number of actual trays required. For the atmospheric pressure case, 10 trays at a 40 % efficiency or a tray efficiency of 4/18 = 22+ % would be needed. This is less than even the minimum 40 % efficiency above, and the design ought to prove satisfactory. If the column of Figure 4 is indeed to be operated at 20 psig, 6/ 0.4 = 15 actual plates at a 40 % efficiency, or 6/ 0.6 = 10 actual plates with a tray efficiency as high as 60 %, would be needed vs. the 18 shown. This latter design is the same as in the atmospheric case if one is willing to accept the higher estimated efficiency of 60 %.

If somehow this design turns out not to be satisfactory, the stripping steam, the number of trays, or both would have to be increased to avoid risking a reduction in aniline removal. Conversely, if it turns out that fewer actual trays are required, one would be able to cut back on the stripping steam and still maintain the desired separation.

Ref. **[57]** indicates that a steam stripper of less than 30 actual trays with steam at 5 % of the treated water should be adequate for most wastewater treatment applications.

SUMMARY

- Aniline is an important industrial chemical, whose chief means of commercial production is by catalytic hydrogenation of nitrobenzene.
- In this process, the water of reaction must be removed from the aniline product.
    - + Liquid aniline-water mixtures separate into an aniline-rich organic layer in equilibrium with a water-rich aqueous layer.
    - + These form a heterogeneous azeotrope upon distillation.
- This wastewater must be treated prior to discharge to minimize aniline contamination.
- Steam stripping reduces the load on downstream wastewater treatment facilities.
- Such a steam-stripper design is contained in a 1973 US EPA development document.
    - + Its basis was not clear, whether theoretical or derived from an operating unit.
    - + Pressure and temperatures listed on stripping-column appear to be inconsistent.
    - + The azeotropic composition appears to be incorrectly stated and the overhead concentrations impossible to achieve; the material balance has been modified.
- Open-literature VLE data are available at atmospheric pressure and several other conditions, but not the elevated pressure shown on the development-document column.
- The present paper employs the aforementioned data along with widely accepted thermodynamic principles to deduce a technical basis for the stripper design.
- For the modified material balance, this analysis has verified that the 18-tray column in the EPA development document can achieve the stated aniline bottoms concentration.
    - + The required number of <u>theoretical</u> plates calculated at 20 psig is greater than at atmospheric pressure.
    - + The number of <u>actual</u> plates needed is the same if one assumes a higher plate efficiency for the 20-psig case.
    - + A conservative range of 40 % to 60 % in overall plate efficiencies has been estimated from laboratory data and published correlations.

ABOUT THE AUTHOR

Robert G. Kunz was an environmental engineering manager at Air Products and Chemicals, Inc., Allentown, PA before retiring after 26+ years of service. He then joined Cormetech, Inc., Durham, NC as Technical Project Manager in support of sales and marketing efforts in the petroleum refining and petrochemical industries. He held engineering positions previously at Esso Research and Engineering Company, Florham Park, NJ and The M.W. Kellogg Company, New York, NY and is currently an independent environmental consultant. "Dr. Bob" has earned a BChE degree in Chemical Engineering from Manhattan College, a PhD in Chemical Engineering from Rensselaer Polytechnic Institute, an MS in Environmental Engineering from Newark College of Engineering, and an MBA from Temple University. He has contributed numerous publications to the technical literature, including topics in phase equilibria, and is a recipient of the Water Pollution Control Federation's Harrison Prescott Eddy Medal for noteworthy research in wastewater treatment. He is author or co-author of two books and holder of one U.S. patent. He is a member of the American Institute of Chemical Engineers (AIChE), the American Chemical Society (ACS), and the Air & Waste Management Association (A&WMA) and is a licensed professional engineer in several states.

ACKNOWLEDGEMENTS

The author is grateful to Roberta Kunz Fox, AIA, of fox^2 design for her assistance with Figures 3 and 4. The author would also like to thank the library staff at North Carolina State University, Raleigh, NC; Lehigh University, Bethlehem, PA; the Linda Hall Library, Kansas City, MO; and the University of Manchester, Manchester, U.K. for assistance in obtaining reference materials. Thanks also to AIChE, Danbury, CT for information on the Student Design Problem.

LITERATURE CITED

1. Kunz, R.G., "Environmental Calculations A Multimedia Approach," pp.41-58, Wiley, Hoboken, NJ (2009).
2. Kunz, R.G., "The Rayleigh Equation Revisited What to Do When Alpha (α) Isn't Constant," Paper 118b presented at the 2012 Spring AIChE Meeting, Houston, TX (Apr. 1-5, 2012).
3. Treybal, R.E., "Mass Transfer Operations," pp.266-267,313 McGraw-Hill, New York (1955).
4. EPA 440/1-73/009, "Development Document for Proposed Effluent Limitations Guidelines and New Source Performance Standards for the Major Organic Products Segment of the Organic Chemicals Manufacturing Point Source Category," pp.211-216, U.S. Environmental Protection Agency, Washington , DC (Dec. 1973).
5. Carlson, H.C. and A.P. Colburn, "Vapor-Liquid Equilibria of Nonideal Solutions Utilization of Theoretical Models to Extend Data," *Ind. Eng. Chem.*, **34**(**5**), 581-589 (May. 1942).
6. NIST Standard Reference Data Program, "Aniline," National Institute of Standards and Technology," Gaithersburg, MD (NIST website accessed on 7/10/2011 and 10/24/2013).
7. Wikipedia Encyclopedia, "Aniline," (http://en.wikipedia.org/wiki/Aniline accessed on 7/10/2011 and 10/24/2013).
8. Weast, R.C., editor, CRC Handbook of Chemistry and Physics, 58th ed., pp.C-112, D-197, and D-207, CRC Press, Inc., Cleveland, OH (1977).
9. Kroschwitz, J.I. and M. Howe-Grant, editors, "Kirk-Othmer Encyclopedia of Chemical Technology," 4th ed., Vol. 2, pp.426-441, 482-503, Wiley, New York (1992).
10. U.S. Dept. of Health and Human Services and U.S. Dept. of Labor, "Occupational Safety and Health Guideline for Aniline," Washington, D.C. (1992).
11. Maxwell, G.R., "Synthetic Nitrogen Products A Practical Guide to the Products and Processes," pp.361-371, Kluwer Academic / Plenum Publishers, New York (2004).
12. Washburn, E.W., editor-in-chief, et al., "International Critical Tables of Numerical Data, Physics, Chemistry and Technology," 1st ed., Vol. 3, p.389, McGraw-Hill, New York (1928).
13. Prutton, C.F. and S.H. Maron, "Fundamental Principles of Physical Chemistry," Revised Edition, pp.168-175, MACMILLAN, New York (1957).
14. McKetta, J.J. and W.A. Cunningham, editors, "Encyclopedia of Chemical Processing and Design," Vol. 3, pp.289-346, MARCEL DEKKER, INC., New York (1977).
15. Treybal, R.E., "Liquid Extraction," 1st ed., pp.5-11, 42-76, McGraw-Hill, New York (1951).
16. Dalos, A., I. Ország, and F. Ratkovics, "Liquid-Liquid and Vapor-Liquid Equilibrium Data for the System Aniline + Water in the Presence of NaCl, NaI, NH4Cl, and NH4I," *Fluid Phase Equilibria*, **11**, 91-102 (1983).

17. Groggins, P.H. editor-in-chief, "Unit Processes in Organic Synthesis," 5th ed., pp.59-67,93, 107-108,112-113,157,78,181,398,445,456-461, McGraw-Hill, New York (1958).
18. U.S. National Bureau of Standards Circular No. 19, "Standard Density and Volumetric Tables," 6th ed., p.46, U.S. Government Printing Office, Washington, DC (1924).
19. Meyer, C.A., R.B. McClintock, G.J. Silvestri, and R.C. Spencer, Jr., "1967 ASME Steam Tables," 2nd ed., pp.86-94 American Society of Mechanical Engineers, New York (1968).
20. Shreve, R.N. "The Chemical Process Industries," 2nd ed., pp.869-876, McGraw-Hill, New York (1956).
21. Shreve, R.N. "The Chemical Process Industries," 3rd ed., pp.813-818, McGraw-Hill, New York (1967).
22. McKetta, J.J. and W.A. Cunningham, editors, "Encyclopedia of Chemical Processing and Design," Vol. 4, pp. 247-288, MARCEL DEKKER, INC., New York (1977).
23. U.S. Environmental Protection Agency, "OPPT Chemical Fact Sheets Aniline Fact Sheet: Support Document (CAS No. 62-53-3)," EPA 749-F-95-002a, U.S. EPA Office of Pollution Prevention and Toxics, Washington, DC (Dec. 1994).
24. U.S. Environmental Protection Agency, "OPPT Chemical Fact Sheets Nitrobenzene Fact Sheet: Support Document (CAS No. 98-95-3)," EPA 749-F-95-015a, U.S. EPA Office of Pollution Prevention and Toxics, Washington, DC (Feb. 1995).
25. Wikipedia Encyclopedia, "Benzene," (http://en.wikipedia.org/wiki/Benzene accessed on 9/25/2011 and 10/24/2013).
26. U.S. Dept. of Health and Human Services and U.S. Dept. of Labor, "Occupational Health Guideline for Nitrobenzene," Washington, D.C. (Sept. 1978).
27. McKetta, J.J. and W.A. Cunningham, editors, "Encyclopedia of Chemical Processing and Design," Vol. 31, pp.146,165-166,182, MARCEL DEKKER, INC., New York (1990).
28. Kroschwitz, J.I. and M. Howe-Grant, editors, "Kirk-Othmer Encyclopedia of Chemical Technology," 4th ed., Vol. 17, pp.133-152, Wiley, New York (1996).
29. Kroschwitz, J.I. and M. Howe-Grant, editors, "Kirk-Othmer Encyclopedia of Chemical Technology," 4th ed., Vol. 4, p.78, Wiley, New York (1992).
30. Chiam, J.T., "A Study of the Steam Distillation of the Aniline-Water System," Master's Thesis, Chemical Engineering Department, University of Manchester, Manchester, U.K. (Oct. 1959).
31. Carey, J.S., J. Griswold, W.K. Lewis, and W.H. McAdams, "Plate Efficiencies in Rectification of Binary Mixtures," *Trans. Amer. Inst.Chem.Engrs.*, **30**, 504-510 (1933-34).
32. Browner, C.M., J.C. Fox, S. Frace, M.B. Rubin, and F. Hund, "Development Document for Final Effluent Limitations Guidelines and Standards for the Pharmaceutical Manufacturing Point Source Category," U.S. EPA Office of Science and Technology, Washington, DC (July 30, 1998).
33. Sittig, M., "Pollution Control in the Organic Chemical Industry," pp.76-79, NOYES DATA CORPORATION, Park Ridge, NJ (1974).
34. Shukla, H.M. and R.E. Hicks, Process Design Manual for Stripping of Organics," EPA-600/2-84-139, PB84-232628, pp. 2-3, 2-14, 2-22 to 2-24, 3-3, 3-9 to 3-10, 3-24, 5-2, 5-5, U.S. EPA Office of Research and Development, Cincinnati, OH (Aug. 1984).

35. Lowenbach, W., J. Schlesinger, and J. King, "Toxic Pollutant Identification: Nitrobenzene/Aniline Manufacturing," EPA 01/79-04, pp. 94-95, 113, U.S. EPA Office of Research and Development, Washington, DC (Jan. 1979).
36. Smith, B.D. "Design of Equilibrium Stage Processes," 3 ed., pp. 14,29-54,384,573-578, McGraw-Hill, New York (1963).
37. American Institute of Chemical Engineers, "Student Contest Problem for 1942," *Trans. Amer. Inst. Chem. Engrs.*, **38**, 653-718 (1942).
38. Wilson, R.Q., W.H. Mink, H.P. Munger, and J.W. Clegg, "Dehydration of Hydrazine by Azeotropic Distillation," *A.I.Ch.E.J.* **1**(**2**), 220-224 (June 1955).
39. LeCat, M. "Tables Azéotropiques: Tome Premier – Azéotropiques Binaires Orthobares," 2nd ed., p.177, UCCLE-BRUXELLES, Brussels, Belgium (July 1949).
40. Horsley, L.H. "Azeotropic Data: Advances in Chemistry Series No. 6," pp.9-10, 313, American Chemical Society, Washington, DC (1952); "Azeotropic Data – II: Advances in Chemistry Series No. 35," (with W.S. Tamplin), pp.12, 95 (1962); Azeotropic Data – III: Advances in Chemistry Series No. 116," pp.28, 597, 608 (1973).
41. Gmehling, J., J. Menke, J. Krafczyk, and K. Fischer, "Azeotropic Data," 2nd ed., Part 1, pp.93-94 and Part 2 List of References, pp.958,978-979,982,992,1034,1036,1057,1062, WILEY-VCH, Weinheim, Germany (2004).
42. Scatchard, G. and W.J. Hamer, "The Application of Equations for the Chemical Potentials to Partially Miscible Solutions," *J. Am. Chem. Soc.*, **57(10)**, 1805-1809 (1935).
43. Griswold, J., D. Andres. E.F. Arnett, and F.M. Garland, "Liquid-Vapor Equilibrium of Aniline-Water," *Ind. Eng. Chem.*, **32(6)**, 878-880 (June 1940).
44. Speakman, J.C., "The Surface Tensions and Partial Vapour Pressures of Aqueous Aniline Solutions," *J. Chem. Soc (London)*, **Part I**, 776-779 (1935).
45. Röck, H. and L. Sieg, "Messungen von Verdampfungsgleichgewichten mit einer modernisierten Umlaufapparatur," *Z. Phys. Chem., Neue Folge,* 355-364 (1955).
46. Shreinemakers, F.A.H., *Z. Phys. Chem. (Leipzig)*, **35**, 459-479 (1900).
47. Horyna, J., "Gleichgewicht Flüssigkeit-Dampf von Wässrigen Anilin-Lösungen unter Erhöhtem Druck," *Collect. Czech. Chem. Commun.*, **24 (10)**, 3253-3259 (1959).
48. Zharikov, L.K., K.S. Krylova, G.M. Kopylevich, *Zh. Prikl. Khim. (Leningrad)*, **48(6)**, 1249-1251 (1975).
49. Kolyuchkina, G.Ya., V.S. Timofeev, L.A. Serafimov, *Uch. Zap. Mosk. Inst. Tonkoi Khim. Tekhnol.*, **1**, 65-70 (1971).
50. Gmehling, J. and U. Onken, "Vapor-Liquid Equilibrium Data Collection Aqueous-Organic Systems," Vol. 1, Part 1, pp. 497-498, DECHEMA Chemistry Data Series, Frankfurt/Main, W. Germany (1977).
51. Gmehling, J., U. Onken, and W. Arlt, "Vapor-Liquid Equilibrium Data Collection Aqueous-Organic Systems (Supplement 1)," Vol. I, Part 1a, pp. 400-401, DECHEMA Chemistry Data Series, Frankfurt/Main, Federal Republic of Germany (1981).
52. Gmehling, J., U. Onken, and J.R. Rarey-Nies, "Vapor-Liquid Equilibrium Data Collection Aqueous-Organic Systems, (Supplement 2)," Vol. I, Part 1b, p. 320, DECHEMA Chemistry Data Series, Frankfurt/Main, Federal Republic of Germany (1988).
53. Dean, J.A., editor, "Lange's Handbook of Chemistry," 11 ed., Table 10-10 (pp. 10-31 to 10-45), McGraw-Hill, New York (1973).

54. Keenan, J.H. and F.G. Keyes, "Thermodynamic Properties of Steam Data for the Liquid and Solid Phases," 1st ed., 34th printing, p.14, Wiley, New York (Feb. 1962).
55. Stull, D.R., "Vapor Pressure of Pure Substances," *Ind. Eng. Chem.*, **39(4)**, Organic Compounds 517-540, Inorganic Compounds 540-550 (April 1947).
56. Bernauer M., V. Dohnal, A.H. Roux, G. Roux-Desgranges, and V. Majer, "Temperature Dependences of Limiting Activity Coefficients and Henry's Law Constants for Nitrobenzene, Aniline, and Cyclohexylamine in Water,"*J. Chem. Eng. Data*, **51**, 1678-1685 (2006).
57. Hwang, Y.L., G.E. Keller, II, and J.D. Olsen, "Steam Stripping for Removal of Organic Pollutants from Water. 1 Stripping Effectiveness and Stripper Design," *Ind. Eng. Chem. Res.*, **31**, 1753-1759 (1992).
58. Souders, M., Jr. and G.G. Brown "Fundamental Design of Absorbing and Stripping Columns for Complex Vapors," *Ind. Eng. Chem.*, **24**(**5**), 519-522 (May 1932).
59. Kunz, R.G. and W.F. Baade, "Predict Contaminant Concentrations in Deaerator-Vent Emissions," *Hydrocarbon Processing* (International Edition), **80**(**6**), 100-A to 100-O (June 2001).
60. McCabe, W.L. J.C. Smith, and P. Harriott, "Unit Operations of Chemical Engineering," 5th ed., p. 516, McGraw-Hill, New York (1993).
61. Smith, J.M., and H.C. Van Ness "Introduction to Chemical Engineering Thermodynamics," 2nd ed., pp.388-389, 392-397, McGraw-Hill, New York (1959).
62. Kunz, R.G. and W.F. Baade, "Vapor-Liquid Activity Coefficients for Methanol and Ethanol from Heat of Solution Data: Application to Steam-Methane Reforming," *Journal of Hazardous Materials*, **88**(**1**), 53-62 (2001).
63. Kister, W.M., "Effects of Design on Tray Efficiency in Commercial Towers," *Chem. Eng. Progr.*, **104**(**6**), 39-47 (June 2008).
64. Murphree, E.V., "Rectifying Column Calculations with Particular Reference to N Component Mixtures," *Ind. Eng. Chem.*, **17**(**7**), 747-750 (July 1925).
65. Lewis, W.K., Jr., "Rectification of Binary Mixtures Plate Efficiency of Bubble Cap Columns," *Ind. Eng. Chem.*, **28**(**4**), 399-402 (April 1936).
66. Perry, R.H., D.W. Green, and J.O. Maloney, editors, "Perry's Chemical Engineers' Handbook," 6th ed., pp.3-111 to 3-112, 3-251 to 3-252, 14-24, 15-15 to 15-17, 18-13 to 18-15, McGraw-Hill, New York (1984).
67. Perry, R.H., D.W. Green, and J.O. Maloney, editors, "Perry's Chemical Engineers' Handbook," 7th ed., pp.14-12, 14-34 to 14-38, McGraw-Hill, New York (1997).
68. D.W. Green and R.H. Perry, editors, "Perry's Chemical Engineers' Handbook," 8th ed., pp. 14-15, McGraw-Hill, New York (2008).
69. Sandler, S.I., "Chemical and Engineering Thermodynamics, pp. 346-348, 354, Wiley, New York (1989).
70. Hougan, O.A., K.M. Watson, "Chemical Process Principles Part II Thermodynamics," 2nd ed., 3rd Printing, pp.601-604, Wiley, New York (Mar. 1964).
71. Smith, J.M., H.C. Van Ness, and M.M. Abbott, "Introduction to Chemical Engineering Thermodynamics," 5th ed., p.86, McGraw-Hill, New York (1996).

APPENDIX A: Safety and Toxicology

Benzene, nitrobenzene, and aniline are dangerous chemicals and must be treated with great caution. Benzene is volatile, flammable, poisonous, a skin irritant, and readily absorbed by the body:[22] acute exposure to benzene vapor at high levels rapidly affects the central nervous system, resulting in severe shortness of breath, tightening of leg muscles, headache, dizziness, confusion, and hysteria; chronic low-level exposure over weeks or months brings on weakness and anemia arising from its effect on bone marrow. It is known to cause leukemia.[25] Maximum allowable workplace concentrations are 1 ppm (8 hr) and (15 min).[25]

Both the nitrobenzene and the aniline prepared from it are easily ingested by humans and animals, resulting in cyanosis and methemoglobinemia.[14,26] Cyanosis is a blue tinge to tongue, lips, and face caused by the body's conversion of the blood's oxygen-carrying hemoglobin to methemoglobin, which is unable to combine reversibly with oxygen. This results in permanent damage from oxygen not reaching the tissues of the body if methhemoglobin reaches too high a level.[9,14] It also affects the central nervous system, leading to headaches, fatigue, heart palpitations, vomiting, unconsciousness, and coma.[27] Cyanosis may be caused by contact with liquid nitrobenzene or aniline or by inhalation of the vapor or its absorption through the pores of the skin.[14] Symptoms usually appear after several hours following exposure to nitrobenzene;[26-28] effects of aniline exposure are usually observed more rapidly.[14]

Both nitrobenzene and aniline are combustible, but neither is classified as a volatile solvent.[14] Chronic exposure to nitrobenzene may cause anemia and damage to the spleen and liver.[27] Work clothes contaminated by nitrobenzene or aniline should be removed immediately, and exposed skin should be washed with soap and water.[10,27]

Aniline is combustible but neither explosive nor spontaneously flammable in air.[11] This chemical is highly toxic and may be fatal if swallowed, inhaled, or absorbed through the skin, lungs, or mucous membranes.[9,11] It is able to cause serious burns and eye irritation. Both vapor and liquid irritate one's eyes, the liquid causing damage to the cornea. Aniline can be absorbed through the skin in lethal amounts.[10] Adequate ventilation should be provided when handling. Skin contact should be avoided. Ref. **[9]** lists recommended personal protective equipment. Aniline is a suspected carcinogen,[11] because of significant evidence of increased bladder cancer in humans.[10]

The AIChE Annual Student Design Problem for 1942[14,37] advised that a plant visitor having an acute sense of smell not be able to detect either nitrobenzene or aniline.[14] Odor (shoe polish)[26] threshold for nitrobenzene in air is 1.9 ppm,[26] perceptible at 3.6 parts per billion (ppb).[24] For aniline, it is 1.1[10] to 1.3,[23] depending on source; the odor (dead fish) of aniline is said to be detectable as low as 0.5 ppm in air.[11] Maximum allowable worker-exposure limits are 1 ppm for nitrobenzene[26] and 2 ppm for aniline[10-11] both averaged over an 8-hr day. The plant must therefore be sealed tightly to prevent fugitive emissions and personnel exposure.

APPENDIX B: Vapor-Phase Non-Idealities Not Significant at This System's Pressure

Because the total pressure in this example case is somewhat elevated, the question arises whether fugacity (*f*) rather than pressure (p) should be used in vapor-liquid equilibrium (VLE) relationships that are valid at or near atmospheric pressure. Fugacity approaches pressure as pressure approaches zero. Depending on the system, it is said that vapor-phase non-idealities can usually be ignored below a pressure of 30 psia (15.304 psig).[36] Above such a pressure, the simple Ideal Gas Law (with standard nomenclature – P, pressure; V, Volume; n, number of moles; R, Universal Gas Constant in consistent units; and T, absolute temperature)

$$P\,V = n\,R\,T \text{ or } (P\,V)\,/\,(n\,R\,T) = 1 \qquad \text{(B-1)}$$

and relationships derived from it may not strictly apply. Fugacity (*f*) and the dimensionless fugacity coefficient (φ) may then have to be used instead to modify the pressure. The fugacity coefficient is that quantity which when multiplied by the pressure gives the fugacity

$$\varphi = f/p \qquad \text{(B-2)}$$

Since a pressure of 20 psig (34.696 psia, 1794.295 mmHg, 2.36 atm) is just beyond the suggested cut-off point, at least a brief check of its applicability to the present case is warranted. Generalized charts and tables for *f*/p [69,70] similar to Compressibility Factor (Z) charts and tables, are available to estimate φ from critical temperature (T_C) and critical pressure (P_C) through the principle of corresponding states.[71] Critical properties for pure aniline and water are listed in Table S-1 below.

TABLE S-1. Critical Constants for Aniline and Water

Constituent	Tc (°C)	Pc (atm)
Aniline	**426**	**52.4**
Water	**374.15**	**218.4**

From Perry's Chemical Engineers' Handbook 6 ed. pp. 3-111 & 3-112 [66]

Critical constants for their vapor-phase mixtures were estimated by the Pseudocritical Constant Method using Kay's Rules [69] of averaging for two components at molar concentrations of x_1 and x_2 (where $x_1 + x_2 = 1$)

$$T_C \text{ or } P_C \text{ (pseudocritical)} = x_1\,(T_{C1} \text{ or } P_{C1}) + (1 - x_1)\,(T_{C2} \text{ or } P_{C2}) \qquad \text{(B-3)}$$

Individual component values of *f*/p for aniline and water obtained from the fugacity coefficient generalized chart in Ref. **[69]** and the corresponding tables in Ref. **[70]** are no less than 0.95 across the entire range of concentrations. Likewise, mixture properties with the worst-case extreme values of Reduced Temperature T_r = (126 + 273.15) (°K)) / 699.15 °K = 0.571 and Reduced Pressure P_r = 2.36 atm/52.4 atm = 0.045 also produce values of *f*/p of 0.95 or higher. As a result, one is led to conclude that pressures rather than fugacities can be employed at 20 psig in thermodynamic relationships for this system without serious error.

APPENDIX C: Estimation of Activity Coefficients from Liquid-Liquid Solubility Data

Activity coefficients can be derived from liquid-liquid mutual solubility data. This is an accepted method to extend one's information.[5,14-15] The procedure makes use of the fact that the single vapor phase and both liquid phases are all in equilibrium with one another. For each component

Aniline Phase   Water Phase

$$y = \gamma_A \, x_A \, p^{vap} / P_T = \gamma_W \, x_W \, p^{vap} / P_T \quad (19)$$

The ratio p^{vap} / P_T, the same for both liquid phases, cancels out, and one is left with

$$y = \gamma_A \, x_A = \gamma_W \, x_W \quad (20)$$

Equation (20) allows the establishment of an activity-coefficient formulation valid at least in the vicinity of the liquid miscibility gap. One then writes such a formulation, e.g. the Margules Equations, for each component in each phase, as exemplified by water (component 1)

(water phase, W) $$\ln \gamma_{1W} = x_{2W}^2 \, [A_{1\text{-}2} + 2\, x_{1W} (A_{2\text{-}1} - A_{1\text{-}2})] \quad \text{(C-1)}$$

(aniline phase, A) $$\ln \gamma_{1A} = x_{2A}^2 \, [A_{1\text{-}2} + 2\, x_{1A} (A_{2\text{-}1} - A_{1\text{-}2})] \quad \text{(C-2)}$$

Now subtract Equation (C-2) from Equation (C-1) to obtain Equation (C-4), remembering that

$$\ln \gamma_{1W} - \ln \gamma_{1A} = \ln (\gamma_{1W}/\gamma_{1A}), \text{ and } \ln (\gamma_{1W}/\gamma_{1A}) = \ln (x_{1A}/ x_{1W}) \quad \text{(from Equation (20))} \quad \text{(C-3)}$$

Therefore $$\ln (x_{1A} / x_{1W}) = x_{2W}^2 \, [A_{1\text{-}2} + 2\, x_{1W} (A_{2\text{-}1} - A_{1\text{-}2})] - x_{2A}^2 \, [A_{1\text{-}2} + 2\, x_{1A} (A_{2\text{-}1} - A_{1\text{-}2})] \quad \text{(C-4)}$$

Similarly, $$\ln (x_{2A} / x_{2W}) = x_{1W}^2 \, [A_{2\text{-}1} + 2\, x_{2W} (A_{1\text{-}2} - A_{2\text{-}1})] - x_{1A}^2 \, [A_{2\text{-}1} + 2\, x_{2A} (A_{1\text{-}2} - A_{2\text{-}1})] \quad \text{(C-5)}$$

Equations are linear in $A_{1\text{-}2}$ and $A_{2\text{-}1}$. Therefore. values of x_{1A}, x_{1W}, x_{2A}, and x_{2W} from the mutual solubility data at a given temperature can be plugged in and terms collected to solve Equations (C-4) and (C-5) simultaneously for $A_{1\text{-}2}$ and $A_{2\text{-}1}$, perhaps using determinants. Values at 100 °C: $x_{1W} = 0.98525326$, $x_{2W} = 0.01474647$; $x_{1A} = 0.37342845$, $x_{2A} = 0.62757155$

Equations to be solved: $$(A_{1\text{-}2}) (-0.1006982) + (A_{2\text{-}1}) (-0.2929304) = -0.9728538$$

$$(A_{1\text{-}2}) (-0.1454620) + (A_{2\text{-}1}) (0.97748306) = 3.75083585$$

from which $A_{1\text{-}2} = -1.0478222$ and $A_{2\text{-}1} = 3.68130933$. The ratios from Equation (20) work out to be $x_{1A} / x_{1W} = \gamma_{1W} / \gamma_{1A} = 0.37800277$, and $x_{2A} / x_{2W} = \gamma_{2W} / \gamma_{2A} = 42.5566381$.

Unfortunately, these coefficients do not fit the Griswold et al. data very well except in the immediate vicinity of the miscibility gap.

Other sets of equations used to solve for Margules coefficients $A_{1\text{-}2}$ and $A_{2\text{-}1}$ and van Laar coefficients $A_{1\text{-}2}$ and $A_{2\text{-}1}$ are given in Ref. **[5]** and Ref. **[15]**.

351335 Kettle Reboiler Hydraulics

Abstract:

Kettle Reboiler Hydraulics

T.J. Cai, M.R. Resetarits* and L. Pless

Fractionation Research, Inc.

424 S. Squires St.

Stillwater, OK 74074

ABSTRACT

In the chemical processing industry, kettle reboilers are often considered troublesome. The problems are often related to excessive pressure drops caused by liquid entrainment from the kettle reboilers. Information and data regarding kettle hydraulics and entrainment is generally lacking. This paper presents pressure drop and entrainment data obtained from commercial size columns, with different hydrocarbon systems. The entrainment data were collected using gamma ray scanning of the kettle's vapor product line. An approach for calculating pressure drops across reboilers, and columns, is presented. Regarding the commercial size column, pressure drops are computed for various reboiler heat duties. Head losses of the liquid phase, from the bottom of the column to the reboiler, are estimated via correlations. All hydraulic calculations are compared with experimental data. A good agreement between the measured data and the calculated values was found except at high entrainment levels. A case study, from an industrial size kettle reboiler that had operational problems, will be discussed. A simple solution to some kettle problems involves an extension of the vapor return nozzle in the bottom of the column.

Key Words: Distillation, reboiler, pressure drop, hydraulics, entrainment

*Corresponding author. E-mail: cai@fri.org , Phone: 1-405-385-0354, Fax: 1-405-385-0357

DISTILLATION FUNDAMENTALS

By

Mike Resetarits

Anand Vennavelli

AY Ogundeji

Fractionation Research Inc.
424 S. Squires Street, Suite 200
Stillwater, OK 74074

2 April 2014

Presented at YP Tutorials: Distillation and Absorption Equipment
AIChE Spring Meeting, New Orleans, Louisiana

UNPUBLISHED

Abstract

In the chemicals processing industry, distillation is the most commonly employed separation technique. Different chemicals have different boiling points. After a simple mixture flash, the vapor product contains more light compounds than does the liquid product. A series of flash vessels can be employed to effect purifications of the light and heavy compounds. Similarly, a series of equilibrium stages arranged vertically in a distillation column can effect simple and difficult separations. Column operating pressure is an important variable as is column pressure drop. A reboiler is required at the bottom of a column to boil the heavy liquid. A condenser is required at the top of a column to condense the light vapor and create a reflux stream that flows down the column and cleans the upflowing vapor. When the reflux flow rate is divided by the top (liquid) product rate, the resultant number is called the reflux ratio. Generally, reflux requirements are a function of the relative volatilities of the light and heavy components. Pumps, pipes, valves, meters, thermocouples and other instrumentation are all essential parts of a distillation unit. Batch, azeotropic and extractive distillations are also employed industrially.

Keywords: Distillation Separation Boiling Flashing

Acknowledgments

This tutorial paper is dedicated to the thousands of global engineers and technicians who design, build, operate and maintain distillation columns.

Introduction

When a liquid mixture is fully boiled, to dryness, the resultant vapor will have the exact same composition as the starting liquid. However, when a liquid mixture is only partially boiled, the lighter compounds (usually those with lower molecular weights) will leave the liquid first. The resultant liquid will be richer in heavier compounds (usually those with higher molecular weights). A partial separation has occurred, but the resultant vapor will not be pure and the resultant liquid will not be pure.

Now imagine a mixing chamber or a vertical column wherein a light liquid stream encounters a heavy vapor stream. Also imagine that the streams are not at compositional equilibrium. Some of the light components in the liquid stream might leave it and enter the vapor stream. Similarly, some of the heavy components in the vapor stream might leave it and enter the liquid stream. If, as the streams encounter each other, compositional equilibrium is achieved, it is said that an "equilibrium stage" or a "theoretical tray" of contacting has occurred. Such a compositional equilibrium will be achieved if the liquid and vapor phases contact each other intimately for a sufficient amount of time. When the equilibrium is attained, further contacting or mixing does not yield additional mass transfer.

Distillation begins with the boiling of a mixture. A distillation column can be viewed as a series of equilibrium stages arranged vertically. A liquid stream flows down from stage to stage and encounters a vapor stream that flows up from stage to stage. As compositional equilibria are attained in each of those stages, the downward-flowing liquid stream becomes heavier and more pure as the upward-flowing vapor stream becomes lighter and more pure. Distillation columns are capable of starting with a mixture and turning it into a very pure top product rich in light compounds, and a very pure bottom product rich in heavy compounds.

Discussion

Boiling Points

The science of distillation begins with the study of boiling points. The temperature at which a liquid boils depends upon the pressure that the top liquid surface is subjected to. When that pressure is atmospheric pressure, the boiling point is referred to as the "normal boiling point." **Table 1** gives the normal boiling points of several compounds of the "aromatics family" of chemicals. All of these compounds exist as liquids at room temperature and pressure. For distillation separations to occur, mixtures of these compounds need to be partially boiled. **Table 2** includes the normal boiling points of the main compounds that are found in a natural gas processing plant. All of these compounds exist as vapors at room temperature and pressure. For distillation separations to occur, mixtures of these compounds need to be partially condensed. Distillation always requires the presence of two fluids, a liquid and a vapor. In both of these tables, higher molecular weight compounds have higher boiling temperatures.

Bubble and Dew Points

The separation of benzene (B) from toluene (T) is an important industrial separation. It is usually effected via distillation. At atmospheric pressure, pure benzene boils at 80°C; pure toluene, at 111°C. A mixture of benzene and toluen e begins to boil across a wide range of temperatures; depending on the mixture composition. The lower curve of **Figure 1** gives the bubble point temperatures. Very unfortunately, taking any mixture of benzene and toluene, and raising its temperature to 80°C or higher will not simply boil off all of the benzene, leaving pure toluene behind. When boiling most mixtures, the lighter component indeed boils first, but the resultant vapor includes very significant quantities of the heavier component.

Referring back to **Figure 1**, regarding B and T, the lower curve represents mixture bubble points. The upper curve represents mixture dew points. The area between the curve is the so-called "2-phase region."

Assume a cold 50/50 BT mixture (Point m_L in **Figure 1**) and assume that its temperature is being raised continuously. It will not begin to boil at 80°C; instead the bubble point will prove to be 92°C (Point b_1 in **Figure 1**). The composition of the first bubble of vapor is represented by Point d_1. It is no surprise that d_1 is high in B content. As the 50/50 2-phase mixture is heated further, when the temperature reaches 95.5°C, rough ly half of the mixture will be vaporized. The vapor composition is represented by d_2; the liquid, b_2. The horizontal lines in **Figure 2** are called "tie lines." At this point, some separation of the two compounds has indeed been achieved, but the two phases are far from being pure. With additional heating of the mixture, the dew point (d_3) will be achieved at a temperature of 99°C. The v ery last droplet of liquid will have the composition represented by b_3.

If the mixture is heated all the way to 110°C, the resultant vapor (Point m_V on the graph) will have the same compositions as the initial liquid (m_L).

As the BT mixture was being heated, regardless of the temperature, there never came a point where the B or the T approached "purity" – not in the liquid phase and not in the vapor phase.

Flash Distillation

Figure 2 summarizes a simple flash distillation, again using the example of a 50/50 B/T liquid mixture. Imagine that the liquid mixture is heated to a temperature of 95.5°C with a subsequent let-down to atmospheric pressure. **Figure 2** shows that a modest amount of separation has occurred. In fact, the after-flash vapor and liquid compositions are represented by Points b_2 and d_2 in **Figure 1.**

Figure 3 shows a series of flash distillations, where the resultant liquids are flashed again and again at ever-increasing temperatures. After 5 flashes, a T purity of 91% has been achieved. (For an industrial benzene column, the T purity target is closer to 99% than 90%.) Most importantly, **Figure 3** shows that a series of operations, in this case flashing, can be used to separate and purify chemical compounds that have different boiling points.

In **Figures 1**, **2** and **3**, boiling was always effected at atmospheric pressures, with boiling temperatures in the neighborhood of 80 to 110℃. Low pressure steam at 2.5 bara (36 psia) and 127℃ could be employed in the boilers (heat ex changers).

Vapor Pressures and Relative Volatilities

Generally, with a pure liquid, molecules are constantly escaping the top liquid surface and are becoming vapor. The pressure that those evaporated molecules exert is called the "vapor pressure" of the pure liquid. When a pure liquid is heated, the vapor pressure of the liquid increases. When the vapor pressure of the pure liquid equals atmospheric pressure, the "normal boiling point temperature" is achieved and evaporation becomes violent, i.e., the pure liquid "boils."

Different liquid compounds have different vapor pressures, and thus, different normal boiling points. Generally, smaller molecules have higher vapor pressures. When a liquid-vapor mixture is enclosed and heated, as the pressure of the vessel rises, the boiling point temperature also rises. If a vacuum is created above that mixture, the boiling temperature of the mixture becomes less than the normal (atmospheric pressure) boiling temperature. As an example, water can boil across a very broad range of temperatures. A very simple laboratory experiment can show how easily water can boil at room temperature. **Table 3** gives details regarding how pressure affects the boiling temperature of water.

Consider a mixture of small(1) and large(2) molecules. When that mixture is heated, and boiled, the smaller molecules will usually have higher vapor pressures and will escape the liquid phase first. The vapor phase composition will have more lighter compounds than the leftover liquid phase; a partial distillation separation will have occurred.

The ratio of the vapor pressures (VP) of different chemical compounds is often called the "relative volatility" and is often represented by "∝". The following simple equations apply:

$$\alpha_{12} = \frac{VP_1}{VP_2}$$

$$y_1 = \frac{\alpha_{12} x_1}{1 + (\alpha_{12} - 1) x_1}$$

$$x_1 + x_2 = 1$$

$$y_1 + y_2 = 1$$

where

x_1 = composition of Compound 1 in liquid phase

x_2 = composition of Compound 2 in liquid phase

y_1 = composition of Compound 1 in vapor phase

y_2 = composition of Compound 2 in vapor phase

For many chemical compounds, their vapor pressures are known across broad ranges of temperatures. These vapor pressures can be employed within the above equations to determine the separations that are achievable by boiling mixtures of compounds.

Figure 4 provides the vapor pressures of benzene and toluene across a broad range of temperatures. Notice how the graphed data converge at higher temperatures. At higher temperatures, the relative volatilities of binary compounds do tend to decrease, i.e., the separations become more difficult. **Figure 5** provides the vapor pressures of propylene and propane.

Figure 6 illustrates the ramifications of relative volatilities (essentially, vapor pressure ratios) on vapor-liquid equilibria. Notice that at α =1.1, the composition of component 1 in the vapor and liquid phases are just about equal. At α =3.0, the composition of the light component in the vapor phase is much higher than it is in the liquid phase.

Column Configuration and Ancillary Equipment

Figure 7 is a diagram of a typical distillation unit including the column and the heat exchangers. Column diameters can be as small as 2 inches or as large as 50 feet. Column heights can range from 12 inches to 350 feet. The inside of the column can be equipped with trays or packings – to mix the countercurrent liquid and vapor streams.

Figure 7 shows a feed stream entering the column about half-way up; this is not unusual. The feed stream can be a liquid, a flashing liquid or a vapor. For the sake of this discussion, it will be assumed that the feed stream is a liquid. The feed enters the column and joins the down-flowing reflux liquid stream. As the combined liquid stream flows downward, mass is exchanged with an upward flowing vapor. At the bottom of the column, most of the liquid stream is boiled in a "reboiler". The resultant vapor stream is called the "boil-up." In **Figure 7**, steam is being used to boil the liquid. A portion of the bottom liquid is not reboiled and is the final "bottoms product." All of the fluids that reach the bottom of the column are rich in heavy components, i.e., chemicals with high boiling points. The vapor leaving the reboiler is similarly heavy. As this heavy vapor flows upwards through the column, it encounters a lighter liquid. There is a compositional disequilibrium between the phases which causes light components to leave the liquid and heavy components to leave the vapor. In fact, as the vapor stream rises past the feed point, it becomes richer in lighter components. By the time that the vapor reaches the top of the column it is quite light. In **Figure 7**, the entire vapor stream enters a condenser which totally converts the vapor stream to a liquid. Some of this liquid is returned to the column as a "reflux" and some becomes the "overhead product." All of the fluids near the top of the column are quite light, including the liquid reflux stream. As this light reflux stream flows down the column it encounters a vapor stream that is relatively heavier. A compositional disequilibrium exists and heavy components in the vapor stream leave the vapor stream. Light components in the liquid stream leave the liquid stream. Some people regard the reflux stream as "clean'; they regard the reflux stream as cleaning the upflowing vapor stream.

In **Figure 7**, cooling water is employed as the heat sink medium in the condenser. Air (in a fin fan condenser) or refrigerants can also be used. Condensers are not actually positioned at the tops of columns. They are positioned about 30 feet above ground level. In fact, reflux drums are typically positioned underneath most condensers to collect condensates. Pumps are required to move the reflux liquids back to the tops of the columns. **Figure 7** depicts a typical distillation unit. There are many variations.

Reflux Requirements

"Reflux ratio" is defined as the reflux flow rate divided by the overhead product rate. Difficult separations, with low relative volatilities, require higher reflux ratios. Higher reflux ratios require more boil-up in the reboilers and higher steam rates and higher cooling water rates in the condensers and, overall, higher energy consumption rates. Additionally, higher reflux ratios lead to increased vapor and liquid traffics inside the column and the column diameter must be increased when the column is being designed. Using the benzene-toluene binary as an example, **Figure 8** shows that taller columns, with more theoretical stages, can lead to reduced reflux ratios. **Figure 8** also shows that as top and bottom purity targets are increased, so are the needs for additional column height and refluxing.

Atmospheric Distillation

Whenever possible, distillation columns are designed for, and operated at, near-atmospheric pressure. **Table 1** provided the atmospheric pressure boiling points of several compounds of the "aromatics family." When these particular chemicals are distilled, low-grade low-temperature steam can be employed in the reboilers. At the same time, cooling water or air can be used in the condensers to effect liquefactions. There are other occasions, and chemicals, where the associated distillation columns need to be run at high pressures or at vacuum conditions. Such columns are more expensive to build and operate. High pressure columns require many extra safety considerations.

Table 1 shows that the boiling points of benzene (B) and toluene (T) are quite different. Indeed, these compounds can be separated and purified via boiling – and distillation. Benzene columns are quite common, globally, and are operated at near-atmospheric pressures. **Table 1** includes other aromatic compounds which will be referred to later.

High Pressure Distillations

Distillation separations cannot occur with chemical compounds that are totally in the liquid state, or totally in the vapor state. Liquid chemicals need to be partially vaporized in reboilers. This is usually "easy" and can be effected using steam, hot water, etc. Vapor chemicals need to be partially or totally condensed in condensers and this often creates "difficulties." Cooling water and air are the most common heat sinks for hot chemicals. Cooling water and air are commonly assumed to be at 100°F = 38°C. The chemical compou nd vapors that they condense must therefore be at a higher temperature, like 130°F. As an alternative, a refrigerant could be employed as the heat sink fluid, but refrigeration is generally expensive and is generally avoided.

See **Table 2**, and in particular, the boiling points at atmospheric pressure, of pure methane and pure ethane. A natural gas plant demethanizer separates methane from ethane using distillation. If such a demethanizer is operated at atmospheric pressure, the methane-ethane mixture would need to be cooled to temperatures lower than -100°C to create a liquid phase. Instead, demethanizers are typically operated at pressures like 500 psia, to facilitate liquefaction. At pressures like 500 psia, the bubble and dew points of methane-ethane are high enough that cooling water and air can be used in the distillation column condenser.

Referring again to **Table 2**, this time noting the (atmospheric pressure) boiling temperatures of propylene and propane. The separation of propylene from propane is an important global process for polypropylene production. If a propylene-propane distillation column were run at atmospheric pressure, a (costly) refrigerant at about -50°C would be required as the heat sink in the overhead condenser.

Figure 5 provides vapor pressure data for these C_3 compounds. Most propylene-propane columns run at pressures like 300 psia. At such pressures, per **Figure 5**, these chemicals condense at temperatures like 130°F. Cooling water or air (at 100°F) can be used to effect these liquefactions in overhead condensers.

Vacuum Distillation

The polymer known as polystyrene is created using relative pure styrene monomer. Plants that produce styrene include distillation columns that separate ethylbenzene (EB) from styrene (S). Refer back to **Table 1** and note the atmospheric-pressure boiling points of pure ethylbenzene and pure styrene, 136 and 146°C, respectively. If an EB/S column were operated at atmospheric pressure, those chemicals would be subjected to such temperatures. Very unfortunately, such temperatures caused styrene monomers to combine repeatedly to form polymer molecules and those molecules solidify. Such atmospheric-pressure EB/S columns would plug up (foul) quickly. Instead, EB/S columns are actually operated at very deep vacuums, like 100 mmHga, to reduce the column temperatures and virtually eliminate the renegade polymerizations and the plugging.

There are other reasons why some distillation columns are operated at sub-atmospheric pressures and relatively low temperatures. At atmospheric pressures and higher temperatures, some molecules crack apart instead of boil. As an example, the second major distillation column in most refineries is typically run at vacuum pressures (and lower temperatures) to prevent large oil molecules from cracking.

Unpractical Distillations

Some distillation separations are quite difficult, but they are performed industrially nevertheless. Refer again to **Table 1**. The boiling points of EB and p-xylene are quite close, and yet, this distillation is performed industrially, but rarely. This is also true of the distillation of p- and m-xylene from o-xylene. Conversely, however, the boiling points of p-xylene and m-xylene are so close (0.8°C, actually) that this distillation is n ever performed industrially. As it turns out, the melting points of these two compounds are amazingly different (+13 and -47°C, respectively) and crystallization can be employed to separate them.

Closing

Thirty years ago, Dr. James Fair of the University of Texas at Austin, predicted that distillation would be a popular separation method for many decades. He was correct!

Figure 1: Benzene - Toluene Bubble and Dew Points; Pressure = 14.7 psia

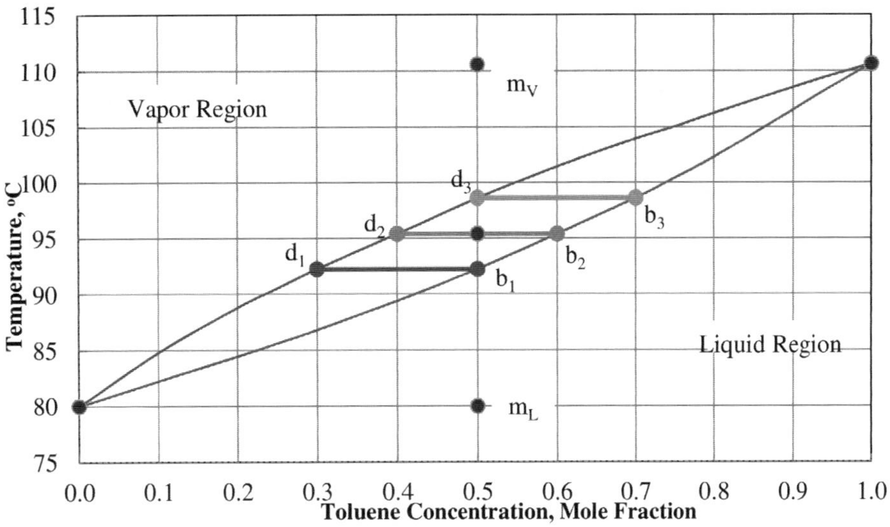

Figure 2: Flash Distillation

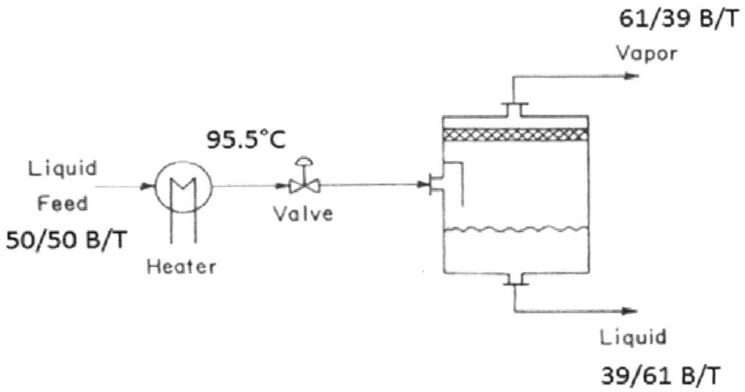

Figure 3: Successive Flash Distillations

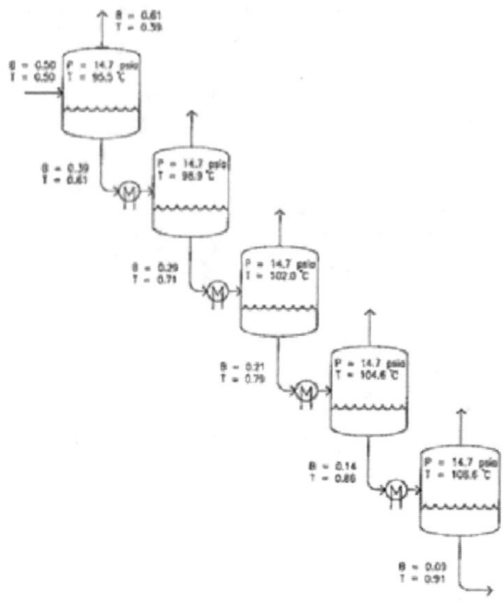

Figure 4: Vapor Pressures of Benzene and Toluene

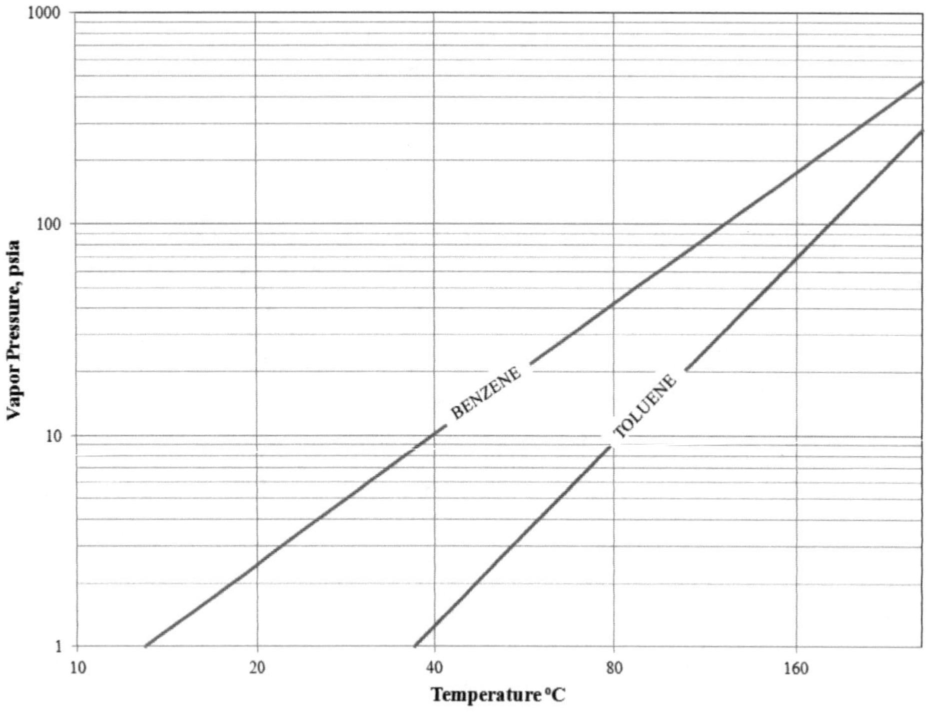

Figure 5

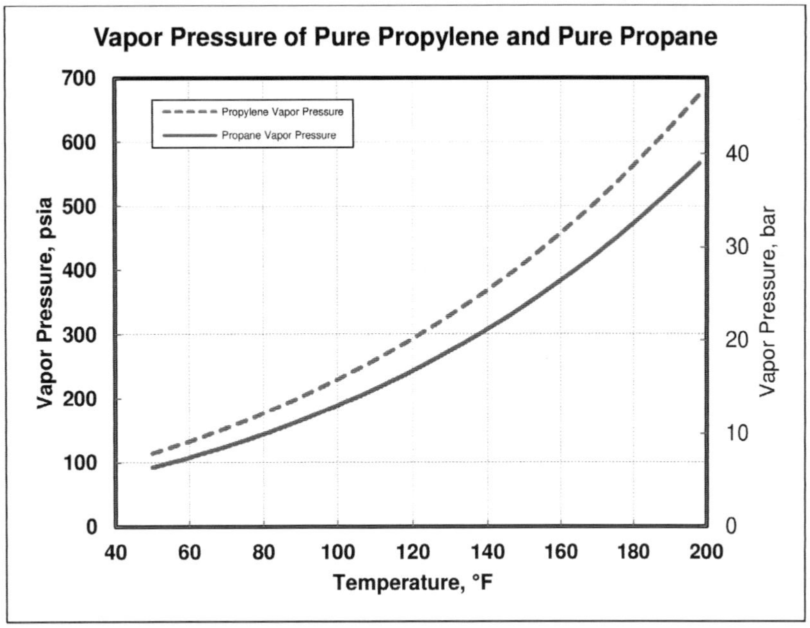

FIGURE 6 Vapor (y_1) versus liquid (x_1) concentration as a function of relative volatility.

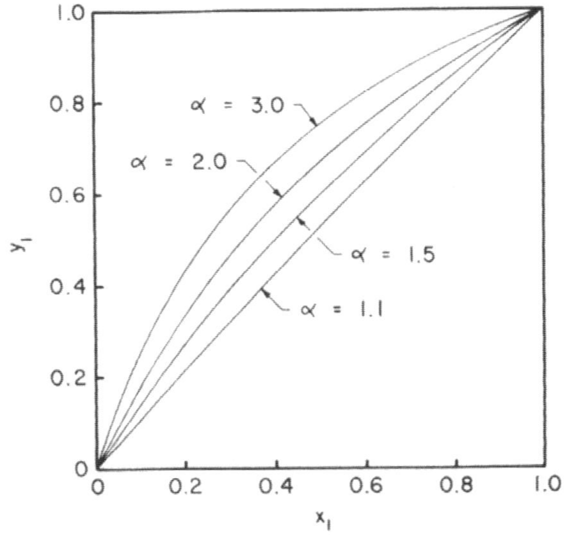

FIGURE 7 Schematic representation of a distillation column.

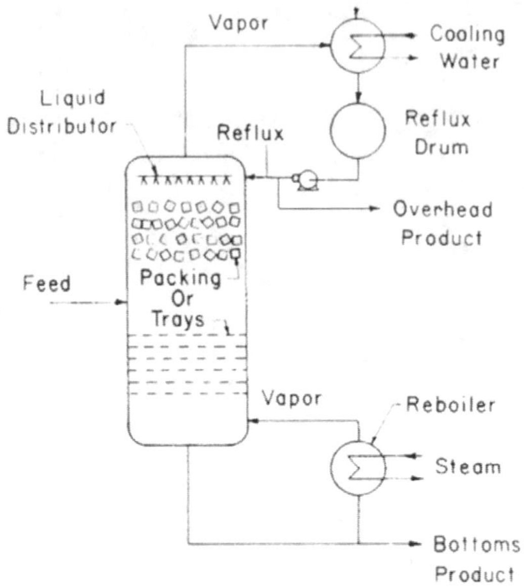

FIGURE 8 Theoretical stages versus reflux ratio (benzene-toluene at 1.0 bar). x_D, mole fraction benzene in overhead; x_B, mole fraction toluene in bottoms.

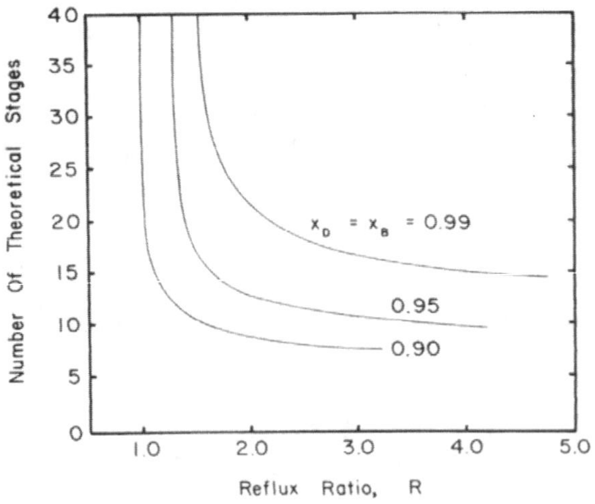

Table 1
Boiling Temperatures at Atmospheric Pressure

	Boiling Point, °C
Benzene (B)	80
Toluene (T)	111
Ethylbenzene (EB)	136
p-Xylene	138
m-Xylene	139
o-Xylene	144
Isopropylbenzene	153
Styrene (S)	146

Table 2
Boiling Temperatures at Atmospheric Pressure

	Boiling Point, °C
Methane	-161
Ethylene	-104
Ethane	-89
Propylene	-48
Propane	-42
Isobutane	-12
n-Butane	-1
n-Pentane	+36

Table 3
Water Boiling Temperatures

Pressure, psia	Temp, °C	Temp, °F
1.0	39	102
14.7	100	212
100	164	328
600	252	486

REFERENCES

1. Kister, H.Z., Distillation Design, McGraw-Hill, NY, NY, 1992.
2. Kister, H.Z., Distillation Operation, McGraw-Hill, NY, NY, 1990.
3. Lockett, M.J., Distillation Tray Fundamentals, Cambridge Univ. Press, NY, NY, 1986.
4. Strigle, R.F., Packed Tower Design and Applications – Random and Structured Packing, 2nd Edition, Gulf Publishing Co., Houston, TX, 1994.
5. Resetarits, M.R., Agnello, J., Lockett, M.J., and Kirkpatrick, H.L., "Retraying Increases C3 Splitter Column Capacity," Oil and Gas Journal (86), June 1988, pp. 54.
6. Resetarits, M.R. and Miller, R.J., "Increase Demethanizer Efficiency and Capacity," Hydrocarbon Processing, October 1991, pp. 79-81.
7. Resetarits, M.R., Navarre, J.L., Monkelbaan, D.R., Hangx, G.W.A. and van den Akker, R.M.A., "Trays Inhibit Foaming," Hydrocarbon Processing, March 1992, pp. 61-64.
8. Resetarits, M.R., Riter, J.B. and Weiland, R.H., "Distillation Technology: Cocurrent Flow Trays in Distillation Columns," Hydrocarbon Engineering, Vol. 7, No. 9, September 2002, pp. 85-88.
9. O'Connell, H.E., Plate Efficiency of Fractionating Columns and Absorbers, Trans. Am. Inst. Chem. Engrs., 42, 741, 1946.
10. Resetarits, M.R., Miller, R.J., Navarre, J.L., Linskeseder, M. and Reich-Rohrwig, P., "New Enhanced Capacity MD Tray Debottlenecks Deethanizer," IChemE Distillation and Absorption Conference, Birmingham, U.K., September 7, 1992.
11. Resetarits, M.R., Monkelbaan, D.R. and Miller, R.J., "Trays in Rocking Services," IChemE Distillation and Absorption Conference, Birmingham, U.K., September 7, 1992.
12. Resetarits, M.R., Urbanski, N.F., and Xu, Z.P., "Column Revamps," AIChE Spring Meeting, Chicago, IL, February 16, 1998.
13. Resetarits, M.R. and Pappademos, N., "Factors Influencing Vapor Crossflow Channeling," AIChE Annual Meeting, Reno, NV, November 8, 2001.
14. Resetarits, M.R. and Lockett, M.J., "Distillation," Encyclopedia of Physical Science and Technology, Vol. 4, 2001.
15. Resetarits, M.R. and Ogundeji, A.Y., "On Distillation Tray Weir Loadings," AIChE Spring Meeting, Tampa, FL, April 27-29, 2009.
16. Resetarits, M.R., Cai, T.J. and Whiteley, J.R., "On Distillation Tray Entrainment", AIChE Spring Meeting, Chicago, IL, March 2011.
17. Chambers, S., Cai, T.J., Vennavelli, A.N., Ogundeji, A.Y., Resetarits, M.R. and Redden, J.C., "Effect of Tray Spacing on Tray Performance," AIChE Spring Meeting, April 3, 2012.

Trays for Distillation and Absorption

By

AY Ogundeji
Mike Resetarits
Tony Cai

Fractionation Research Inc.
424 S. Squires Street, Suite 200
Stillwater, OK 74074

2 April 2014

Presented at YP Tutorials: Distillation and Absorption Equipment,
AIChE Spring Meeting, New Orleans, Louisiana, April 2, 2014

Abstract

Distillation columns are essential to the chemical processing industries. They need to be properly designed and operated. Column internals, such as trays and packings, are essential to column performance. If column internals malfunction, the entire plant may suffer. Among column internal, trayed columns are widely used in distillation and absorption applications. This presentation will discuss the ranges of the liquid and vapor loads that a trayed column can handle. It will review the systems where trayed columns are particularly applicable. A brief history of trays and trayed columns will be presented. New generations of high performance trays, including augmented crossflow, counter current flow and cocurrent flow trays, will be discussed. The performance differences between trays and packings will be summarized. Column designs that provide high energy efficiencies will be briefly discussed. Extensive video footage of tray operations will be shown, from commercial size columns with various distillation systems ranging from deep vacuum to high pressure.

Keywords: Distillation Absorption Tray Flooding

OVERVIEW

Distillation is a process of separating mixtures into two or more products based on differences in their boiling points in a boiling mixture. The distillation process has been around for hundreds of years and, it is one of the most widely used processes in the industries.

Absorption is used to separate gas mixtures, remove impurities, or recover valuable chemicals. In absorption, there is a transfer of one or more species from the gas phase to a liquid absorbent. The species transferred to the liquid absorbent are referred to as solutes or absorbates.

Distillation and absorption employ trays and packings as column internals. This paper focuses only on tray internals. These column internals are used as phase contacting devices. There are different types of trays. The table below shows a rough comparison of the common tray types.

Type	**Sieve Tray**	**Valve Tray**	**Bubble-Cap Tray**	**Dual-Flow Tray**
Capacity	High	High	Moderate	Very high
Efficiency	High	High	Moderate	Lower than others
Pressure drop	Moderate	Moderate	High	Low to moderate
Turndown	About 2:1	About 3:1	Highest	Low
Cost	Low	About 20% higher than sieve	High. About 2-3 times more than sieve	Low

GENERAL

For distillation, a liquid phase flows countercurrently with a <u>vapor</u> phase, while for absorption, a liquid phase flows countercurrently with a <u>gas</u> phase. This is shown in **Figure 1**.

Figure 2 shows a drawing of a simple distillation column with trays. The figure shows that the column contains 20 trays. Feed and reflux streams are introduced into Tray 10 and Tray 20, respectively. Vapor is introduced into the bottom of the column via the "From Reboiler" nozzle.

The vapor rises and mixes with the liquid flowing down the column. The liquid flowing downward exits the column via the "To Reboiler" piping while the upward flowing vapor exits the column via the "To Condenser" nozzle.

Figure 3 shows the different tray configurations. Multi-pass trays are usually employed when liquid rates are too large for a 1-pass tray.

GENERATION OF TRAYS

There are different generations of trays. The generations include crossflow, augmented crossflow, countercurrent and cocurrent flow trays. The first generation is simple crossflow trays which include bubble-cap, sieve and valve trays, and which have the least capacities. The cocurrent flow trays (Shell Consep, Koch-Glitsch UltraFrac and UOP SimulFlow trays), which are the latest generation and have the highest capacities.

TRAY PARTS

Figure 4 shows a drawing of a 1-pass crossflow tray. It shows different parts of a tray. Liquid flows across each tray deck, over an outlet weir, and into a downcomer. The outlet weir purpose is to build a liquid head on the tray deck. A receiving pan collects and redirects the liquid onto the next deck through the downcomer clearance. The vertical distance between successive trays is called "spacing". The inner side of the downcomer is a vertical metal plate called the curtain. The column wall is also known as the shell.

Figure 5 shows a simplification of 1-pass crossflow tray hydraulics. Liquid enters the tray deck from the downcomer of the tray above by gravity. The liquid on the tray is aerated with vapor rising through the deck perforations to form a froth. The froth flows across the tray deck, over the outlet weir and into the downcomer where the vapor disengages. The distance between the top of the outlet weir and the top surface of the liquid flowing over the weir is called crest.

Figures 4 & 6 can be used to compare conventional and augmented crossflow trays, respectively. Augmented crossflow trays have truncated downcomers. That is, the downcomers have bottoms, and the bottoms are perforated. Augmented crossflow trays also have perforated decks below the bottoms of the downcomers as shown in Figure 6. Conventional crossflow trays do not have truncated downcomers. The higher open area of augmented crossflow trays yields higher capacities than conventional crossflow trays.

Figure 7 shows different types of tray decks. The choice of deck usually depends on the application. Vapor-liquid handling capacities and pressure drops of decks differ. For example, bubble-cap decks turn down better than sieve and valve decks. Bubble cap trays have higher pressure drops than sieve and valve trays. Moving valve decks have larger rangeabilities than simple sieve decks. Push valves reduce froth heights and increase capacities.

FLOODING

The performance of trayed columns is determined by many factors that include flooding, weeping and dumping. Flooding is excessive accumulation of liquid inside a trayed tower. When trays flood, there is sudden increase in column differential pressure and decrease in mass transfer efficiency. There are many different mechanisms that cause trays to flood. These mechanisms include choking, downcomer backup, spray entrainment and froth height flooding. More details about the mechanisms are as follows:

Choke Flooding

It occurs when the liquid flow rate increases to a point where the velocity of the biphase in the downcomer exceeds a certain limit. At this point, the biphase cannot even fit into the downcomer mouth and the biphase cannot be transported to the tray below. It happens when the downcomer top area (mouth) is too small to accommodate the aerated liquid above the outlet weir. A sketch is provided in **Figure 8**.

Downcomer Backup Flooding

It occurs because of high tray pressure drop, high liquid height on the tray and large downcomer clearance losses. When the backup of biphase in the downcomer exceeds the tray spacing (plus the outlet weir height), liquid accumulates on the tray above.

Spray Entrainment Flooding

It happens at low liquid rates when vapor velocity is increased to a point where the liquid on the tray is dispersed by the rising vapor into liquid droplets which are entrained to the tray above.

Froth Height Flooding

It occurs at high liquid and vapor rates. If the tray spacing is 24 inches or less, as vapor rate increases, a point is reached where the froth height equals the tray spacing. Liquid accumulates on the tray above.

Weeping

This phenomenon occurs at low vapor rates. At low vapor rates, when the pressure exerted by the vapor is insufficient to hold up the liquid on the deck, liquid starts to leak through the deck perforations. This liquid backmixing yields reduced mass transfer efficiency.

Dumping

It occurs at very low vapor rates. Excessive weeping will lead to dumping. With dumping, all the liquid on the deck will weep through the perforations. In this condition, no liquid flows over the outlet weir.

CLOSING

Roughly half of old and new columns contain trays; the other half, packing. Fundamental research regarding these contacting devices has been performed at Fractionation Research Inc. (FRI) since 1954.

REFERENCES

- H.Z. Kister, Distillation Design, McGraw-Hill, 1992.
- J.D. Seader, E.J. Henley and D.K. Roper, Separation Process Principles, John Wiley & Sons, 2011.
- M.J. Lockett, Distillation Tray Fundamentals, Cambridge University Press, N.Y., N.Y. 1986.

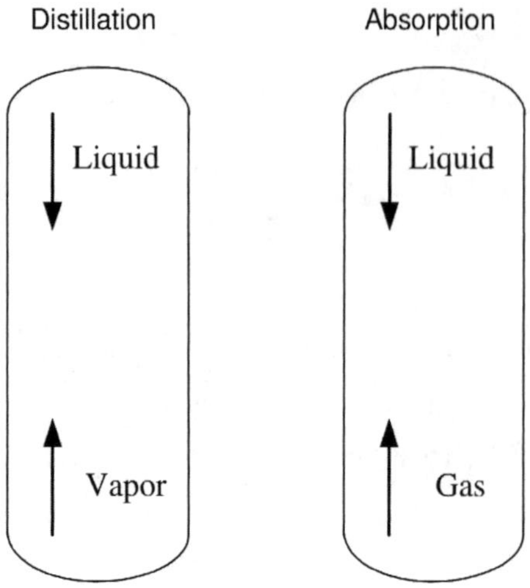

FIGURE 1.

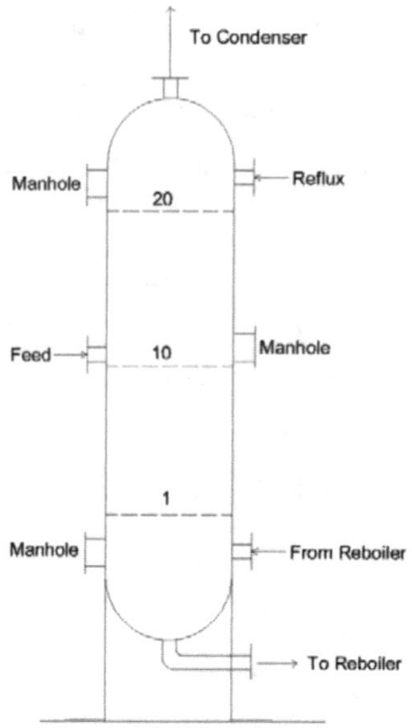

FIGURE 2. Simple Distillation Column with Trays

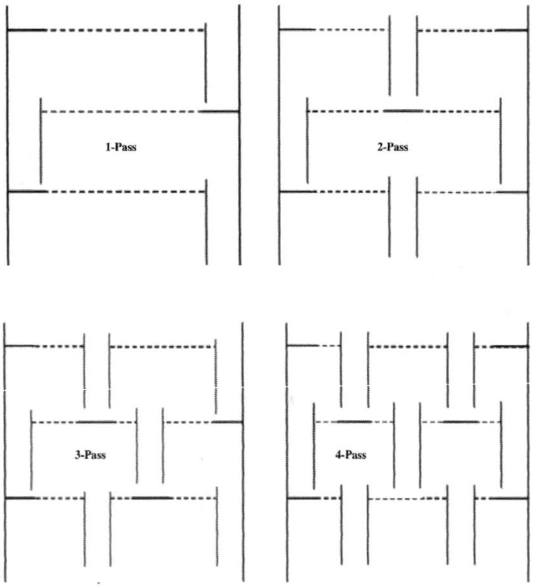

FIGURE 3. Crossflow Trays

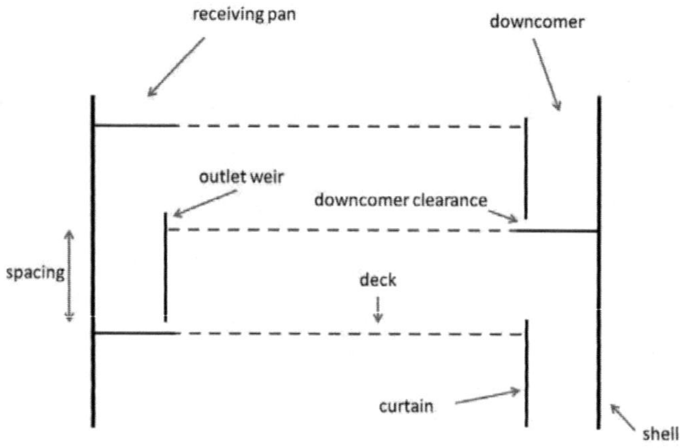

FIGURE 4. 1-Pass Crossflow Trays

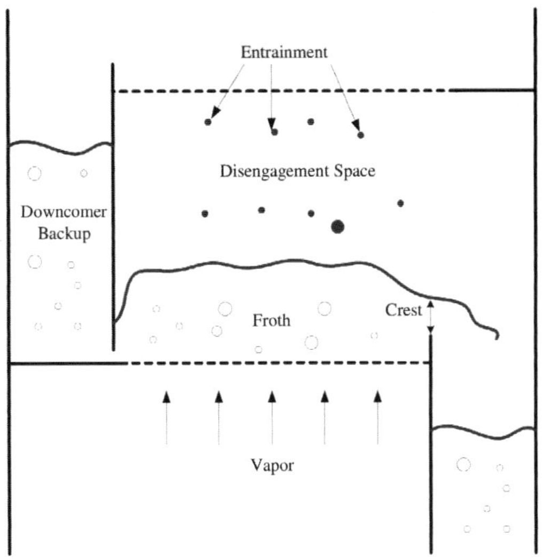

FIGURE 5. Flow Pattern on 1-Pass Crossflow Tray

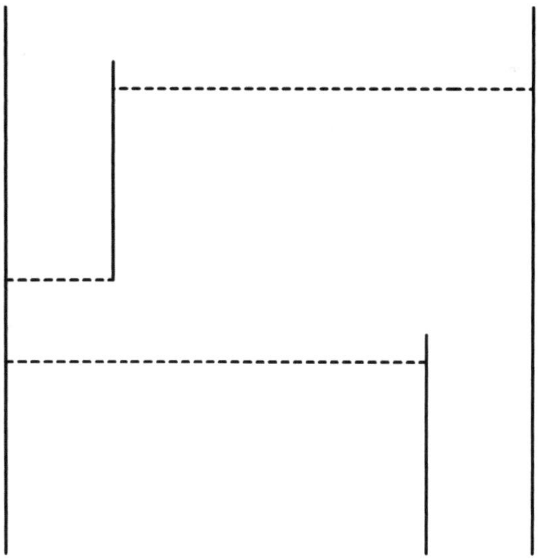

FIGURE 6. Augmented Crossflow Trays

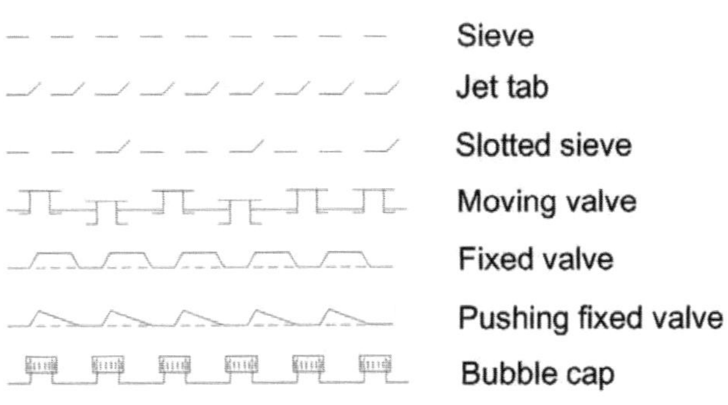

FIGURE 7. Tray Deck Types

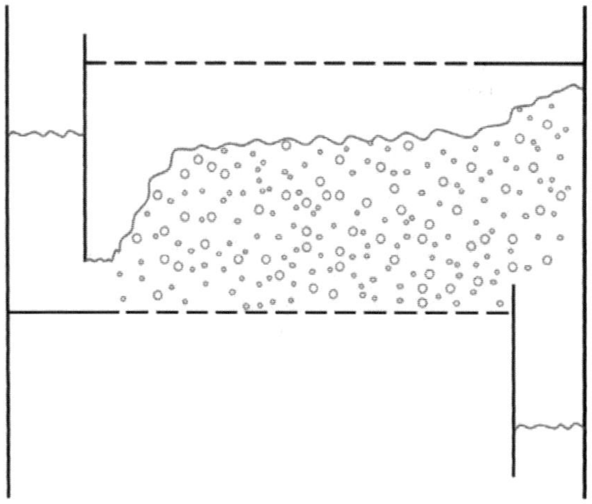

FIGURE 8. Choke Flooding

Structured Packings for Distillation and Absorption

By

Simon Chambers
Anand Vennavelli

Fractionation Research Inc.
424 S. Squires Street, Suite 200
Stillwater, OK 74074

2 April 2014

Presented at YP Tutorials: Distillation and Absorption Equipment,
AIChE Spring Meeting, New Orleans, Louisiana, April, 2014

Abstract

Structured packings used in industrial practice can be broadly classified into three generations: wire gauze, standard, and high capacity corrugated sheet metal. Standard structured packings were a popular choice in the industry for many years. Newer high capacity structured packings, which comprise either "half-S" or "full-S" crimp angle shapes, can operate at much higher load points or maximum useful capacities (MUC's), and considerably lower pressure drops than standard structured packings.

Structured packings are favored in high vapor rate, low liquid rate distillation services and in absorption. They are particularly advantageous in vacuum columns, where low pressure drops are sometimes critical. For example, in Ethylbenzene-Styrene monomer columns, low pressure drops yield low liquid temperatures and, consequently, reduced renegade polymerization.

In this paper, structured packing physical characteristics, installation, and applications are discussed. In addition, basic performance characteristics, such as HETP, pressure drop, and capacity, and liquid distribution quality are reviewed. A 20-minute video was prepared by FRI to illustrate the basics of structured packings.

Keywords: Packings Distillation Absorption Low Pressure Drop

1 Overview

Structured packings are often the choice of internals for low-pressure distillation and absorption. Structured packings provide good efficiencies, low liquid hold-ups, and low pressure drops.

Structured packings are frequently used in:

- Chemicals and fine chemicals (e.g., Ethylbenzene/styrene distillation)
- Refining (e.g., vacuum and atmospheric crude oil distillation)
- Petrochemicals

Structured packings are best suited for vacuum to moderate pressure applications.

2 Generations of Structured Packings

Structured packings can be broadly classified into three generations – wire gauze, standard corrugated sheet metal, and high capacity corrugated sheet metal.

2.1 First Generation: Wire Gauze Structured Packings

First introduced in the late 1950's to early 1960's, these packings are most suitable for difficult separations and when there is a low pressure drop requirement. However, they are relatively expensive to fabricate. Examples include Sulzer Wire Gauze BX and CY, Koch Glitsch Goodloe, Montz A3-500, AMACS Woven Gauze. Figure 1 shows the photographs of some wire gauze packings.

2.2 Second Generation: Standard Structured Packings

Because of the high fabrication costs associated with the wire gauze packings, more cost-effective corrugated sheet metal structured packings were introduced in the late 1970's. These packings have since become the standard of the industry. Examples are shown in Figure 2.

Because these packings have a sharp change in angle between the layers, the vapor and the liquid have to change direction abruptly at the interface of the packing layers. Such a sudden change in the direction of the liquid and vapor flow can limit the performance of the packings.

2.3 Third Generation: High Capacity Structured Packings

The third generation, or high capacity structured packings, are very similar to the standard packings, except that they provide smooth transitions between layers for the liquid and vapor streams. Examples are shown in Figure 3.

3 Anatomy of a Corrugated Structured Packing

Structured packing comprises metal sheets that are textured, crimped, and often crimped – and assembled together as blocks.

3.1 Texture and Perforation

Texturing can be seen, for example, in the photograph of the Mellapak packing of Figure 2. Texturing promotes liquid film spreading while perforating minimizes vapor channeling and promotes vapor-liquid contacting. Texturing and perforating can increase the packing efficiency.

3.2 Blocks and Layers

Structured packings are usually built in layers of heights 200-300 mm, which are further segmented into multiple blocks. Figure 4 shows a full bottom layer of the packing with a few blocks of another layer placed on top of the bottom layer.

3.3 Crimps

The crimps are an important physical feature of a structured packing, and effectively determine the surface area and flow patterns. Referring again to Figure 2, the crimps of most of the structured packings are easily seen.

3.3.1 Crimp Angles

Based on the crimp angle from the vertical axis, structured packings are classified into X (30°), Y (45°), and Z (60°) packings. More-vertical crimps exhibit lesser hold-ups, lesser pressure drops, lower efficiencies, and higher capacities

3.3.2 Crimp Depth

Large crimp depths provide a low surface area, and conversely, small crimp depths provide a larger surface area. Obviously, to create more surface area, more metal is required - more metal means more cost, but also a higher efficiency. With more surface area, liquid hold-ups are higher, pressure drops are higher, and capacities are lower.

4 Installation of structured packings

First, the packing support grid is installed, and then, packing elements are lowered down by means of a belt or a rope to prevent damage to the packing. Metal guide plates or shoes are employed to facilitate installation of packing. The installer stands on a board to prevent damage to the packing sheets. As shown in Figure 4, each layer contains wall wiper bands that are bent out so they touch the column wall so that any liquid wall flow will be redirected back into the bulk of the packing. Each layer is usually rotated 90 degrees to minimize seam alignments of the packing blocks, and to prevent short-circuiting of vapor liquid traffic. Careful installation of the structured packing is vital for the column to work.

5 Performance Characteristics of Structured Packings

The key performance characteristics are HETP, Capacity or often, the Maximum Useful Capacity (MUC), and pressure drop.

5.1 Pressure Drop

Pressure drops of structured packings are lower than those of random packings, and much lower than those of trays. The pressure drop across a structured packing increases gradually with increasing liquid and vapor rates; the pressure drop increases sharply close to the flood point, as liquid begins to stack up in the bed.

5.2 HETP (Height Equivalent to a Theoretical Plate)

Structured packings use the concept of a "Height Equivalent to a Theoretical Plate", or HETP, to describe the performance of the packings. The HETP is defined as

$$\text{HETP}=\frac{\text{Packing Bed Depth}}{\text{Number of theoretical plates}} \tag{1}$$

The number of theoretical plates[1] required for a given separation is usually determined by a process simulation. An easy separation requires few theoretical plates; a difficult separation requires many theoretical plates.

For the same number of theoretical plates, a lower HETP results in a lesser packed bed depth, and consequently, a shorter column. Therefore, the lower the HETP, the more efficient the column is.

[1] Number of theoretical plates is also sometimes referred to as the number of theoretical stages, or the number of equilibrium stages.

For example, a highly efficient structured packing might achieve an HETP of about 300 mm (12 inches); a less efficient structured packing might achieve an HETP of 500 mm (20 inches).

5.3 Capacity or Maximum Useful Capacity (MUC)

The highest vapor rate at which the HETP of the column is good is called the MUC. A further increase in rate from the MUC will cause the HETP to rise rapidly, as the column progresses into a flooded condition.

5.4 Effect of packing specific surface area

Specific surface area, determined by the crimp characteristics, is an important factor affecting performance characteristics of structured packings.

High surface area structured packings exhibit high pressure drops and vice-versa. Typical pressure drop curves are shown in Figure 5.

High surface area structured packings exhibit low HETP's and vice-versa. Typical HETP curves are shown in Figure 6.

High surface area structured packings exhibit low MUC's and vice-versa. In Figure 6, the MUC point is indicated for both high surface area and low surface area packings. .

6 REFERENCES

Some of the best sources of information regarding structured packing are listed below:

Kister, H.Z., **Distillation Design**, McGraw-Hill, 1992

Kister, H.Z., **Distillation Operation**, McGraw-Hill, 1990

Perry's Chemical Engineer's Handbook, 8th edition, "Chapter 14 – Equipment for Distillation and Gas Absorption"

Strigle, R.F., **Packed Tower Design and Applications**, 2nd edition, Gulf Publishing Co., Houston, TX, 1987.

Vendor Resources:[2]

http://www.sulzer.com/en/Products-and-Services/Separation-Technology/Structured-Packings
http://www.koch-glitsch.com/masstransfer/pages/structured_packing.aspx
http://www.amacs.com/packing/structured/
http://amacs.com/wp-content/uploads/2012/09/AMACS-Montz-Brochure.pdf
http://raschig.com/-Structured-Packings

[2] The listing here is not comprehensive. The list of references is provided for the benefit of the reader. FRI does not endorse or recommend any particular vendor product over another, including those listed in this document.

7 FIGURES

Figure 1: Wire Gauze Structured Packings

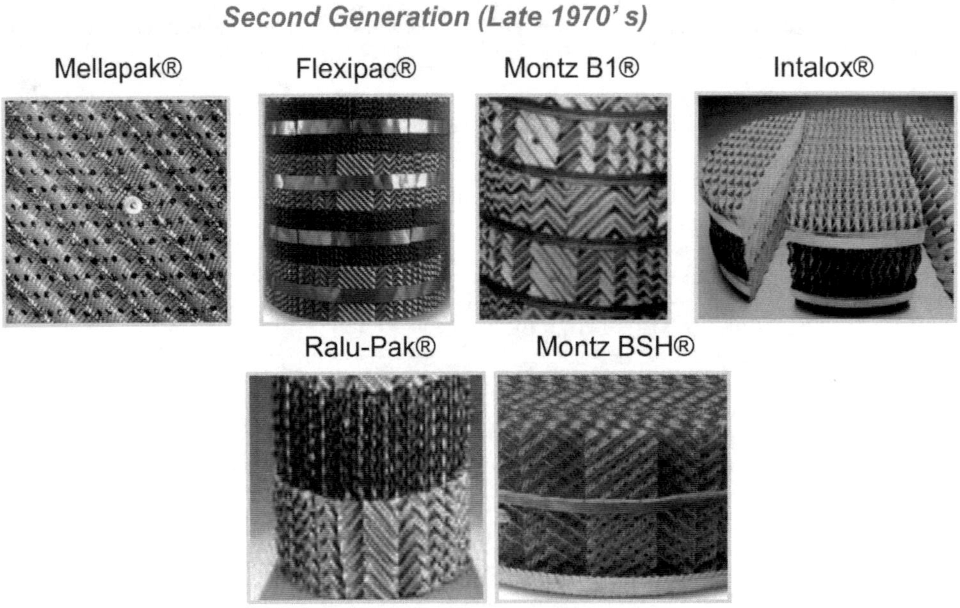

Figure 2: Standard Sheet Metal Structured Packings

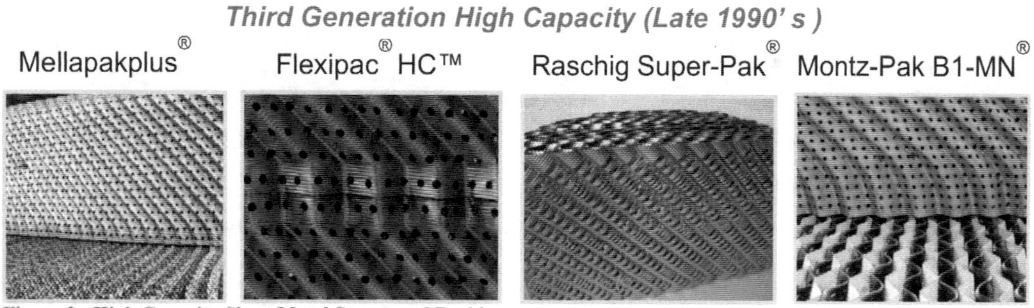

Figure 3: High Capacity Sheet Metal Structured Packings

Figure 4: Sulzer Mellapak 250.Y Packing - Packing Layer and Blocks

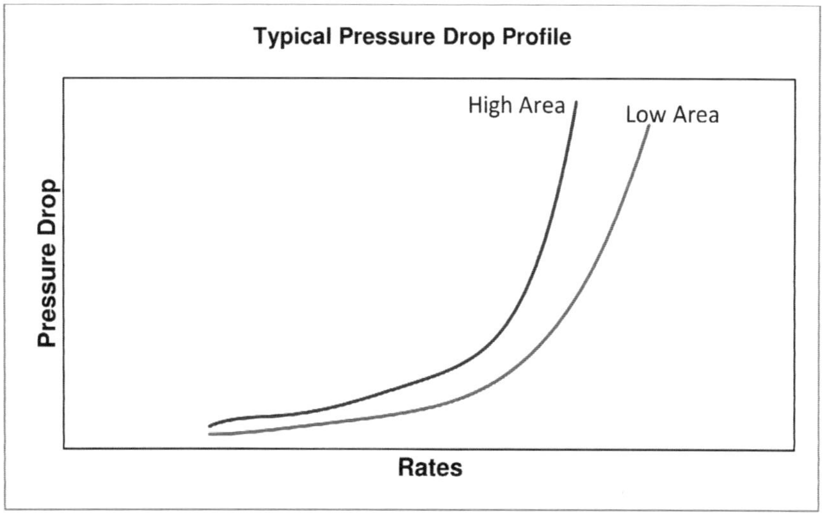

Figure 5: Typical Pressure Drop Curves

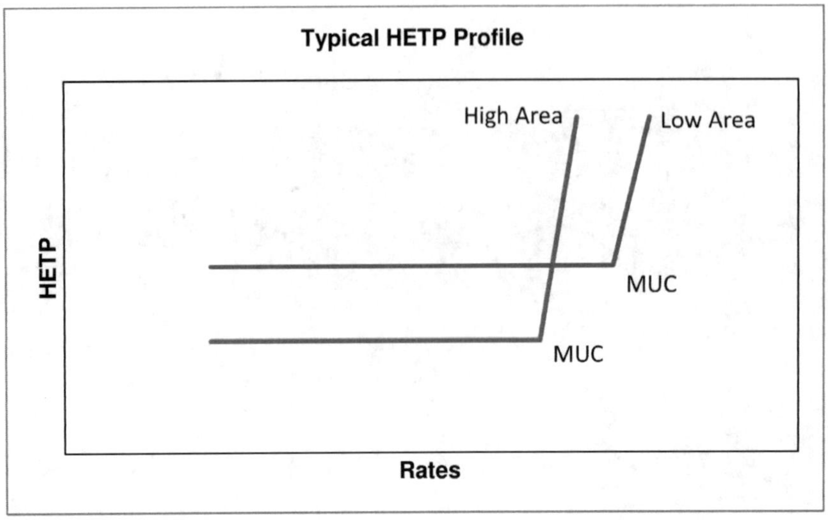

Figure 6: Typical HETP Curves

Liquid Extraction
Processes and Equipment

By

Mike Resetarits
AY Ogundeji
Tony Cai

Fractionation Research Inc.
424 S. Squires Street, Suite 200
Stillwater, OK 74074

Presented at YP Tutorials: Distillation and Absorption Equipment, AIChE Spring Meeting, New Orleans, Louisiana, 2 April, 2014

UNPUBLISHED

Abstract

What is the best way to remove C_6, C_7 and C_8 compounds from a feed stream containing C_6, C_7 and C_8 compounds? How do you take a lubricating oil and remove certain compounds and improve its viscosity index (so that its viscosity does not change excessively with the oil's temperature)? What is propane deasphalting? The answers to all of these questions is "liquid extraction processes."

What technology was supported and funded by the actor Kevin Costner to help clean up the large 2010 Gulf of Mexico oil spill? What are Karr reciprocating plate columns, Oldshue-Rushton contactors and Podbielniak centrifuges? What the heck is a Graesser Raining Bucket? The answer to all of these questions is "liquid extraction equipment."

Imagine heavy and light fluids flowing vertically past each other with only a very small density difference. Imagine rotating equipment that works better without rotating. Liquid extraction is different – very different. There are very few experts.

Keywords: Extraction Aromatics Ternary Equilibria

Acknowledgment

The authors acknowledge the many contributions of Dr. Frank Seibert and Dr. Bruce Eldridge (both of the University of Texas at Austin) to the field of liquid-liquid extraction.

OVERVIEW

Some people say that chemical engineering is 50% Reactions and 50% Separations. In the field of Separations, distillation is the technique that is employed the most often, but liquid-liquid extraction, absorption, adsorption and membranes also play important roles.

Liquid-liquid extraction has many, and diverse, applications. Extraction devices can employ agitation to hasten mass transfer; there are many types of agitated devices. There are also many types of devices that do not employ agitation. Mass transfer calculations usually begin with liquid-liquid equilibrium (thermodynamic) data. Hydraulic calculations usually begin with force balances around droplets, and interfacial tension measurements.

GENERAL

As shown in **Figure 1**, when distillation is employed, a liquid phase and a vapor phase flow countercurrently. With extraction, a heavy liquid phase and a light liquid phase move countercurrently. With extraction, those two phases must be generally immiscible in each other. Chemical compounds in one phase, however, diffuse into the other phase. Very often, one liquid phase is aqueous; the other liquid phase is hydrocarbon. (Many people know that water and oil do not mix.) **Figure 2** shows a swarm of light liquid droplets (maybe "oil") rising through a heavier liquid (maybe "water"). At the top of the photograph, the light liquid droplets are seen to be coalescing.

Extraction columns usually have two feed streams. One is called the "Feed" and the other is called the "Solvent." The Feed stream of **Figure 3** is comprised of just two chemicals, A and B. The solvent stream is designated by "S". "A" is the "Solute" which ultimately transfers from the Feed stream to the solvent stream. After the Feed stream loses component A, the residual liquid, which is mostly component B, is often called the "Raffinate." (Generally, in the extraction world, the word "raffinate" refers to a low-value product stream.) In Figure 3, the product stream that is designated by "S + A" usually goes to a solvent recovery column that separates S from A. The resultant solvent stream is returned to the extractor. "A" is usually the final target product. The solvent recovery column is usually some form of distillation column, e.g., an extractive distillation column.

In trayed distillation and absorption columns, the vapor phase is dispersed as droplets in the downward-flowing liquid stream. As shown in Figure 3, extraction processes offer four different possibilities as follows:

1. Solvent = heavy phase; feed = light phase; heavy phase is dispersed
2. Solvent = heavy phase; feed = light phase; light phase is dispersed
3. Solvent = light phase; feed = heavy phase; heavy phase is dispersed
4. Solvent = light phase; feed = heavy phase; light phase is dispersed

Generally, the heavy phase is an aqueous phase and the light phase is a hydrocarbon phase. Most extraction experts believe that the more voluminous phase should be dispersed.

APPLICATIONS

A partial list of extraction applications is provided below:

- Pharmaceutical – production of penicillin
- Food – decaffeination of coffee
- Metals extraction
- Chemical and petrochemical
  - Aromatics recovery
  - Lube oil extraction
  - Acetic acid extraction
  - Desulfurization
- Liquid sweetening

BTX aromatics extraction is a particularly popular process, especially in refineries. A product stream from a catalytic reformer contains hexane, benzene, heptane, toluene, octane, xylenes and heavier aromatic and non-aromatic compounds. The aromatic (BTX) compounds are far more valuable than the non-aromatic compounds. The former cannot be separated easily from the latter via distillation columns.

Figure 4 shows a light liquid feed stream (C_6-C_8) entering the side of an extractor. A heavy liquid solvent stream enters the top of the extractor. There are at least two popular solvents – sulfolane and tetraethylene glycol. Both of these solvents have a strong tendency to absorb (or "extract") the benzene, toluene and xylenes, leaving behind the hexane, heptane and octane, i.e., the non-aromatic raffinate stream. Rich Solvent leaving the bottom of extractor goes to a (vapor-liquid) Solvent Recovery Column which separates the solvent from the BTX – the final product. That Recovery Column also generates a light reflux stream that is returned to the extractor. The BTX product stream goes to a series of distillation columns that separate and purify the benzene, toluene and xylenes.

Extraction can also be used to remove heavy aromatics compounds from a lubricating oil feed stream. The most popular solvents are furfural, phenol and NMP. The final product from the extractor is a lubricating oil that has an improved Viscosity Index, i.e., its viscosity is less dependent on temperature.

CONTACTORS

There are many types of liquid-liquid contacting devices. Some employ agitation, e.g., mixers, to hasten mass transfer. Others do not require or employ agitation. For those, in fact, agitation can be harmful, creating a totally-emulsified contactor wherein the light and heavy phases never separate out. The interface between any two liquids has an intermolecular "skin." The strength of that skin is called the "interfacial tension." Many extraction experts believe that liquid-liquid systems and interfacial tensions higher than 15 dynes/cm have a tendency towards large droplets and low interfacial areas for phase-to-phase mass transfer. Such systems require agitation to create more, smaller droplets. Liquid-liquid systems with interfacial tensions less than 15 dynes/cm have a tendency to easily produce many small droplets and a large interfacial mass transfer area. Agitation is not required with such systems; efficiencies are acceptable without it.

Generally, liquid phase diffusivities (diffusion coefficients) are much lower than vapor phase diffusivities. With a liquid-liquid extractor, diffusing solute compounds move slowly through one liquid phase, pass through the interface, and then move slowly through the second liquid phase. As a result of these slow mass transfers, extractor efficiencies are generally low. For example, a typical trayed extractor might have a tray efficiency of only 20%.

A partial list of agitated and non-agitated extraction devices is given below:

Agitated
- Mixer-settlers
- Podbielniak centrifuges
- Costner Centrifugal
- Rotating Disc Contactors
- Scheibel
- Karr reciprocating plate
- Kuhni

Non-agitated
- Packing
- Conventional trays
- Seal Deck and Rain Deck (UOP) trays
- Multiple Upcomer (UOP) trays

Figure 5 shows the UOP Multiple Upcomer (MU) tray. With this tray, the heavy phase is the dispersed phase. Each tray deck has many circular perforations which create a swarm of heavy phase droplets. As the droplets fall through the continuous light phase, mass is transferred. Those droplets coalesce on the tray below. The cycle is repeated again and again. As shown in Figure 5, the light liquid phase snakes its way up the column as the continuous phase.

There are many proprietary agitated extraction devices. Rotating disc contactors, such as the one shown in **Figure 6**, have often been used in Sulfolane aromatics extraction units. The Costner Centrifugal Extractor has been considered for off-shore oil-water separations. Podbielniak extractors are popular in pharmaceutical separations.

HYDRAULICS

For some liquid-liquid systems, at some temperatures, interfacial tensions can be measured easily with a ring tensiometer such as the one shown in **Figure 7**. Initially, the ring is lowered to the bottom of the containment cylinder. Then, the ring is pulled upwards, by a tin wire, until the interface between the two liquid phases is impacted. At that point, the interfacial "skin" resists any other upward ring movement. Eventually, the ring breaks through the skin with a force that is measured. That force is easily converted to an interfacial tension in dynes/cm. Thereafter, if the ring is pulled upwards through the light liquid phase, the liquid-vapor interface will soon be impacted. The force that will be required to pull the ring through that interface is a direct measure of the light liquid surface tension. For liquid-liquid systems at higher temperatures and pressures, interfacial tensions can be determined indirectly from the sizes of droplets that are created by sieve plates.

THERMODYNAMICS

Distillation calculations employ the concept of "equilibrium stages" to determine column heights. Vapor-liquid equilibrium laboratory data are employed to determine the number of equilibrium stages that are required to effect specified distillation separations.

Similarly, the extraction technology employs the concept of "equilibrium stages," and liquid-liquid equilibrium data, to determine staging requirements and contactor heights (sizes). Unlike distillation, extraction often employs ternary (triangular) diagrams to represent liquid-liquid equilibria. **Figure 8** provides a ternary diagram for the BTX example of Figure 4. The three apexes of the diagram represent pure components. Every position within the triangle represents a ternary composition. Every position under the curved umbrella represents a non-equilibrium mixture that sooner or later becomes two immiscible liquid phases that are in equilibrium with each other. The exact middle of the triangle represents a non-equilibrium mixture of 33% toluene and 33% heptane and 33% sulfolane. In a BTX extractor, Sulfolane is the solvent, toluene is the solute and heptane is the raffinate. Regarding this diagram, the feed stream is a toluene/heptane mixture.

In Figure 8, Points a, c and e represent light liquids that are high in heptane concentrations, low in toluene concentrations and almost absent of sulfolane. Points b, d and f represent heavy liquids that are high in sulfolane concentrations, high in toluene concentrations and almost absent of heptane. The almost-horizontal lines are called "tie lines." The points at the ends of each tie line represent light and heavy phases that are in equilibrium with each other. For example, Points a and b are in equilibrium with each other. The "Plait Point" represents the unique situation where the light and heavy liquid phases have the exact same compositions. Ternary diagram equilibrium data can be employed to determine the number of equilibrium phases that are required to effect target separations. Mass transfer models and computer programs exist for determining tray and packing efficiencies. More often than not, new extractors are designed using the efficiencies that were observed with similar extractors that are already operating.

REFERENCES

Some of the best sources of information regarding the science of liquid-liquid extraction are listed below:

- Treybal, R.E., **Liquid Extraction**, McGraw-Hill 1963.
- Lo, T.C., Baird, M.H.I., and C. Handson, **Handbook of Solvent Extraction**, John Wiley and Sons 1983.
- Laddha, G.S. and T.E. Degaleesan, **Transport Phenomena in Liquid Extraction**, McGraw-Hill 1978.
- Arlt, W., Macedo, M. Rasmussen, P., and J.M. Sorensen, **Liquid-Liquid Equilibrium Data Collection, vol. V, Parts 1-4,** DECHEMA Chemistry Data Series, Frankfurt 1997.
- **Perry's Chemical Engineer's Handbook**, 7th edition, "Liquid-Liquid Extraction Operations and Equipment", 1997. (revision in 2006)
- Walas, S.M., **Phase Equilibria in Chemical Engineering**, Butterworth Publishers 1985.
- Maddox, R.N., **Gas and Liquid Sweetening**, Campbell Petroleum Series, 1977.

FIGURES

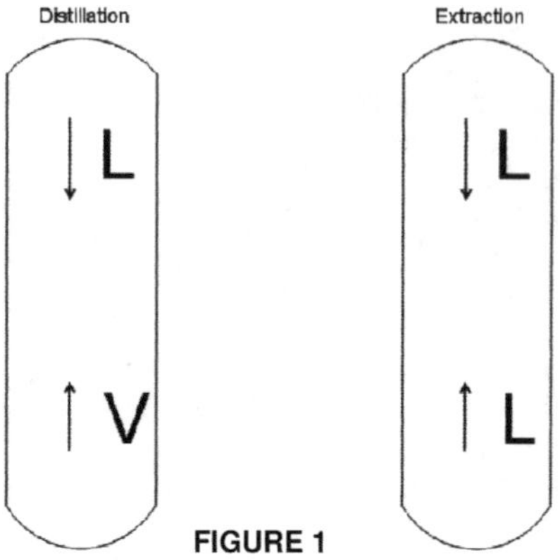

FIGURE 1

FIGURE 2

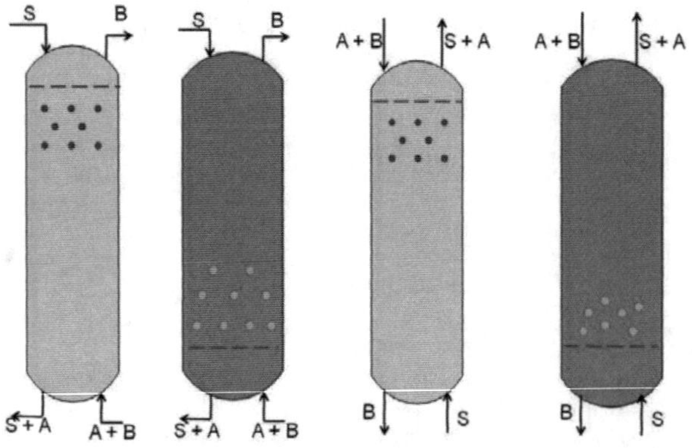

FIGURE 3. Extraction Options

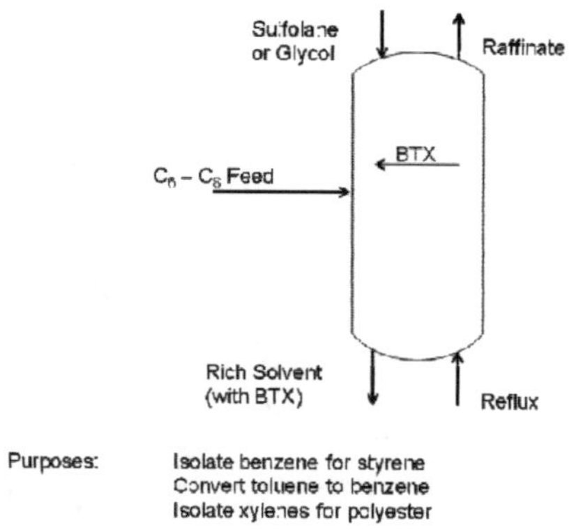

FIGURE 4. Aromatics Extraction

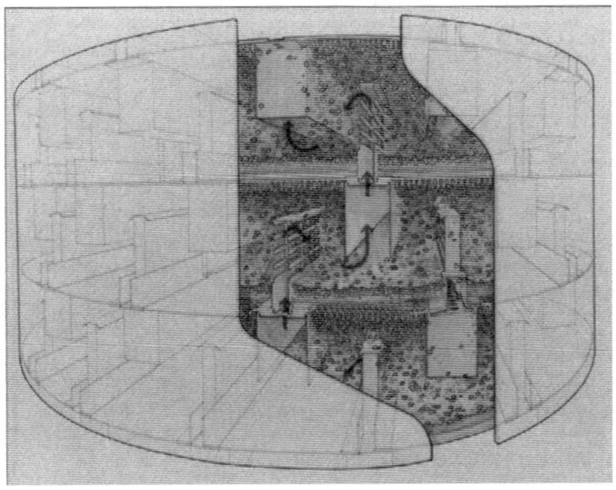

FIGURE 5. MU Tray Hydraulics

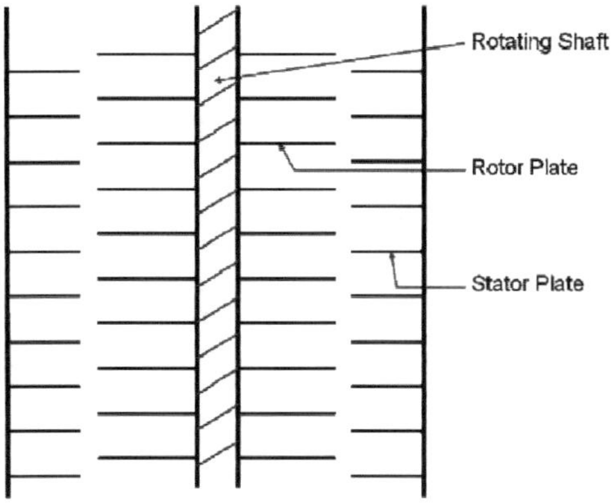

FIGURE 6. Rotating Disc Contactor

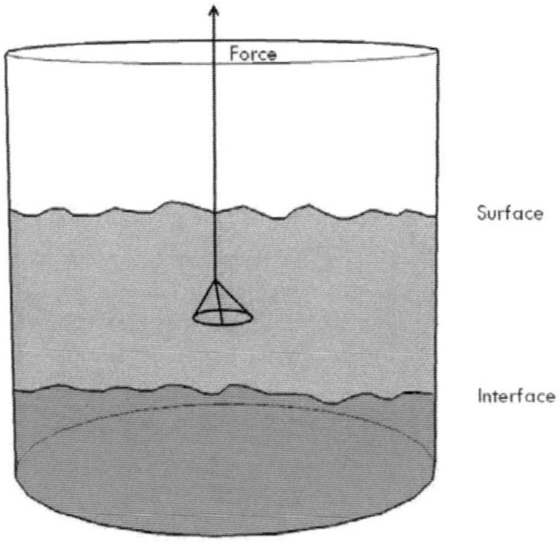

FIGURE 7. Ring Tensiometer

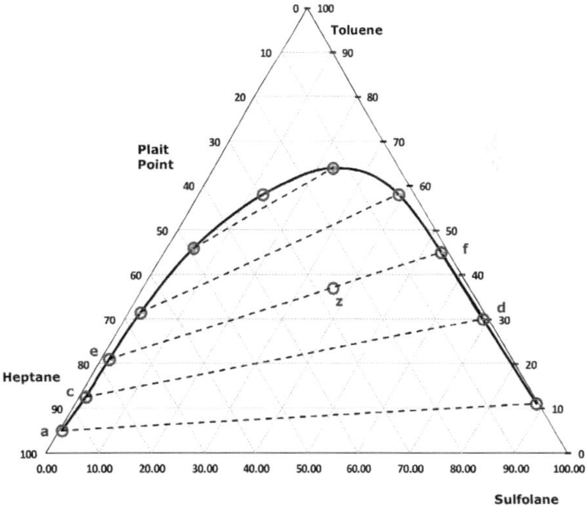

FIGURE 8. Heptane + toluene + sulfolane at 25℃, a type I system. [*Data taken from De Fre and Verhoeye*, J. Appl. Chem. Biotechnol., ***26,*** *pp. 1-19 (1976).*]

Acid Gas Absorption Fundamentals

By

Clint P. Aichele
James. R. Whiteley
Yash Tamhankar

School of Chemical Engineering
Oklahoma State University

Mike Resetarits

Fractionation Research Inc.
424 S. Squires Street, Suite 200
Stillwater, OK 74074

2 April 2014

Presented at YP Tutorials: Distillation and Absorption Equipment,
AIChE Spring Meeting, New Orleans, Louisiana, April 2014

ABSTRACT

Absorption is a standard unit operation that has been studied by chemical engineering students for decades. One of its primary uses is for the removal of acid gases, like CO_2 and H_2S, from natural gas streams. Historically, aqueous amines and hot carbonate solutions have been used to perform these absorptions. Many different amines, solution strengths, activators and passivators have been tested and employed industrially. Approximately half of the industrial towers are trayed; half, packed. Foaming adversely affects the hydraulic performance in these systems.

Acid gas absorption is receiving appreciable new attention as concerns rise regarding CO_2 in the atmosphere. In addition, the United States has recently experienced significant growth opportunities in shale gas production which require improvements in gas treating technologies. This paper reviews acid gas absorption and regeneration fundamentals. A PFD will be discussed at length, to illustrate important concepts. New absorbent solutions will be discussed, including ionic liquids. Flue gas applications will be discussed, including the difficulties associated with column pressure drops and very large column diameters. Rate-based modeling approaches will also be reviewed.

INTRODUCTION

Subsequent to the Industrial Revolution, global carbon dioxide emissions have increased from virtually zero to 9 billion total metric tons. These emissions originated from the natural gas, petroleum, coal and cement industries. Carbon dioxide, CO_2, emissions increase annually, as an increasing percentage of the global population consumes gas/oil/coal energy and drives cars. Carbon dioxide is a "greenhouse gas" and many people contend that such gases are contributing to the warming of the planet. As CO_2 emissions continue to increase, the technologies that reduce those emissions receive increasing attention from academia, industry and global governments.

For many years, and now, acid gas absorption units have been the primary means whereby CO_2 emissions have been reduced. Such units are – everywhere – and are relatively easy to understand. Several excellent books describe such units. [1,2,3]

DISCUSSION

Natural Gas Processing

When it first comes from the ground, natural gas contains many compounds, as shown in **Table 1**. Methane burns nicely in a furnace, but CO_2 does not burn at all. Therefore, the consumers who purchase natural gas do not want to receive, and pay for, CO_2. Typically, natural gas production companies must reduce the CO_2 content of natural gas down to 2% or less.

Raw natural gas sometimes contains hydrogen sulfide, H_2S, which is particularly-dangerous poison. Natural gas production companies must reduce the H_2S content of natural gas down to 4 ppmv or less.

Very recently hydraulic fracturing or "fracking" is yielding gases that are very high in ethane, propane and butane. These "natural gas liquids" or "NGL's" are very valuable in that they can easily be converted to ethylene and propylene. As fracking increases in the U.S., more raw natural gas streams are being created and all of those streams need the acid gases (CO_2 and H_2S) to be removed.

Flue Gases

"Flue gas" often refers to a combustion exhaust gas from the chimney of a power plant. Its composition depends on what is being burned, e.g., natural gas or coal. **Table 2** provides typical flue gas compositions. Globally, the CO_2 content of flue gases is receiving increasing attention, but the removal of that CO_2 is particularly challenging. Flue gases are at very low (near atmosphere) pressure and need to be compressed to flow (through absorbers). Many flues create huge volumes of such gases and those gases contain particulates and oxygen. Perhaps the largest challenge of all: What can be done with the CO_2 once we capture it? "Sequestration" is an umbrella word that includes all of those natural and unnatural places where CO_2 can be stored or utilized. For example, compressed CO_2 can be used for Enhanced Oil Recovery (EOR).

Solvents

Water can be used as a solvent for the acid gases in a natural gas stream, but water is very ineffective. The Benfield hot carbonate process employs aqueous potassium carbonate as the solvent. CO_2 and H_2S are removed from natural gas streams via the following overall chemical reactions:

$$K_2CO_3 + CO_2 + H_2O = 2KHCO_3$$

$$K_2CO_3 + H_2S = KHS + KHCO_3$$

Globally, for acid and gas removals, aqueous amine solutions are probably the most popular. Forty years ago, MEA and DEA were the workhorses. Then, MDEA became popular especially because of its ability to remove H_2S quickly and totally, without over-absorbing CO_2. Most recently, piperazine is being employed regularly. **Figure 1** is a partial list and classification of amines. When CO_2 is absorbed into an aqueous amine solution, many chemical reactions are involved. These chemical reactions are very exothermic. Amine solutions are very corrosive. Passivators are sometimes employed to reduce attacks on the metals of acid gas processing units.

Typical Unit

Figure 2 shows a typical amine absorption unit. Two processing columns are employed – an absorber and a solvent regenerator.

Referring to **Figure 2**, the Feed Gas enters the bottom of the absorber. Typical pressures of that gas range from 100 to 1200 psia. Absorber shells can be as thick as 3 inches. As the gas proceeds upwards through the column, it encounters a down-flowing liquid solvent stream. Lean liquid solvent enters the top of the absorber. This lean solvent stream must be low in acid gases (CO_2 and H_2S) or gases issuing from the lean solvent stream will contaminate the gas product stream. It is also important that the lean solvent stream be relatively cool to remove CO_2 to the lowest possible levels. Additionally, a too-warm lean solvent will flash off its acid gases too easily. The treated gas product stream that leaves the top of the absorber often goes to a train of distillation columns that separates and purifies the methane, ethane, propane, isobutane and n-butane.

Still referring to Figure 2, as the liquid solvent stream flows down through the absorber, the exothermic reaction associated with the chemical absorption causes the solvent to increase appreciably in temperature. Most of the temperature increase occurs near the bottom of the absorber. After the rich solvent leaves the bottom of the absorber, it often goes to a heat exchanger that increases the rich solvent temperature further. The hot rich solvent goes to the regenerator, where a massive flash occurs especially because the regenerator operates at near-atmospheric pressure. Some regenerators have two in-series flash chambers in their tops to allow additional flashing off of the acid gases. All of the vapors leaving the top of the regenerator (steam, CO_2 and H_2S) go to a condenser that liquefies the steam and returns the water to the top of the regenerator.

After flashing, the rich solvent flows all the way down the regenerator until it reaches a reboiler. That heat exchanger boils off a large portion of the water that is in the solvent. Very little amine is boiled off. The resultant steam flows upwards through the regenerator, and strips the acid gases out of the down-flowing solvent.

The lean solvent that leaves the bottom of the regenerator is very hot (depending on the reboiler temperature) and is low in pressure. A pump is required to increase the pressure of the lean solvent to a pressure somewhat higher than the absorber pressure.

A high-temperature lean solvent would be very ineffective at acid gas removal, especially due to vapor-liquid equilibrium (thermodynamic) considerations. The rich solvent can be used to decrease the temperature the lean solvent. The lean solvent temperature can also be reduced using a cooling water exchanger, as shown in Figure 2.

There are many variations of absorption unit processing schemes. Examples follow:

1. A semi-lean solution can be pulled from the side of the regenerator and fed to the side of the absorber
2. The bulk of the lean solution can be fed to the side of the absorber, with only a slip stream of the lean solution going to the cooling water exchanger, and the top of the absorber

Regarding the acid gases that leave the top of the regenerator, the H_2S can be converted to elemental sulfur in a Claus Unit. Historically, the CO_2 has been released into the atmosphere. It can, however, be compressed and utilized for Enhanced Oil Recovery (EOR).

Other Solvents

Aqueous amines and potassium carbonate solutions have historically been the most commonly employed absorption solvents. Some amine solutions contain a mixture of primary, secondary and/or tertiary amines. Some potassium carbonate solutions employ amines as “activators.” Generally speaking, H_2S is absorbed quickly into amine solutions; CO_2 slowly. When an excess of CO_2 is absorbed by the lean solution, an excess of regeneration (and energy) is required. A perfect amine solution formulation would simultaneously exactly achieve the targeted CO_2 and H_2S product gas compositions.

Solvents other than aqueous amines and hot carbonate solutions have been employed and are being investigated

Physical solvents, ionic liquids, and enzymes have been explored as possible CO_2 capture solvents to replace aqueous amine absorption for flue gas applications[4]. These approaches have been investigated primarily for their ability to reduce the size and overall cost of acid gas absorption applications. Physical solvents require the physical solubility of the acid gas into the solvent; therefore, physical solvents are commonly used for high pressure applications such as synthesis gas scrubbing or natural gas absorption. Because physical solvent processes don't rely on chemical reactions to absorb the acid gases, the solvent regeneration energy is typically less than that associated with conventional aqueous amine absorption. Ionic liquids have received considerable attention for their potential ability to reduce the regeneration energy of the CO_2 removal process[5]. However, their large-scale use for CO_2 removal applications has not been achieved. Enzymes show promise as a way to enhance the reaction rate of CO_2 into solution, thereby reducing the size of the absorber. However, large scale demonstration of this technology at competitive costs remains to be seen. Despite these advances in solvent development, aqueous amine absorption remains the most competitive technology for CO_2 removal from flue gas, and it continues to be a technology of choice for natural gas treating and refinery scrubbing applications.

Column Internals

Historically, most hot carbonate columns have employed packings, especially random packings. For many years, most amine units have employed 20 trays in the absorbers and 20 trays in the regenerators. In recent years, however, packings have sometimes been employed in amine columns, especially when the lower pressure drops of packings have yielded processing benefits.

Foaming

Acid gas absorbers are known "foamers," i.e., the 2-phase mixtures on the trays or in the packed beds bubble excessively. These foams cause column capacities to be reduced. FRI has demonstrated that packings do not break foams, but the impact of the foaming is less with packings than it is with trays. On trays, foaming can be reduced, somewhat, via short froths, large deck holes and designing nearer the spray regime (as compared to the froth regime)[6] .

With time, aqueous amine solutions degrade and begin to carry particulates. It is generally held that these particulates cause foaming to increase. With operating untis, old amine solutions need to be filtered – and augmented regularly with fresh amine solution.

Equilibrium Stages and Process Simulations

With amine units and hot carbonate units, the absorber trays/packings and the regenerator trays/packings do not need to yield very many equilibrium stages (or "theoretical trays") to achieve the targeted absorption and regeneration separations.

McCabe-Thiele graphical constructions show huge areas between the operating and equilibrium curves. Step-offs between those curves show that a typical absorber only requires 1-3 equilibrium stages; likewise for a typical regenerator. At the very same time, in these mass transfer services, tray and packing efficiencies are low and difficult to calculate because of the chemical reactions and equilibria that are involved. Typical tray efficiencies are around 20%; some are as low as 5%. Because of these unusual circumstances and difficulties, rate-based theoretical calculations are increasingly preferred for unit simulations. For example, the ProTreat computer program of Optimized Gas Treating[7] is very versatile and can be used to calculate/predict column performances without using the concepts of equilibrium stages or tray/packing efficiencies.

Safety Considerations

Amine solutions are very corrosive. They can dissolve metal and burn skin. Acid gas absorbers are operated at very-high pressures, as high as 1200 psi. Pressure relief systems are, of course, required on such units.

In 1984, welding work was performed inside an amine absorber in Illinois. Following that work, Post-Weld Heat Treatment (PWHT) work was not performed. A weld failed due to stress corrosion cracking. The gas that escaped from the operating column ignited. The column ended up landing 1 kilometer away. Seventeen workers were killed. There was approximately $100 million in property damage. All pressure vessels must be properly designed and maintained.

Current Research

Acid gas absorption/regeneration is a very popular research area in 2014. At the November 2013 AIChE meeting in San Francisco, there were 22 sessions with "CO_2" or "carbon dioxide in their titles. Those sessions included 130 presentations! The presenters represented governments, academia and industry.

Many aspects of acid gas treating are now being studied globally. Two examples follow:

1. Dr. David Van Wagener (Ph.D. from the University of Texas with Dr. Gary Rochelle) focused on 2-stage flashing to reduce regeneration energy consumption[8].
2. Oklahoma State University is studying spray columns for the reduction of the capital cost associated with flue gas treating (see **Figure 3**)[9].

There are many other examples of on-going research work.

CLOSING

Acid gas removal from gas streams is not new. For many years, global gas plants have been removing CO_2 and H_2S from natural gas streams, especially using aqueous amines and hot carbonate solutions.

During recent years, acid gas removal has received appreciable fresh attention. Planet Earth is coming out of a relatively recent ice age, and so, it is generally held that Earth is warming. Many people believe that the warming is being exacerbated by human actions, especially the combustions associated with automobiles and power plants. Like the metabolism reaction, the combustion reaction yields CO_2 as a product. CO_2 is regarded as a "medium" strength" greenhouse gas. The reduction of global CO_2 emissions is correctly on many people's minds, these days, and in many scientists' laboratories.

REFERENCES

1. Maddox, R.N., Gas and Liquid Sweetening, 2nd printing, John M. Campbell, Norman, OK, 1977.
2. Humphrey, J.L. and Keller II, G.E., Separation Process Technology, McGraw-Hill, New York, NY, 1997.
3. Seader, J.D., Henly, E.J. and Roper, D.K., Separation Process Principles: Chemical and Biochemical Operations, 3rd Ed., John Wiley & Sons, Inc., Hoboken, NJ, 2011.
4. Figueroa, J.D., Fout, T., Plasynski, S., McIlvried, H., Srivastava, R.D., "Advances in CO_2 Capture Technology – The U.S. Department of Energy's Carbon Sequestration Program," International Journal of Greenhouse Gas Control, 2, 9-20, 2008.
5. Brennecke, J.D. and Gurkan, B.E., "Ionic Liquids for CO_2 Capture and Emission Reduction," The Journal of Physical Chemistry Letters, 1, 3459-3464, 2010.
6. Resetarits, M.R., Navarre, J.L., Monkelbaan, D.R., Hangx, G.W.A., and van den Akker, R.M.A., "Trays Inhibit Foaming", Hydrocarbon Processing, March 1992, p. 61-64.
7. Weiland, R.H. and Hatcher, N.A., Advanced Gas Treating: The Engineering Science, Optimized Gas Treating Inc., Houston, TX, 2011.
8. Van Wagener, D.H., "Stripper Modeling for CO_2 Removal by Monoethanolamine and Piperazine Solvents," PhD thesis, Univ. Texas Austin, 2011.
9. Tamhankar, Y., Whiteley, J.R., Resetarits, M., and Aichele, C.P., "Spray Droplet Characterization Inside a Glass Column Through Dense Wall Flow," Atomization and Sprays, 24, 115-128, 2014.

Table 1
Typical Natural Gas Concentrations

Gas	Feed	Product Specification
CH_4	70-95 %	-
C2-C4 (NGLs)	0-20 %	-
CO_2 (no heating value)	0-20 %	< 2 % for pipeline, < ~100 ppmv for cryogenic
H_2S (lethal)	0-5 %	< 4 ppmv
O_2	0-1 %	0.4 %
N_2	0-5 %	-
Ar, He, Ne, Xe	trace	-
H_2O	Equilibrium	4-7 lb/MMscf

Table 2
Typical Flue Gas Concentrations

Gas	Feed	Product Specification
CO_2	10-15 %	-
O_2	0-5 %	< 10 ppmv
N_2	70-75 %	-
H_2O	Saturated	< 1 %
SOx	~50 ppmv	-
NOx	~25 ppmv	-

Figure 1
Amine Chemistry

Monoethanolamine (MEA)

Primary amine
Fast reaction rates
Limited capacity

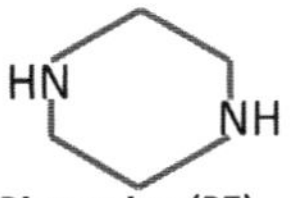

Piperazine (PZ)

Secondary amine
Faster reaction rates
Enhanced capacity

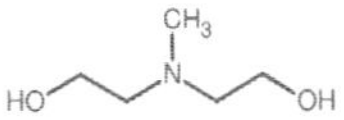

Methyldiethanolamine (MDEA)

Tertiary amine
Slow reaction rates
No carbamate formation
High capacity

Figure 2
Amine Absorption

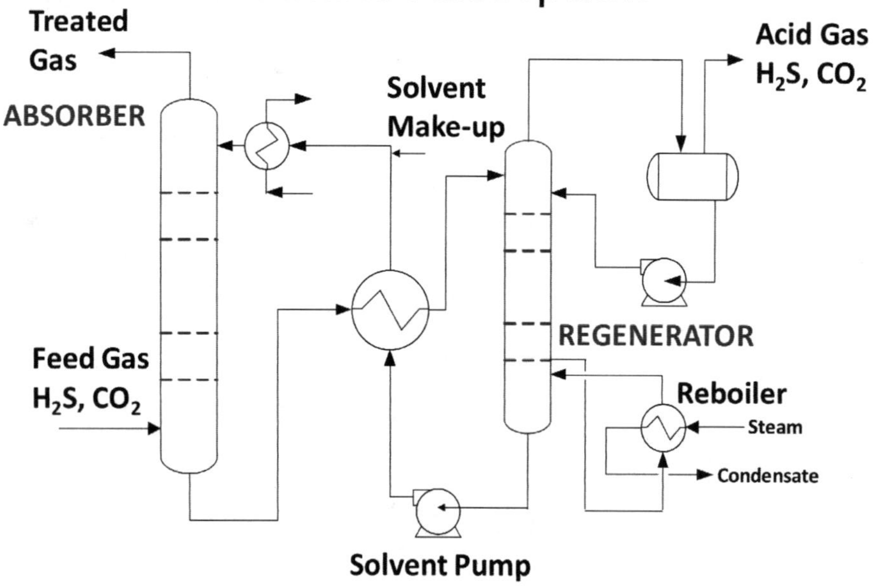

Figure 3
Oklahoma State University
Acid Gas Absorption Apparatus

350235 Radioisotope Applications for Obtaining Detailed Process

Abstract:

In all chemical plants and refineries there generally is not enough instrumentation in the right places in our processes. During the design phase, decisions were made between knowing more about the process, which means buying more instruments, and saving capital funds. When a problem appears with the operation of a vessel, there are always gaps in the knowledge of what is happening due to a lack of instruments in the places where knowledge is not routinely required. Process engineers and operations personnel need to know what is happening so they can take corrective action. When knowledge is missing, they often take actions that make matters worse.

For distillation and separation processes Chemical Engineers have historically relied upon plant process measurements such as flow rates, temperatures, and pressures and model or simulation results for data with which to solve troubleshooting or re-design problems. An array of on-line diagnostic services are available that can provide additional data offering real-time information on how pieces of process equipment are actually operating.

The most common application is Gamma Scanning. This test is primarily applied, but not necessarily limited to, distillation or separation columns. Gamma scans provide a density profile of the internal process of operating distillation columns and other process vessels. The density profile can be used to diagnose the hydraulic operating conditions of mass transfer devices such as damage to internals, flooding, degree of entrainment or weeping, liquid levels on trays and distributors, liquid distribution through packed beds, etc. The presenter will show case studies where scanning revealed vital process information that helped solve an operating problem, or helped make a revamp/re-design successful.

Additional derivatives from gamma scanning of detecting flow distribution patterns will be demonstrated. The CAT-Scan technique is a specialized horizontal gamma scan used to generate a topographic profile of the internal cross-sectional density of process equipment. This profile is useful for the detailed study of liquid flow distribution through packed columns as well as having applications beyond distillation columns. A hybrid application involving tracers and scanning on fixed bed catalytic reactors will also be discussed.

Another technique involving radioactive materials is the Tracer Test, in which a radioactive tracer is introduced into the process for investigating the process flows. External radiation detectors are placed at strategically chosen locations outside vessels and piping to monitor the tracer flow through the process. Either liquid, gas or a solid tracer can be utilized depending on process conditions and the fluid to be studied. This tracer technique can identify liquid or gas maldistribution in fixed or fluidized bed reactors. The residence time through a tank or reactor can be measured and compared with the calculated residence time. Observed differences can relate directly to the degree of mixing or channeling inside the reactor. These tracer techniques have also been used in identifying leakage of heat exchangers and valves.

AUTHOR INDEX

AbdulWahhab, Hamad.....124
Adewole, Jimoh.....97
Agrawal, Rakesh.....166, 168
Aichele, Clint P.....42, 288
Al-Ahmed, Amir.....97
Al-Harthi, MA.....124
Alshaaer, Mazen.....124
AlShoaibi, Ahmed.....120
Anneken, David.....119
Bai, Chuansheng.....90
Banholzer, William.....1
Bearrow, Mike.....69
Beebe, Dustin.....54
Bernard, Andre.....206
Bluck, David.....176
Bondy, Ronald W.....176
Burleigh, Andy.....298
Bustam, Mohd.Azmi.....165
Cai, T.J.....42, 281
Cai, Tony.....49, 251, 267
Cai, Zhiyong.....74
Cao, Xijia.....121
Carlson, Ron.....129
Chambers, Simon.....49, 129, 274
Chikamatsu, Nobuyasu.....84
Clancy-Jundt, Brian.....201, 203
Clavelle, Eric.....206
Coppella, Steven.....91
Dejanovic, Igor.....169
Delgass, W.N.....90
Dixit, Ravi.....170
Donahue, Mike.....93
Duvvuri, Subbarao.....165
Elsharkawy, Adel.....75
Faqir, Naim.....124
Fedor, Patrick.....77
Fellner, Joseph P.....119
Fidkowski, Zbigniew T.....191
Fu, Jinxia.....125
Graeme, Ellis.....126
Gummadi, Srividya.....176
Haider, Mohammad Faruq.....105
Haidl, Jan.....138
Hanson, Daryl.....2
Hennigan, Sean.....205
Honda, Kazunori.....84
Hua, Naomi.....93
Huff, Joshua.....166
Jansen, Helmut.....137, 169
Johnson, Darin.....77
Kannan, Pravin.....120
King, Alan.....127
Kister, Henry Z.....2, 201, 203
Krueger, Artur W.....89
Kunz, Robert G.....223
Laine, Patrick.....92
Lee, SounHo.....154
Linek, Vaclav.....138
Liu, Guangxin.....121
Maharaj, Nilia.....31
Marsh, Michael.....48
Miller, Bob.....48
Miller, Jeffrey T.....90
Miller, Randy.....201, 203
Misra, Rahul.....76
Mohammad, Nurhidayah.....165
Mohanty, Sanat.....76
Morita, Mitsuo.....84
Mutalib, Mohamed I. A.....165
Nagaoka, Tadayoshi.....150
Nolan, Tom.....54
Obayashi, Shuji.....84
Ogundeji, A. Y.....253, 267
Ogundeji, A.Y.....281
Okita, Atsushi.....84
Olujic, Zarko.....137, 169
Oyama, Koji.....84
Pazmino, Jorge.....90
Pendergast, John.....170
Pilling, Mark W.....41
Pless, Lowell.....29, 129, 251, 298
Praderio, Attilio J.....31
Ramapriya, Gautham Madenoor.....168
Rejl, Frantisek J.....138, 143
Resetarits, M.R.....253
Resetarits, Mike.....251, 267, 281, 288
Ribeiro, Fabio H.....90
Rice, James W.....125
Richardson, Kevin J.....48
Rietfort, Thomas.....137
Sandhu, Sarwan S.....119
Sandström, Emma.....125
Santos, Jesse.....31
Saunders, Veronica R.....206
Sawant, Mahesh.....170
Schultes, Michael.....31
Scott, Ethenia.....73, 118
Shawabkeh, Reyad.....124
Shiau, Lie-Ding.....123
Smith, Robert.....126
Sultan, Abdullah S.....97
Summers, Daniel R.....41, 201, 203
Sun, Jinsheng.....121
Suuberg, Eric M.....125
Takahashi, Jumpei.....84
Tamhankar, Yash.....288
Tawarmalani, Mohit.....166
Teng, Hou-Guo.....123
Trauth, Daniel.....170
Valenz, Lukas.....138

AUTHOR INDEX

Vennavelli, A.N. 253, 274
Wang, Wenping..121
Weiland, Ralph..31
Whiteley, James R...42, 288
Wiesman, Daryl..41
Yan, Qiangu...74
Zaidi, S. M. Javid..97
Zhang, Jilei 74
Zich, Egon..137